NONLINEAR ELECTROMAGNETICS

Academic Press Rapid Manuscript Reproduction

NONLINEAR ELECTROMAGNETICS

Edited by

PIERGIORGIO L. E. USLENGHI

Communications Laboratory
Department of Information Engineering
University of Illinois at Chicago Circle
Chicago, Illinois

ACADEMIC PRESS
A Subsidiary of Harcourt Brace Jovanovich, Publishers
New York London Toronto Sydney San Francisco **1980**

ACADEMIC PRESS, INC.
111 Fifth Avenue, New York, New York 10003

United Kingdom Edition published by
ACADEMIC PRESS, INC. (LONDON) LTD.
24/28 Oval Road, London NW1 7DX

LIBRARY OF CONGRESS CATALOG CARD NUMBER: 80-20413

ISBN 0-12-709660-4

PRINTED IN THE UNITED STATES OF AMERICA

80 81 82 83   9 8 7 6 5 4 3 2 1

Contents

CONTRIBUTORS

Numbers in parentheses indicate the pages on which the authors' contributions begin.

Frank S. Barnes (391), Department of Electrical Engineering, University of Colorado, Boulder, Colorado 80309

Ioannis M. Besieris (87), Department of Electrical Engineering, Virginia Polytechnic Institute and State University, Blacksburg, Virginia 24061

F. W. Crawford (239), Institute for Plasma Research, Stanford University, Stanford, California 94305

Giorgio Franceschetti (303), Istituto Elettrotecnico, Università di Napoli, Napoli, Italy

Chia-lun J. Hu (391), Department of Electrical Engineering, University of Colorado, Boulder, Colorado 80309

F. Kaiser (343), Institut für Theoretische Physik, Universität Stuttgart, Stuttgart, Federal Republic of Germany

Nathan Marcuvitz (117), Polytechnic Institute of New York, New York, New York 11735

David W. McLaughlin (65), Department of Mathematics and Program in Applied Mathematics, University of Arizona, Tuscon, Arizona 85721

George C. Papanicolaou (253), Courant Institute, New York University, New York, New York 10012

Innocenzo Pinto (303), Istituto Elettrotecnico, Università di Napoli, Napoli, Italy

A. G. Ramm (263), Department of Mathematics, The University of Michigan, Ann Arbor, Michigan 48109

Alwyn C. Scott (35), Department of Electrical and Computer Engineering, University of Wisconsin, Madison, Wisconsin 53706

A. B. Shvartsburg (189), IZMIRAN, Academgorodok, Moscow Region, USSR

Piergiorgio L. E. Uslenghi (1), Communications Laboratory, Department of Information Engineering, University of Illinois at Chicago Circle, Chicago, Illinois 60680

PREFACE

The preparation of this book began at a symposium on nonlinear electromagnetics sponsored by several commissions of the U.S. National Committee of the International Union of Radio Science (URSI) and by the Antennas and Propagation Society of the Institute of Electrical and Electronics Engineers (IEEE). This symposium was held as a part of the IEEE/URSI INTERNATIONAL SYMPOSIUM at the University of Washington, Seattle, in June 1979. At the conclusion of that symposium, several participants agreed to provide me with expanded versions of the papers they had just presented, to be published in a book on nonlinear electromagnetics. Eight of the thirteen chapters in the book originated in this manner, whereas five chapters were submitted by other authors at my invitation. The final result is a collection of reviews and original research results that span several important areas of nonlinear phenomena in electromagnetism.

The first two chapters are introductory, and provide the unfamiliar reader with a survey of some mathematical methods for nonlinear problems and with a detailed historical perspective on studies of nonlinear waves and, in particular, on the concepts of soliton. These are followed by nine chapters on wave phenomena in a variety of nonlinear media, networks and antennas. The book concludes with two chapters on nonlinear phenomena in biological materials and systems. Although some omissions are evident (such as nonlinear optics, which would require a separate volume), the reader will find comprehensive coverage of many nonlinear research areas of interest to the applied physicist and electrical engineer. The treatment is at the advanced level, often at the forefront of current research, and incorporates many new results. Thus, I believe that this book will be valuable to all researchers and students of nonlinear phenomena in physics and engineering.

I am grateful to the authors for their enthusiasm and cooperation, to Drs. W. Ross Adey and Frank S. Barnes for their help in organizing the nonlinear symposium at Seattle, to Ms. Lellie Buford for her expert typing, and to Academic Press for their patient collaboration. Completion of this project was made possible by the financial support of the Air Force Office of Scientific Research, of the Office of Naval

Research, and the Department of Information Engineering in the University of Illinois at Chicago Circle.

Chicago, June 1980 *Piergiorgio L. E. Uslenghi*

AN INTRODUCTION TO SOME MATHEMATICAL TECHNIQUES FOR NONLINEAR PROBLEMS[1]

Piergiorgio L.E. Uslenghi

Communications Laboratory
Department of Information Engineering
University of Illinois at Chicago Circle
Chicago, Illinois

I. INTRODUCTION

This chapter is intended for those readers who are unfamiliar with the mathematical techniques used in dealing with nonlinear problems, and may be skipped by all who, having some knowledge of the subject, wish to delve at once into the specialized treatments contained in the remainder of the book.

A comprehensive discussion of methods in nonlinear mathematics would fill several volumes; thus, this introduction is necessarily selective. The methods and examples discussed were chosen with several criteria in mind: firstly, they are rather simple and can be presented without a laborious and sophisticated mathematical apparatus; secondly, they are dictated by important practical problems and provide some historical perspective on the development of the field; thirdly, they lead the reader to more specialized and updated bibliography. The overall aim is to arouse the interest of the engineer, applied mathematician

[1]*The preparation of this survey was supported in part by the Air Force Office of Scientific Research under grant AFOSR-77-3253.*

ISBN 0-12-709660-4

and applied physicist in nonlinear phenomena, by providing him with a glimpse into the beauty and relevance of the results obtained thus far and with a motivation for further study.

The focus of this chapter is on nonlinear ordinary differential equations and on some aspects of nonlinear partial differential equations, but some bibliographical references are provided for other topics as well (nonlinear integral, integro-differential, difference and functional equations). The material presented is borrowed from a vast array of books and journal articles; despite the efforts to reference the original sources, inadvertent omissions cannot be avoided, and the bibliography is necessarily selective rather than comprehensive.

II. ORDINARY DIFFERENTIAL EQUATIONS

II.1. The Fundamental Theorem

The ordinary (linear or nonlinear) differential equation of order n in the normal form

$$y^{(n)} = f(x,y',y'',\ldots,y^{(n-1)}) \;, \tag{1}$$

where $y^{(\ell)} = \frac{d^\ell y}{dx^\ell}$ and $y' = y^{(1)}$, $y'' = y^{(2)}$, is equivalent to a system of n first-order equations in normal form. In fact, if we let $y = y_1$, $y^{(1)} = y_2$, ..., $y^{(n-1)} = y_n$, then Eq. 1 may be replaced by the system

$$\begin{cases} y_1' = y_2 \;, \\ \cdot\;\;\cdot\;\;\cdot\;\;\cdot \\ y_{n-1}' = y_n \;, \\ y_n' = f(x,y_1,y_2,\ldots,y_n) . \end{cases} \tag{2}$$

Conversely, the system

$$y_j' = f_j(x,y_1,\ldots,y_n) \quad , \quad (j = 1,2,\ldots,n) \tag{3}$$

of which (2) is a particular case, yields an n-order differential equation in normal form for the unknown function $y_h(x)$, $(h = 1,2,\ldots,n)$ as follows: first we differentiate each of the n Eqs. (3) $(n - 1)$ times with respect to x, thus obtaining n^2 equations overall; then we eliminate the $n^2 - 1$ unknowns y_j, y'_j, ..., $y_j^{(n)}$, $(j = 1,2,\ldots,n; j \neq h)$ from the n^2 equations, and are left with one nth-order equation in normal form (i.e., explicitly solved with respect to $y_h^{(n)}$) for the unknown $y_h(x)$.

In the following we consider the system (3) together with the initial conditions

$$y_h(x_0) = y_{0h} \quad , \quad (h = 1,2,\ldots,n) \tag{4}$$

where y_{0h} are given arbitrary constants. We recall that a function f_j satisfies Lipschitz' condition with respect to the variable y_h in a given domain D of y_h if a positive constant C exists such that

$$|f_j(x,y_1,y_2,\ldots,y_{h1},\ldots,y_n) - f_j(x,y_1,y_2,\ldots,y_{h2},\ldots,y_n)| \leq C|y_{h1} - y_{h2}|$$

for any two values y_{h1} and y_{h2} of y_h in D.

Theorem. Given the system (3) with the initial conditions (4), if we can find two positive numbers α and β such that in the domain D defined by the inequalities

$$x_0 \leq x \leq x_0 + \alpha \;,\; y_{0h} - \beta \leq y_h \leq y_{0h} + \beta \;,\; (h = 1,2,\ldots,n)$$

the functions f_j, $(j = 1,2,\ldots,n)$ are continuous and satisfy Lipschitz' condition with respect to the n variables y_h, then:

(i) We can find a number γ, $0 < \gamma \leq \alpha$, such that in the interval $x_0 \leq x \leq x_0 + \gamma$ there exists one and only one solution to the system (3) that satisfies the initial conditions (4).

(ii) If K is a positive number such that $|f_j| \leq K$, $(j = 1,2,\ldots,n)$ everywhere in the domain D, then γ is the smaller of the two numbers α and β/K.

(iii) The one and only solution of (3-4) in $x_0 \le x \le x_0 + \gamma$ satisfies the system of n (nonlinear) Volterra integral equations

$$y_h(x) = y_{0h} + \int_{x_0}^{x} f_h(\xi, y_1(\xi), \ldots, y_n(\xi))\, d\xi \,, \qquad (h = 1,2,\ldots,n) \tag{5}$$

(iv) Successive approximations to the exact solution $y_h(x)$ are obtainable via the recursion formula

$$y_{h,\ell+1}(x) = y_{0h} + \int_{x_0}^{x} f_h(\xi, y_{1,\ell}(\xi), \ldots, y_{n,\ell}(\xi))\, d\xi \,, \qquad (\ell = 0,1,2,\ldots) \tag{6}$$

where $y_{h,0}(x) = y_{0h}$ and $\lim_{\ell\to\infty} y_{h,\ell}(x) = y_h(x)$.

Extensive discussions on the above fundamental theorem may be found in several books (see, e.g., [1]). Here we limit ourselves to a few remarks:

(a) If the f_j are continuous but do not satisfy Lipschitz' condition with respect to the y_h, then a solution to (3-4) still exists but is, in general, no longer unique.

(b) If the f_j are continuous functions of some parameters, then also the y_h are continuous functions of these parameters.

(c) Picard's method of successive approximations (5) allows for an explicit, although laborious, solution of (3-4).

(d) Global existence and uniqueness theorems are available for systems such as (3-4) as well as for implicit systems (i.e., with equations not explicitly solved with respect to y_j'); see, e.g., [2-3].

I.2. Solutions of Some First-Order Equations

1. The linear equation $y' = f(x)$ has the solution $y = \int f(x)\,dx + A$, where A is an arbitrary constant.

2. The equation $y' = \phi(y)$ has the implicit solution

$$x = \int \frac{dy}{\phi(y)} + A \ .$$

3. The equation $y' = f(x)\phi(y)$ has the solution

$$\int \frac{dy}{\phi(y)} = \int f(x)\,dx + A \ .$$

4. The homogeneous equation $y' = f(y/x)$ is solved by letting $y/x = z$. Then $y(x) = xz(x)$ and $z(x)$ is given by

$$\int \frac{dz}{f(z) - z} = \ln x + A \ .$$

5. The equation

$$y' = f\left(\frac{ax + by + c}{\alpha x + \beta y + \gamma}\right)$$

reduces to case 4 if $c = \gamma = 0$. Let $x = \xi + x_0$, $y = \eta + y_0$ and impose the conditions

$$ax_0 + by_0 + c = 0 \quad , \quad \alpha x_0 + \beta y_0 + \gamma = 0$$

on x_0 and y_0. If $a\beta \neq \alpha b$ we are back to case 4. If instead $a\beta = \alpha b$, set $a/b = \alpha/\beta = \delta$ and $z = \delta x + y$, obtaining

$$\frac{dz}{dx} = \delta + f\left(\frac{bz + c}{\beta z + \gamma}\right) ,$$

which is of the type considered in case 2.

6. The equation $y = f(x,y')$, with the substitution

$$\frac{dy}{dx} = \tau \tag{7}$$

becomes

$$y = f(x,\tau) \ , \tag{8}$$

wherefrom

$$\frac{dx}{d\tau} = \frac{\dfrac{\partial}{\partial \tau} f(x,\tau)}{\tau - \dfrac{\partial}{\partial x} f(x,\tau)} \ ;$$

this by integration yields $x(\tau)$, which is then substituted into (8) to find $y(\tau)$. Thus, the solution is obtained in parametric form. Note that the choice (7) excludes solutions

τ = constant; that is, possible solutions $y = Ax + B$ should be attempted separately. As an example, consider the Lagrange equation

$$y = xF\left(\frac{dy}{dx}\right) + G\left(\frac{dy}{dx}\right) ,$$

for which

$$\begin{cases} y = xF(\tau) + G(\tau) \\ \dfrac{dx}{d\tau} = \dfrac{xF'(\tau) + G'(\tau)}{\tau - F(\tau)} \end{cases} ;$$

however, separate solutions $y = Ax + B$ also occur if there exist values of A such that $A = F(A)$, in which case $B = G(A)$.

7. Another class of first-order equations which may be solvable in parametric form is $x = F(y,y')$, which with substitution (7) becomes

$$x = F(y,\tau) , \tag{9}$$

wherefrom

$$\frac{dy}{d\tau} = \frac{\dfrac{\partial}{\partial\tau} F(y,\tau)}{\dfrac{1}{\tau} - \dfrac{\partial}{\partial y} F(y,\tau)} .$$

Integration of this last equation yields $y(\tau)$, which substituted into (9) gives $x(\tau)$. The above remark on additional solutions of type $y = Ax + B$ applies here.

8. Bernoulli equations are of the type

$$y' + a(x)y + b(x)y^n = 0 , \tag{10}$$

where we may assume $n \neq 1$ (nonlinear case), because for $n = 1$ the solution is obtained at once by separation of variables (case 3). The general solution of (10) is:

$$y(x) = e^{-\int a(x)\,dx}\left[A + (n - 1)\int b(x)\exp\left\{-(n - 1)\int a(x)\,dx\right\}dx\right]^{-\frac{1}{n-1}} ,$$

where A is an integration constant. Particular cases of Eq. 10 are important in applications; thus, for example, if

$n = 2$ and $a(x)$ and $b(x)$ are constants, we have the Verhulst (or logistic) equation of ecology; similar to it is the Landau equation of fluid dynamics.

II.3. Second-Order Equations Reducible to First Order

1. If in the equation $f(x,y',y'') = 0$ we perform the substitution (7), we obtain $f(x,\tau,\tau') = 0$, which is a first-order equation. If it can be integrated to yield $\tau(x,A)$ with A an integration constant, then from (7):

$$y(x) = \int \tau(x,A)\,dx + B \ .$$

2. With substitution (7), the equation $f(y,y',y'') = 0$ becomes

$$f(y,\tau,\tau \frac{d\tau}{dy}) = 0 \ ;$$

if this can be integrated to yield $\tau(y,A)$, then from (7):

$$x = \int \frac{dy}{\tau(y,A)} + B \ .$$

3. By letting

$$y'/y = \xi \tag{11}$$

in the equation $f(x,y'/y,y''/y) = 0$, we obtain

$$f(x,\xi,\xi' + \xi^2) = 0 \ ;$$

if this can be solved to yield $\xi(x,A)$, then

$$y(x) = B \exp\left\{\int \xi(x,A)\,dx\right\} \ .$$

As an example, consider the equation

$$y'' + a(x)y' + c(x)y = 0$$

which by (11) becomes

$$\xi' + a(x)\xi + \xi^2 + c(x) = 0 \ ;$$

this is an inhomogeneous version of Eq. 10 with $n = 2$ and $b(x) = 1$, and is known as Riccati's equation.

II.4. Some Notable Second-Order Nonlinear Equations

1. Consider the equation

$$y'' = f(x,y)\,g(x) \tag{12}$$

in the domain $0 \leq x < \infty$, $0 \leq y < \infty$ with the boundary conditions

$$y(0) = y_0 \geq 0 \quad , \quad y(+\infty) = 0 \ . \tag{13}$$

Theorem [4]. If:

(a) $f(x,y)$ is continuous with respect to x and y, positive for x and y positive, and monotonically increasing with respect to y;

(b) $f(x,0) = 0$, $\forall x \geq 0$;

(c) for any fixed $y > 0$, $f(x,y)$ has a positive lower bound $\forall x \geq 0$;

(d) $g(x)$ is continuous; $g(x) > 0\ \forall x > 0$; $g(x)$ is integrable over every finite interval $0 \leq x \leq x_0$, and $\int_0^\infty g(x)\,dx = +\infty$;

then there exists one and only one solution to Eq. 12 satisfying boundary conditions (13).

A particular case of Eq. 12 is the famed Fermi-Thomas equation [5]:

$$y'' = x^{-1/2} y^{3/2} \ ,$$

which arises in the study of heavy atoms. Another important case is the Emden-Fowler equation of astrophysics [6]:

$$y'' + x^{-m} y^{n} = 0 \quad , \quad (m > 2) \ .$$

2. Milne's equation of damped vibrations [7] may be written in the form:

$$y'' + [a + f(y')]y' + by = 0 \tag{14}$$

with $a \geq 0$, $b > 0$, $f(0) = 0$, $f(y') = f(-y') > 0$ for $y' \neq 0$, and $f(y')$ continuous and satisfying Lipschitz' condition in any finite interval. Under these hypotheses, there exists one and only one solution of (14) in $x_0 \leq x < +\infty$ that satisfies the initial conditions $y(x_0) = y_0$, $y'(x_0) = y'_0$.

3. Liénard's equation of relaxation oscillations is [8]:

$$y'' + f(y)y' + y = 0 \ . \tag{15}$$

A particular case of (15) is the celebrated van der Pol equation, for which $f(y) = C(y^2 - 1)$ with C a constant [9]. More general versions of (15) are

$$y'' + f(y)y' + g(y) = 0 \tag{16}$$

and

$$y'' + f(y,y')y' + g(y) = 0 \ , \tag{17}$$

studied by Graffi [10] and by Levinson and Smith [11]. Equations 15-17 are of great importance in mechanical, electrical and biological systems, especially insofar as they allow for the existence of periodic solutions ("limit cycles" in phase space). An elementary account of phase-space techniques is provided in the following section of this chapter, while an exhaustive review of applications to biosystems is given in the chapter by Kaiser. Here we limit ourselves to stating some basic results.

Theorem 1. For $-\infty < x < +\infty$, there exists only one solution of Eq. 15 satisfying the initial conditions

$$y(x_0) = y_0 \quad , \quad y'(x_0) = y'_0 \ , \tag{18}$$

provided that

$$y_0^2 + y_0'^2 \neq 0 \tag{19}$$

and that $f(y)$ is continuous and bounded for $-\infty < y < +\infty$.

Theorem 2. If x_1 and $x_2 > x_1$ are two zeros of a solution of (15), then $x_2 - x_1 \geq \pi$.

The boundedness condition on $f(y)$ does not allow us to apply Theorem 1 to the van der Pol equation; however, we may relax this condition as follows:

Theorem 3. If $f(y)$ is continuous in $-\infty < y < +\infty$ and a constant δ exists such that $f(y) \geq 0$, $\forall |y| \geq \delta > 0$, then in $x_0 \leq x < +\infty$ there exists one and only one solution of Eq. 15 that satisfies the initial conditions (18-19).

Theorem 4. If $f(y)$ is continuous for $-\infty < y < +\infty$, if $f(y) \lesseqgtr 0$ for $|y| \lesseqgtr \delta$ where δ is a positive constant, and if either

$$\int_0^\infty f(y)\,dy = +\infty \quad \text{or} \int_{-\infty}^0 f(y)\,dy = +\infty ,$$

then there exists one and only one *periodic* solution of Eq. 15.

4. Forced oscillations of mechanical systems with a nonlinear restoring force are often described via the Duffing equation [12]:

$$y'' + Ay' + f(y) = B \cos x ,$$

which is extensively studied in a number of books (see, e.g., [13-14]).

5. Discussions of the above and other nonlinear equations may be found, for example, in the books [3] and [13-17]. See also the monumental work by Cesari [18] and the two volumes by Bellman [19].

II.5. Phase-Plane Analysis

II.5.1. Introduction. In many applications, the mathematical formulation of the physical problem leads to the (nonlinear) ordinary differential equation of the first order

$$\frac{dy}{dx} = f(x,y) \quad , \tag{20}$$

whose solutions, when represented geometrically in the *phase-plane* (x,y), are curves called *trajectories*. The fundamental theorem of Section I.1 assures us that one and only one trajectory passes through each point (x,y), at which $f(x,y)$ is continuous and lipschitzian; however, these conditions on $f(x,y)$ may be violated at some points (x,y), called *singular points* (or *critical points*) of Eq. 20 .

Equation 20 may be rewritten in parametric form as the system:

$$\begin{cases} \dfrac{dx}{d\tau} = F(x,y) \ , \\ \dfrac{dy}{d\tau} = G(x,y) \ , \end{cases} \qquad (21)$$

where $f(x,y) = G(x,y)/F(x,y)$ and τ is the parameter. The system (21) is slightly more general than Eq. 20, because it includes solutions for which $F(x,y) = 0$, i.e., trajectories parallel to the y-axis. Systems of type (21) arise quite frequently in biological, ecological and socio-economical studies, as well as in problems of nonlinear control theory; for a comprehensive study, see, e.g., [20].

In the following, we assume that at the singular points of (21) both $F = 0$ and $G = 0$, i.e., at a singular point (x_0,y_0) the system (21) has the constant solution $x = x_0$, $y = y_0$. We also assume that each singular point (x_0,y_0) is isolated, i.e., that there exists a neighborhood of (x_0,y_0) in the phase plane wherein no other singular point is contained. Finally, we assume that the singular point is located at the origin of the phase plane; this can always be achieved by moving from the (x,y) plane to the (ξ,η) plane via the substitution $x = x_0 + \xi$, $y = y_0 + \eta$.

II.5.2. The linear case. Assume that F and G in (21) are linear functions:

$$F(x,y) = Ax + By \quad , \quad G(x,y) = Cx + Dy \quad , \qquad (22)$$

where A, B, C and D are constants with

$$AD - BC \neq 0 \ . \qquad (23)$$

It can then be proven that the following types of singularities may occur at the origin $x = y = 0$. Let

$$H = (A - D)^2 + 4BC \quad , \qquad (24)$$

then:

a)	If $H > 0$ and:	$AD - BC > 0$:	improper node;
		$AD - BC < 0$:	saddle.
b)	If $H < 0$ and:	$A + D \neq 0$:	spiral;
		$A + D = 0$:	center.
c)	If $H = 0$:	improper node (proper node only if $A = D$ and $B = C = 0$).	

General sketches of the various types of trajectories bearing the above names are illustrated in Fig. 1.

The arrow on each trajectory of Fig. 1 indicates how a point (x,y) moves along the trajectory as τ increases. If all trajectories approach the singular point as $\tau \to \infty$, the point is called "asymptotically stable" (see, e.g., Figs. 1a, b, d); if the trajectories neither approach the singular point nor move away to infinity as $\tau \to \infty$, as in the case of the center of Fig. 1e, the point is "stable"; finally, if at least one trajectory moves to infinity as $\tau \to \infty$, the point is "unstable" (this is the case of Fig. 1c, and also of Figs. 1a, b and d if the arrow directions are reversed). As τ is often identified with time, these connotations of asymptotic stability, stability and instability are intuitive; the stability condition is determined by the roots of the equation

$$\eta^2 - (A + D)\eta + AD - BC = 0 . \qquad (25)$$

If both roots of (25) are distinct and real positive (negative), then we have an unstable (asymptotically stable) improper node; if one positive and one negative, we have a saddle point (unstable); if equal and positive (negative), we have an unstable (asymptotically stable) node. If the roots are pure imaginary we have a center (stable); if they are complex with a positive (negative) real point, we have an unstable (asymptotically stable) spiral.

Finally, we observe that if $AD - BC = 0$, the system (21-22) does not have a singular point at the origin.

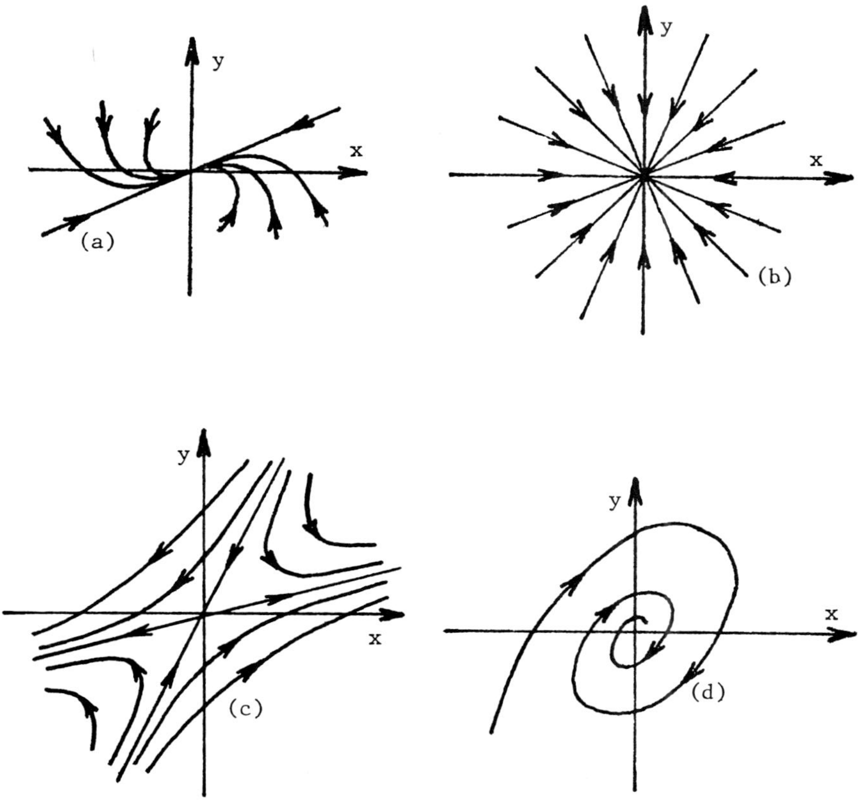

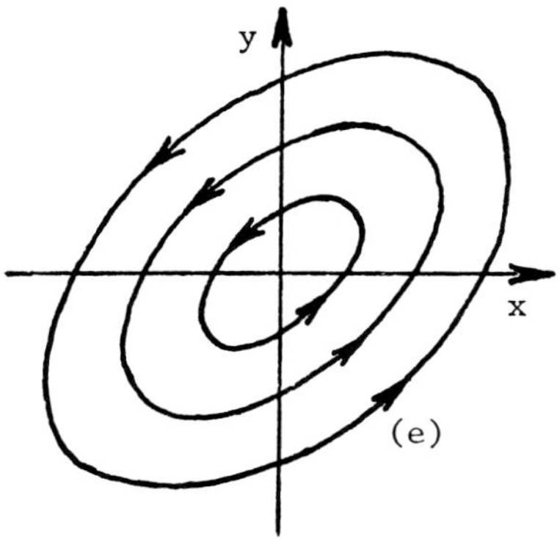

FIGURE 1. Types of trajectories near an isolated singular point: (a) improper node, (b) proper node, (c) saddle, (d) spiral, (e) center.

II.5.3. The piecewise-linear case. If the phase plane (x,y) can be subdivided into regions corresponding to different values of the constants A, B, C, D of Eqs. 22, then the global trajectories for this special type of nonlinear system (called piecewise-linear) may be obtained as follows: (i) in each region of the (x,y) plane in which the system(21-22) is linear, determine the position and type of the singular points; (ii) draw the trajectories in each region, and connect the trajectories of neighboring regions to obtain the entire phase-plane trajectory picture.

II.5.4. The nonlinear case. The results obtained for the linear case may be extended to the general nonlinear case under not very restrictive conditions for the functions F and G of (21). Most works on this subject are based on the classical paper by Bendixson [21]; especially notable are those by Lonn, Frommer (see references under [21]) and Tricomi [1].

The following theorem is due to Bendixson and Tricomi.

Theorem. If $F(x,y)$ and $G(x,y)$ in (21) are continuous and lipschitzian in a domain containing the origin $x = y = 0$; if

$$F(x,y) = P_1(x,y) + f(x,y) \quad , \quad G(x,y) = P_2(x,y) + g(x,y) \tag{26}$$

where $P_{1,2}$ are homogeneous polynomials of the same degree $m \geq 1$ having no common factor of the type $Ax + By$ with A and B real; if

$$\rho^{-m} f(x,y) \to 0 \quad , \quad \rho^{-m} g(x,y) \to 0 \tag{27}$$

in the limit $\rho = \sqrt{x^2 - y^2} \to 0$; if the homogeneous polynomial of degree $m + 1$:

$$xP_2(x,y) - yP_1(x,y) = M(x,y) \tag{28}$$

is not identically zero; then any trajectory of the system (21), for which either $\lim_{\tau\to+\infty} \rho = 0$ or $\lim_{\tau\to-\infty} \rho = 0$, ends on the origin either as a spiral or with a specified tangent forming an angle α with the x-axis that satisfies the equation (which is an algebraic equation of degree $m + 1$):

$$M(\cos\alpha, \sin\alpha) = 0 \ . \tag{29}$$

An analysis of the above theorem in the case $m = 1$ shows that the trajectories of (21)-(26) about the origin $x = y = 0$ exhibit, in general, the same behavior of the trajectories of the linear case (21-22), with two exceptions:

(a) $H < 0$, $A + D = 0$

with H given by Eq. 24, in which case the linear system (21-22) has a center, whereas the nonlinear system (21) - (26) has a spiral or some other singularity;

(b) $H = 0$,

in which case a node is assured for the nonlinear system (21-26) only if Eqs. 27 with $m = 1$ are replaced by the more restrictive conditions

$$F(x,y) = Ax + By + O(\rho^{1+\varepsilon}) \quad , \quad G(x,y) = Cx + Dy + O(\rho^{1+\varepsilon}) \ ,$$

where ε is any positive constant.

The case $AD - BC = 0$ has been excluded from the above considerations; if this case occurs, new types of singularities may appear. Consider, for example, the equation [1]:

$$\frac{dy}{dx} = \frac{3x^2}{2y}$$

whose general solution is $y^2 = x^3 + K$, with K a constant. The trajectory for $K = 0$ has a cusp at the origin and is the only trajectory reaching the origin (Fig. 2).

Stability considerations are easily extended to the nonlinear case; the interested reader is referred to specialized treatises, such as [18], [20] and [22]. Here we only point out that the types of singular points and of stability for the nonlinear case (21)- (26) with $m = 1$ are those of the linear case (21-22), with two exceptions: (i) if Eq. 25 has a double real root, then (21) - (26) with $m = 1$ may have either a node (proper or improper) or a spiral at the origin, but the stability is still

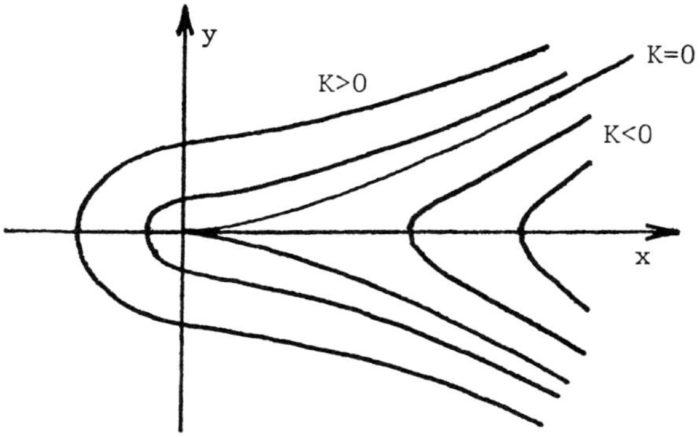

FIGURE 2. A cusp singularity.

that of the linear case; (ii) if the roots of (25) are pure imaginary, then (21)-(26) with $m = 1$ may have either a center or a spiral at the origin, and the stability of the system is indeterminate.

The above treatment is based on the behavior of trajectories near a simple singular point. However, non-simple singularities may also occur in which case, as illustrated in the example of Fig. 2, the behavior of the trajectories may become rather complicated; the interested reader is referred to the articles by Aggarwal [23] and the book by Lefschetz [24].

II.5.5. Limit cycles. A limit cycle is an isolated closed trajectory. Isolated means that no other closed trajectory exists in a neighborhood of the limit cycle. Thus, a limit cycle represents an isolated periodic solution for the nonlinear system; no such solution can exist for a linear system. The van der Pol equation [9] is a well-known example of an equation which allows for a unique periodic solution; a great amount of research has been devoted to finding under what hypotheses for f and g, the more general equations (15-17) admit limit cycles.

Limit cycles may be stable or unstable; in the former case, all trajectories approach the limit-cycle trajectory as τ increases; in the latter case, all trajectories move away from the limit cycle as τ increases. A given system may have several limit cycles. Consider the example [1]:

$$\begin{cases} \dfrac{dx}{d\tau} = y - x\rho \sin \dfrac{1}{\rho} , \\ \dfrac{dy}{d\tau} = -x - y\rho \sin \dfrac{1}{\rho} , \end{cases} \tag{30}$$

where $\rho = \sqrt{x^2 + y^2}$; its solution is

$$\rho = \frac{1}{2 \arctan (e^{\theta - \theta_0})} \tag{31}$$

where $\tan \theta = y/x$ and θ_0 is an arbitrary constant. Closed trajectories are circles, for which $d\rho/d\theta = 0$; these occur at

$$\rho_n = \frac{1}{\pi n} \qquad (n = 1,2,3,\ldots) . \tag{32}$$

Thus, the system (30) has an infinite number of limit cycles, which are the concentric circles with radii given by Eq. 32 and accumulation point at the origin $\rho = 0$. All other trajectories are spirals which approach two neighboring limit circles asymptotically as $\theta \to \pm\infty$.

In studying periodic solutions, it is useful to associate to the system (21) a vector field

$$\underline{P} = \hat{x}F(x,y) + \hat{y}G(x,y) \tag{33}$$

where $\hat{x}$ and $\hat{y}$ are unit vectors along the positive x and y directions, respectively. At a non-singular point (x,y) in the phase plane, the vector $\underline{P}$ is tangent to the trajectory passing through that point, and has Cartesian components F and G; at a singular point, $\underline{P}$ is zero.

Consider a Jordan closed curve in the phase plane, i.e., a curve which can be reduced topologically (i.e., by continuous deformations) to a circle, and that is oriented in the positive (counter-clockwise) direction; such a curve is called a "scroc." For a vector field $\underline{P}$, we define the *Poincaré index* p of a given scroc which does not pass through any singular point of (33) as follows: p is the ratio of the increment of the angle (in radians) that $\underline{P}$ forms with a fixed direction as we go once around the scroc, to 2π radians. The following properties hold:

(a) The index p is an integer number.

(b) The index of a scroc which contains no singular points on it or in its interior is zero.

(c) The index of a scroc containing in its interior an isolated singular point is the same for all scrocs containing that isolated point only. We can thus assign an index to each singular point.

(d) The index of a scroc containing in its interior several singular points is the sum of the indexes pertaining to each point. Thus, the index of an arbitrary scroc is known if the indexes of all singular points of the vector field are known.

(e) The index of the singular point $x = y = 0$ of the quasi-linear system (21) - (26) with $m = 1$, i.e., with $P_1(x,y) = Ax + By$, $P_2(x,y) = Cx + Dy$, is equal to the index of $x = y = 0$ of the linear system (21-22), provided that $AD - BC \neq 0$.

(f) The index of a saddle is -1; the index of a spiral, a center or a node is +1.

(g) The index of a limit cycle, considered as a scroc in its own vector field, is +1. Thus, a periodic solution in the phase plane encloses at least one singular point; if it encloses one singular point only, this cannot be a saddle.

The following two theorems are useful in investigating the presence of limit cycles.

Theorem 1 (Bendixson [21]). A simply connected domain D of the phase plane, in which

$$\nabla \cdot \underline{P} = \frac{\partial F}{\partial x} + \frac{\partial G}{\partial y}$$

is either positive or negative, contains no limit cycles. This theorem gives a necessary condition for the existence of a limit cycle in *D*: $\nabla \cdot \underline{P}$ must change sign in *D*.

Theorem 2 (Poincaré [25] and Bendixson [21]). If *D* is a doubly connected domain of the phase plane that contains no singular points, and if at both boundaries of *D* the vector $\underline{P}$ is everywhere pointing either into *D* or outward from *D*, then there exists in *D* at least one limit cycle. For the proof of this theorem, which establishes a sufficient condition for the existence of periodic solutions, see, e.g., [1] or [26].

The successful application of Theorem 2 depends on the construction of the two curves which define the annular domain *D*; this problem is discussed in detail by Cesari [18]. Extensions of Theorem 2 to higher-order systems, i.e., to systems comprising three or more differential equations of type (21), have been carried out with mixed success. A fruitful application to a third-order system is found in Zubov [27].

II.6. Approximation Methods

Considerations of geometric type in the phase plane such as those developed in the previous section, and considerations of stability (not specifically addressed here; see, e.g., [18-20], [22], [27-29]) yield important qualitative information on the solution of nonlinear systems of the type (3), for which exact general solutions with global validity are, in general, unknown.

Quantitative results may be obtained by the use of either approximation methods or computational techniques (analog or digital). A review of these results is outside the scope of this chapter; in this and the following section we simply present some introductory results following the treatment in [20], and refer the reader to some specialized works.

Most approximation methods belong to one of two groups: small-parameter methods, and singular perturbation methods. In either case, the differential equations which describe the system under study contain a small parameter ε; the goal of an approximation method is the determination of the behavior of the system when $\varepsilon \neq 0$ from the (known) behavior of the same system when $\varepsilon = 0$.

II.6.1. Small-parameter methods. Consider a system of type (3):

$$y_j' = f_j(x, y_1, \ldots, y_n; \varepsilon) \quad , \quad (j = 1, 2, \ldots, n) \tag{34}$$

where the dependence of f_j on a small parameter ε is explicitly noted. For brevity, we write (34) as

$$y' = f(x, y; \varepsilon) \tag{35}$$

where y is an n-vector. We wish to infer the behavior of (35) from the behavior of the system

$$y' = f(x, y; 0) \quad , \tag{36}$$

under the assumption that the n-vector f is continuously differentiable with respect to its arguments x, y and ε.

A particular category of systems (35) is that of *quasi-linear systems*:

$$y' = Ay + \varepsilon\phi(x, y, \varepsilon) \quad , \tag{37}$$

where $\varepsilon \ll 1$ and A is a constant $n \times n$ matrix; for $\varepsilon = 0$, (37) reduces to the linear system $y' = Ay$. We have the following:

Theorem [20]. If the eigenvalues of A are in the left half-plane and ϕ is periodic in x with period x_0, then the system (37) has one and only one periodic solution with period x_0, and this solution is asymptotically stable.

The main small-parameter methods for quasi-linear systems are listed below.

(a) Perturbation methods

Following Poisson, we assume a Taylor series solution for small ε:

$$y(x,\varepsilon) = \sum_{\ell} \varepsilon^{\ell} y_{\ell}(x) \ , \tag{38}$$

substitute (38) into (37) and equate the coefficient of each power of ε to zero, thus obtaining differential equations for $y_{\ell}(x)$. The main difficulty with Poisson's method is the possible occurrence of *secular terms* in the solutions $y_{\ell}(x)$, i.e., terms which diverge as $x \to \infty$. This difficulty is avoided by modifying the method as suggested by Lindstedt: the independent variable x is replaced by a new variable $\xi = \omega x$ such that $y(\xi) = y(\xi + 2\pi)$, then we set

$$y(\xi,\varepsilon) = \sum_{\ell} \varepsilon^{\ell} y_{\ell}(\xi) \quad , \quad \omega(\varepsilon) = \sum_{\ell} \varepsilon^{\ell} \omega_{\ell} \ , \tag{39}$$

substitute (39) into the original system and equate powers of ε; for details, see Cesari [18]. Another version of the perturbation method is due to Poincaré [30], who was the first to use small-parameter techniques.

(b) Averaging methods

If the solution for the linear system ($\varepsilon = 0$) is

$$y(x) = A \cos (\omega x + B) \tag{40}$$

where ω, A and B are constants, it is assumed that the solution of the quasi-linear system is still of the type (40) but with A and B slowly varying functions of x, namely:

$$y(x) = A(x) \cos \left[\omega x + B(x)\right] \ , \tag{41}$$

$$y'(x) = -\omega A(x) \sin \left[\omega x + B(x)\right] \ , \tag{42}$$

with

$$\frac{A'(x)}{A(x)} = B'(x) \tan \left[\omega x + B(x)\right] \tag{43}$$

to make (42) consistent with (41). From (43) and the quasi-linear system we obtain expressions for $A'(x)$ and $B'(x)$ that are then expanded in Fourier series of $\omega x + B(x)$; this technique has been systematized by Bogoliubov and Krylov.

A similar technique was introduced by van der Pol, who set

$$y(x) = A(x) \cos \omega x + B(x) \sin \omega x$$

$$y'(x) = -\omega A(x) \sin \omega x + \omega B(x) \cos \omega x$$

with

$$A'(x) \cos \omega x + B'(x) \sin \omega x = 0 ,$$

and then proceeded as Bogolinbov and Krylov.

To first order, the above two methods yield the same results.

(c) Other methods

The equivalent linearization method [20] is better suited than the previous ones for third- and higher-order systems. Other more general results are to be found in books [18], [31] and [32].

II.6.2. Singular perturbation methods. Consider a system of the type

$$\begin{cases} y' = f(x,y,z;\varepsilon) , \\ \varepsilon z' = g(x,y,z;\varepsilon) , \end{cases} \tag{44}$$

where $y(x,\varepsilon)$ is an n-vector, $z(x,\varepsilon)$ is an m-vector, and $\varepsilon \ll 1$. We compare (44) to the system:

$$\begin{cases} y' = f(x,y,z;0) , \\ 0 = g(x,y,z;0) , \end{cases} \tag{45}$$

obtained from (44) by setting $\varepsilon = 0$, under the hypothesis that f and g are continuously differentiable with respect to x, y, z and ε.

A major advantage in going from (44) to (45) is that (44) is of order $(n+m)$, whereas (45) is of order n plus an algebraic condition. A major disadvantage is that when $\varepsilon \to 0$ we lose the derivatives in the second of (44); it is then intuitive that, in general, it may not be possible to infer the behavior of (44) from that of (45). Perhaps the simplest and most famous case of this type is the van der Pol equation [9]. Singular perturbation methods are treated in detail in the papers by Pontryagin and Miščenko [33] and in the book by Halanay [34], among others. Finally, all readers interested in perturbation techniques should refer to the book by Nayfeh [35].

II.7. Digital Computations

The importance of numerical results in nonlinear problems, for which analytical results, both qualitative and quantitative, are few and difficult to establish, is dramatically illustrated by two historical examples. The existence of a periodic solution for the van der Pol equation was revealed by drawing the vector field corresponding to numerical results; this discovery led to a flurry of investigations into limit cycles and relaxation oscillations. In 1955, Fermi, Pasta and Ulam published a numerical study of a one-dimensional dynamical system of 64 particles interacting via nonlinear forces, that they modeled by means of a system of 64 second-order ordinary differential equations [36]. Despite computational limitations, it was quite clear that the system showed no tendency toward equipartition of energy among the degrees of freedom as time increased. This unexpected result spurred a good portion of the research that has led to our present knowledge of nonlinear wave theory.

An advantage of digital computations with respect to analytic methods is that there is generally no need to distinguish between linear and nonlinear problems; that is, numerical methods already developed for linear problems apply equally well to nonlinear problems. Three types of numerical methods are commonly used: (i) Picard's iteration method, already presented in the fundamental theorem of Section II.1; (ii) polynomial approximation methods, based on the fact that on a finite interval, an arbitrarily close fitting to a continuous function may be achieved by an appropriate polynomial (Weierstrass approximation theorem); (iii) methods of the Runge-Kutta type, that are based on a Taylor series expansion of the unknown function. Details may be found in several books; see, e.g., [20], [37] and [38].

III. PARTIAL DIFFERENTIAL EQUATIONS

III.1 *Introduction*

A partial differential equation is called *linear* if it is linear in the unknown function and its derivatives with coefficients that depend only on the independent variables; otherwise, it is called *nonlinear*. A nonlinear equation is called *quasilinear* if it is linear in the highest-order derivative of the unknown function.

Consider, for example, the second-order equation

$$f(x_i, u, u_{x_i}, u_{x_i x_j}) = 0, \quad (i, j = 1,2,\ldots,n) \tag{46}$$

where u is an unknown function of the n independent variables x_i, and the subscripts indicate partial differentiation, i.e.

$$u_{x_i} = \frac{\partial u}{\partial x_i}, \qquad u_{x_i x_j} = \frac{\partial^2 u}{\partial x_i \partial x_j} .$$

Equation 46 can be classified in an invariant way even if it is nonlinear, by rules similar to those adopted for linear equations. However, this classification is often not very useful because it is based on the signs of the eigenvalues of a matrix $\{a_{ij}\}$ with elements

$$a_{ij} = \frac{\partial f}{\partial u_{x_i x_j}} ;$$

since the a_{ij} now contain u and its derivatives, the type of equation depends not only upon the independent variables x_i, but also upon *u*. Despite this fact, it is often possible to reduce the equation to a useful canonical form if it depends on two variables only, as is the case for one-dimensional wave equations (the two variables being a spatial coordinate and the time).

The bibliography on partial differential equations is enormous; see, for example, the books [39] and [40]. Non-linear equations are addressed in a number of monographs and symposia proceedings; see, e.g., [16], [41] and [42]. Here our main interest is in partial differential equations which arise in studies of nonlinear wave propogation. Among the many books on this subject, we point out the one by Jeffrey and Taniuti [43] for its discussion of the method of characteristics and of conservation laws as applied to magnetohydrodynamics; that on nonlinear waves edited by Leibovich and Seebass [44]; the excellent (and, unfortunately, out-of-print) monograph by Scott [45] on nonlinear waves in electronics; the volume by Karpman [46] on nonlinear waves in dispersive media; and the encyclopaedic work by Whitham [47] that has become the standard reference on this subject.

One of the most fascinating aspects of nonlinear wave propagation is the appearance of *solitons*. The widespread interest that this new concept has aroused in such varied fields as elementary particle theory, plasmas, lasers, hydrodynamics, transmission lines, biophysics and so on is evident in the following chapters and in the references therein. The beginner will find the review article by Scott et. al. [48] and several chapters in [49] especially illuminating.

In the following we briefly consider the concept of soliton, then list some important partial differential equations which admit solitons as solutions.

III.2 *Solitons*

Consider the one-dimensional linear wave equation

$$u_{xx} - \frac{1}{c^2} u_{tt} = 0 \ , \tag{47}$$

which is satisfied by arbitrary functions of argument $\xi = x - ct$, representing traveling waves which propagate with velocity c in the $\pm x$ direction (depending on whether c is positive or negative). A *solitary wave* is a traveling wave which is localized in the variable ξ, i.e. the transition of u from a constant value at $\xi \to -\infty$ to a (possibly different) constant value at $\xi \to +\infty$ essentially takes place along a finite (and sufficiently short) segment of the ξ-axis.

If two solitary waves collide with each other, and emerge from the collision with unchanged shapes and velocities (having suffered, at the most, a phase shift), they are called *solitons*. For the linear equation 47, all solitary waves are solitons; however, this is the exception rather than the rule in the case of nonlinear wave equations, because the nonlinear interaction during collision is expected to modify both shape and velocity of each wave. In general, three physical phenomena may be present: nonlinearity, dispersion and dissipation. We limit our consideration to lossless cases, in which dissipation is nonexistent or negligible, i.e. to conservative systems; these possess a Lagrangian density function from which the equations of motion (such as eq. 47) can easily be derived (see, e.g., [50]). Since the nonlinearity sharpens the wave while the dispersion smoothes it, it is conceivable that these two phenomena may compensate each other, thus allowing for solitary wave existance. On the other hand, if the medium is either linear and dispersive, or nonlinear and dispersionless, solitary waves (and therefore solitons) cannot exist.

III.3 *Some Nonlinear Wave Equations with Soliton Solutions.*

(a) *Korteweg-de Vries (KdV) equation.*

The KdV equation

$$u_t + \alpha u u_x + u_{xxx} = 0 \,, \quad \alpha \text{ constant,} \tag{48}$$

arises in the mathematical description of a large number of physical phenomena; a comprehensive survey up to 1976 was given by Miura [51]. Equation (48) can be derived from the Lagrangian density function

$$L = \frac{1}{2} U_x U_t + \frac{\alpha}{6} U_x^3 + U_x V_x + \frac{1}{2} V^2 \tag{49}$$

where $U_x = u$ and $U_{xx} = V$. An explicit solution of (48) that represents two interacting solitons is [48]:

$$u(x,t) = \frac{216+288\cosh(2x-8t)+72\cosh(4x-64t)}{\alpha[3\cosh(x-28t)+\cosh(3x-36t)]^2} \;; \tag{50}$$

notice that here u is not a function of $\xi = x-ct$. If, however, t is large, then (5) is approximated by the superposition of two solitons:

$$u(x-c_j t) = \frac{3}{\alpha} c_j \operatorname{sech}^2\left\{\frac{1}{2}\sqrt{c_j}\,(x-c_j t) + A_j\right\}, \qquad (j=1,2), \tag{51}$$

where A_j are constants and $c_1=4$, $c_2=16$.

A procedure for constructing a solution which represents the interaction of N solitons is discussed in detail by Whitham ([47], section 17.2). Firstly we change dependent variable by letting

$$U = \frac{12}{\alpha} \frac{\partial}{\partial x} \ln F \tag{52}$$

in (48-49), then we prove that an exact solution exists in the form

$$F = 1 + \sum_{\ell} f_\ell + \sum_{\ell \neq m} C_{\ell m} f_\ell f_m + \sum_{\ell \neq m \neq n} C_{\ell m n} f_\ell f_m f_n + \ldots.$$

$$\ldots. + C_{12\ldots N} f_1 f_2 \ldots. f_N \,, \tag{53}$$

where the f_j products contain no repeated subscripts, the C's are constants and

$$f_j = \exp\,[-\beta_j(x-x_{oj}) + \beta_j^3 t] \ , \tag{54}$$

with β_j and x_{oj} constants. Thus, the trick consists in transforming the original nonlinear equation 48 into another equation for which soliton solutions appear as simple exponentials.

(b) *Sine-Gordon equation*

The equation

$$u_{tt} - u_{xx} + \sin u = 0 \tag{55}$$

arises in studies of elementary particles, ferromagnetic materials, crystal dislocations, Josephson junctions, and lipid membranes [48]. It can be derived from the Lagrangian density function

$$L = \frac{1}{2}(u_x^2 - u_t^2) - \cos u \quad . \tag{56}$$

The quantity u often has the physical meaning of an angle (e.g., in relation to a rotating vector). Thus, u varies by 2π when x goes from $-\infty$ to $+\infty$ in the solutions

$$u_{\pm}(x-ct) = 4\tan^{-1}\left\{\exp\left(\pm\frac{x-ct}{\sqrt{1-c^2}}\right)\right\} \tag{57}$$

of eq. 55; u_+ and u_- are called soliton and anti-soliton, respectively; their interactions are described in [48].

(c) *Born-Infeld equation*

The equation

$$(1-u_t^2)\,u_{xx} + 2\,u_x u_t u_{xt} - (1+u_x^2)\,u_{tt} = 0, \tag{58}$$

first proposed by Born in relation to a nonlinear modification of Maxwell's equations, can be derived from the Lagrangian density function

$$L = (1+u_x^2 - u_t^2)^{\frac{1}{2}} \ . \tag{59}$$

The surprising fact is that (58) admits as exact solutions arbitrary functions of argument $x \pm t$ (the velocity is here normalized to unity), and that these solutions conserve their shapes after collision, just as for the linear case of eq. 47.

(d) *Other equations*

Soliton solutions are possible for other nonlinear equations, such as the Boussinesq equation, the Toda lattice equations, the nonlinear Schroedinger equation, as well as generalized versions of the KdV equation, and the Hirota equation which contains some of the above equations as particular cases. For details see, e.g., [47] and [48].

For many of these nonlinear wave equations, a general method exists, called the "inverse scattering transform method", which allows us to find the exact solution to an initial-value problem as the sum of a discrete set of solitons plus a radiation field. This method is a nonlinear generalization of the Fourier transform method, and its description lies outside the scope of this Introduction; see, e.g., [48] and several contributions in [49], as well as the chapter by McLaughlin. Perturbation analyses of nonlinear physical problems have been performed by several authors (see [52]).

IV. OTHER NONLINEAR METHODS

For nonlinear integral equations and their applications, see the books by Saaty [16] and by Krasnosel´skii [53], the symposium proceedings edited by Anselone [54], and references therein. The work of Saaty also discusses nonlinear functional, difference, delay-differential, integro-differential and stochastic differential equations. For functional differential equations, see Hale [55].

The difficult topic of bifurcation theory has not been addressed herein; an excellent reference is the book by Keller and Antman [56].

Another topic not addressed here is catastrophy theory, which has become of ever increasing importance in recent years. Unfortunately, sweeping generalizations and applications have been carried out without a rigorous basis, by some of its more enthusiastic proponents, and these have met with justifiable criticism. The interested reader should consult the books by Lu [57] and by Poston and Stewart [58], as well as the article [59].

Finally, one last reference: the special issue [60] provides insight into up-to-date applications of computational techniques to some nonlinear problems.

REFERENCES

1. Tricomi, F.G., "Equazioni differenziali," Einaudi, Turin (1953).
2. Carathéodory, C., "Vorlesungen über Reelle Funktionen," zweite aufl., Berlin, 665-674 (1927).
3. Sansone, G., "Equazioni differenziali nel campo reale," Zanichelli, Bologna, Vol. II, Ch. 8, 140-150 (1949).
4. For a proof of this theorem, see, e.g., [3], Vol. II, Ch. 12, and references therein.
5. Fermi, E., *Rend. Acc. Naz. Lincei (6) 6,* 602-607 (1927); Thomas, L.H., *Proc. Cambridge Phil. Soc. 23,* 542-548 (1927).
6. Emden, R., "Gaskugeln," Leipzig (1907); Fowler, R.H., *Quart. J. Math. (Cambridge Series) 45,* 289-350 (1914); ibid. *(Oxford Series) 2,* 259-288 (1931). See also: Eddington, A.S., "The Internal Constitution of the Stars," Cambridge, England, Ch. 4 (1926).
7. Milne, W.E., "Damped Vibrations," University of Oregon (1923).
8. Liénard, A., "Étude des oscillations entretenues," *Revue Général de l'Eléctricité 23,* 901-912, 946-954 (1928).
9. van der Pol, B., "The Nonlinear Theory of Electric Oscillations," *Proc. IRE 22,* 1051-1086 (1934); see also, *Phil. Mag. (7) 2,* 978-992 (1926).
10. Graffi, D., *Mem. Accad. Sci. Bologna (9) 7,* 121-129 (1940); ibid. *9,* 83-91 (1942).
11. Levinson, N. and Smith, O.K., "A General Equation for Relaxation Oscillations," *Duke Math. J. 9,* 382-403 (1942). See also Minorsky, N., "Introduction to Nonlinear Mechanics," Edwards, Ann Arbor, Michigan (1947).
12. Duffing, G., "Erzwungene Schwingungen bei veränderlicher Eigenfrequenz und ihre technische Bedentung," Vieweg, Braunschweig (1918).
13. Stoker, J.J., "Nonlinear Vibrations," Interscience, New York (1950).

14. Hayashi, C., "Nonlinear Oscillations in Physical Systems," McGraw-Hill, New York (1964).
15. McLachlan, N.W., "Ordinary Nonlinear Differential Equations in Engineering and Physical Sciences," Second Edition, Oxford University Press (1955).
16. Saaty, T.L., "Modern Nonlinear Equations," McGraw-Hill, New York (1967).
17. Minorsky, N., "Nonlinear Oscillations," Van Nostrand, Princeton, NJ (1962).
18. Cesari, L., "Asymptotic Behavior and Stability Problems in Ordinary Differential Equations," Springer, New York (1963).
19. Bellman, R., "Methods of Nonlinear Analysis," Academic Press, New York, Vol. 1 (1970); Vol. 2 (1973).
20. Aggarwal, J.K., "Notes on Nonlinear Systems," Van Nostrand, New York (1972).
21. Bendixson, L. "Sur les courbes définies par des équations différentielles," *Acta. Math 24,* 1-88 (1901). See also Lonn, E.R., *Math. Z. 44,* 507-530 (1938); Frommer, M., *Math. Ann. 99,* 222-272 (1928); ibid. *109,* 345-424 (1934).
22. Bellman, R., "Stability Theory of Differential Equations," Dover, New York (1969).
23. Aggarwal, J.K., "Singular Points of Planar Ordinary Differential Systems," *J. Diff. Eqs. 3,* 203-213 (1967); "On Non-elementary Singular Points," *J. Franklin Inst. 281,* 41-50 (1966).
24. Lefschetz, S., "Differential Equations: Geometric Theory," Interscience, New York (1963).
25. Poincaré, H., "Mémoire sur les curbes définies par une équation différentielle," *J. de Mathématique 7,* 375-422 (1881); ibid., *8,* 251-296 (1882).
26. Coddington, E.A. and Levinson, N., "Theory of Ordinary Differential Equations," McGraw-Hill, New York (1955).
27. Zubov, V.I., "Mathematical Methods for the Study of Automatic Control Systems," Macmillan, New York (1963).
28. LaSalle, J. and Lefschetz, S., "Stability by Liapunov's Direct Method with Applications," Academic Press, New York (1961).
29. Nayfeh, A.H. and Mook, D.T., "Nonlinear Oscillations," Wiley, New York (1979).
30. Poincaré, H., "Leçons de Mecanique Céleste," in three volumes, Gauthier-Villars, Paris (1905, 1907, 1910).
31. Hale, J.K., "Oscillations in Nonlinear Systems," McGraw-Hill, New York (1963).
32. Bogoliubov, N.N. and Mitropolsky, Y.A., "Asymptotic Methods in the Theory of Nonlinear Oscillations," Gordon and Breach, New York (1961).

33. Pontryagin, L.S., "Asymptotic Behavior of the Solutions of Systems of Differential Equations with a Small Parameter in the Higher Derivatives," *Trans. AMS (2) 18,* 295-319 (1961). Miščenko, E.F., "Asymptotic Calculation of Periodic Solutions of Systems of Differential Equations Containing Small Parameters in the Derivatives," *Trans. AMS (2) 18,* 199-230 (1961).
34. Halanay, A., "Differential Equations, Stability, Oscillations, Time Lags," Academic Press, New York (1966).
35. Nayfeh, A.H., "Perturbation Methods," Wiley, New York (1973).
36. Fermi, E., Pasta, J. and Ulam S., "Studies of Nonlinear Problems I," Report LA1940, Los Alamos Scientific Laboratory, New Mexico (1955). Reprinted in: Newell, A.C., ed., "Nonlinear Wave Motion," Lectures in Applied Mathematics, Vol. 15, American Mathematical Society, Providence, RI, 143-156 (1974).
37. Henrici, P., "Discrete Variable Methods in Ordinary Differential Equations," Wiley, New York (1962).
38. Ralston, A., "A First Course in Numerical Analysis," McGraw-Hill, New York (1965).
39. Courant, R., and Hilbert, D., "Methods of Mathematical Physics", vol. II, Interscience, New York (1961).
40. Garabedian, P.R., "Partial Differential Equations", Wiley, New York (1964).
41. Finn, R., editor, "Applications of Nonlinear Partial Differential Equations in Mathematical Physics", American Mathematical Society, Providence, R.I. (1965).
42. Ames, W.F., editor, "Nonlinear Partial Differential Equations", Academic Press, New York (1967).
43. Jeffrey, A., and Taniuti, T., "Non-linear Wave Propagation", Academic Press, New York (1964).
44. Leibovich, S., and Seebass, A.R., "Nonlinear Waves", Cornell University Press, Ithaca, New York (1974).
45. Scott, A., "Active and Nonlinear Wave Propagation in Electronics", Wiley-Interscience, New York (1970).
46. Karpman, V.I., "Non-linear Waves in Dispersive Media", Pergamon Press, New York (1975).
47. Whitham, G.B., "Linear and Nonlinear Waves", Wiley-Interscience, New York (1974).
48. Scott, A.C., Chu, F.Y.F., and McLaughlin, D.W., "The soliton: a new concept in applied science", *Proc. IEEE 61,* 1443-1490 (1973).
49. Scott, A.C., and Lonngren, K.E., editors, "Solitons in Action", Academic Press, New York (1978).
50. Goldstein, H., "Classical Mechanics", Addison-Wesley, Reading, Mass. (1950).

51. Miura, R.M., "The Korteweg-deVries equation: a survey of results", *SIAM Review 18,* 412-459 (1976).
52. McLaughlin, D.W., and A.C. Scott, "A multisoliton perturbation theory", in [49], 201-256.
53. Krasnosel´skii, M.A., "Topological Methods in the Theory of Nonlinear Integral Equations", Macmillan, New York (1964).
54. Anselone, P.M., editor, "Nonlinear Integral Equations", University of Wisconsin Press, Madison, Wisconsin (1964).
55. Hale, J., "Theory of Functional Differential Equations", Springer, New York (1977).
56. Keller, J.B., and Antman, S., "Bifurcation Theory and Nonlinear Eigenvalue Problems", Benjamin, New York (1969).
57. Lu, Y.-C., "Singularity Theory and an Introduction to Catastrophe Theory", Springer, New York (1976).
58. Poston, T., and Stewart, I.N., "Taylor Expansions and Catastrophes", Pitman, London (1976).
59. Golubitsky, M., "An introduction to catastrophe theory and its applications", *SIAM Review 20,* 352-387 (1978).
60. Computational methods in nonlinear problems in mechanics and engineering science", special issue, *Intern. J. Eng. Science 18,* no. 2 (1980).

THE BIRTH OF A PARADIGM

Alwyn C. Scott

Department of Electrical and Computer Engineering
University of Wisconsin
Madison, Wisconsin

There is no better, there is no more open door by which you can enter into the study of natural philosophy than by considering the physical phenomena of a candle.

Michael Faraday

In the foundations of any consistent field theory, the particle concept must not appear in addition to the field concept. The whole theory must be based solely on partial differential equations and their singularity-free solutions."

Albert Einstein

A funny thing happened to solitary wave research over the past decade: it became respectable. No longer is it possible for all soliton buffs to meet in a small room; nor can one now read the important papers in a few weeks. The early, innocent days are gone, and (as Fig. 1 shows) soliton research output has entered a period of exponential growth with a doubling time of about 18 months. The solitary wave concept has emerged as a widely accepted paradigm for exploring and modeling the dynamics of the real world.

More than simple curiosity leads us to ask about the gestation process preceding this explosive event. We do not wish it to be said of us: "If we seem to see farther than others, it is because we are standing on the faces of giants."

ISBN 0-12-709660-4

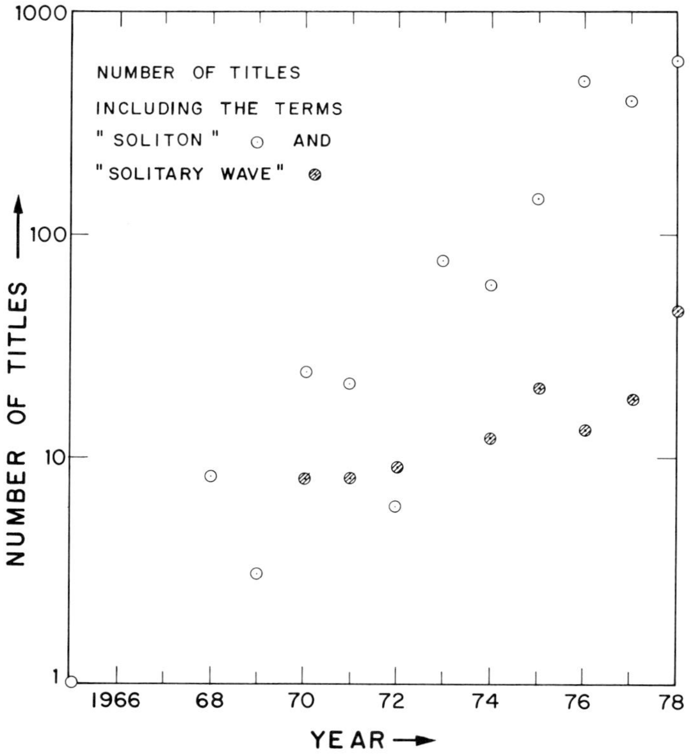

FIGURE 1. Number of scientific journal articles including the term soliton and solitary wave for each year (data from the "Permuterm" Subject Index).

I. FROM THE BEGINNING TO THE GREAT WAR

Hydrodynamic solitary waves (like Lorentz contraction, universal gravitation and the survival of the fittest) must have been around since the dawn of time; but the *noticing* of them began only with that accidental observation so fluently recorded by John Scott Russell (see Fig. 2) in his "Report on Waves" to the British Association in 1844 [1].

> I was observing the motion of a boat which was rapidly drawn along a narrow channel by a pair of horses, when the boat suddenly stopped - not so the mass of water in the channel which it has put in motion; it accumulated round the prow of the vessel in a state of violent agitation, then suddenly leaving it behind, rolled forward with great velocity, assuming the form of a large solitary elevation, a rounded, smooth and well-defined heap of water, which continued its course along the channel apparently without change of form or diminution of speed. I followed it on horseback, and overtook it still rolling on at a rate of some eight or nine miles an hour, preserving its original figure some thirty feet long and a foot to a foot and a half in height. Its height gradually diminished, and after a chase of one or two miles I lost it in the windings of the channel. Such, in the month of August 1834, was my first chance interview with that singular and beautiful phenomenon.....

The "narrow channel" was (and still is, thanks to the efforts of local canal enthusiasts [2]) the Union Canal linking Edinburg with Glasgow, and the "rapidly drawn boat" was not there by accident. It was part of a careful series of experiments through which Russell was studying the force-velocity characteristics of variously shaped boat hulls in order to determine design parameters for conversion from horse- to steam-power. Extensive wave-tank measurements in 1834 and 1835 established the following important properties of hydrodynamic solitary waves [1]:

1) these waves are independent and localized dynamic entities that move with fixed shape and velocity;

2) a wave of height h travels in water of depth d with velocity

FIGURE 2. Photograph of John Scott Russell (ca. 1860).

$$u = \sqrt{g(d+h)} \quad , \tag{1}$$

where g is the acceleration of gravity;

3) a sufficiently large initial mass of water will produce two or more solitary waves as indicated in Fig. 3;

4) solitary waves of depression are not observed;

5) solitary waves cross each other "without change of any kind."

Russell's understanding of hydrodynamic solitary waves provided a scientific basis for his "wave line hull" (or "solid of least resistance") which profoundly altered the contemporary tendency to build a ship's bow "with the shape of a duck's breast" and led him to become one of the great naval architects of the nineteenth century [3]. Although subsequent scientific discussions of hydrodynamic solitary waves involved Airy, Boussinesq, Rayleigh and Stokes (to name a few of the most prominent [4]), attention was focused upon the correctness rather than the

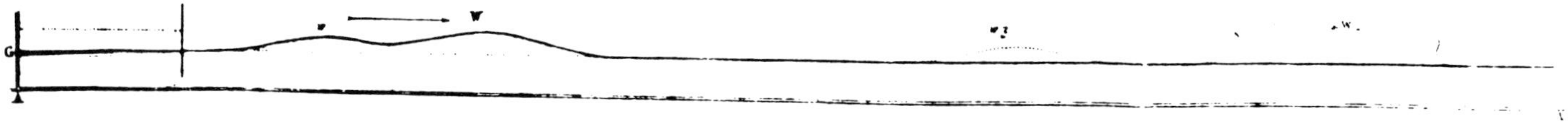

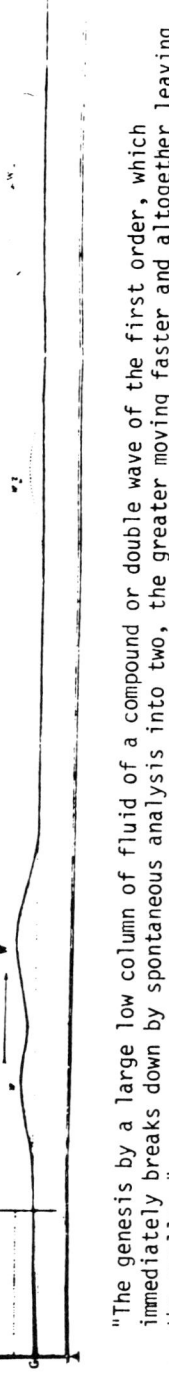

"The genesis by a large low column of fluid of a compound or double wave of the first order, which immediately breaks down by spontaneous analysis into two, the greater moving faster and altogether leaving the smaller."

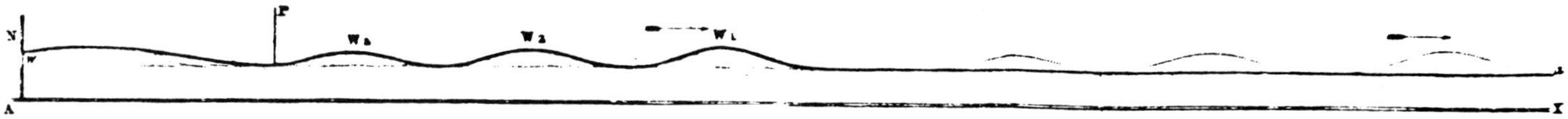

"Instead of genesis of a compound wave, the added mass sends off a series of single waves, the first being the greatest: these however do not remain together, but speedily separate as shown in the dotted lines, and become the further apart the longer they travel."

"The genesis of a compound wave by impulsion of the plate with a variable force and velocity, which variations have produced corresponding variations on the wave form. After propagation the wave breaks down by spontaneous analysis, so that after the lapse of a considerable period the compound wave is resolved into single separate waves."

FIGURE 3. Multiple solitary waves generated in the tank experiments of John Scott Russell. [J. Scott Russell, B.A.R. 14, 331 (1844)].

significance of Russell's observations. The collective judgement of nineteenth century science is quantitatively recorded in Lamb's opus on hydrodynamics which allots only 3 of 730 pages to the solitary wave [5]. Evidence that Russell himself retained a much broader appreciation of the ultimate importance of his discovery is provided by an almost unnoticed posthumous work in which (among many provocative ideas) he correctly estimated the height of the earth's atmosphere from Eq. (1) and the fact that the sound of cannon fire travels faster than the command to fire it [6].

In 1895 Korteweg and deVries presented a study of hydrodynamic wave motion that, because it reduced the problem to its essential elements, was to play a key role in subsequent theoretical developments [7]. Their results can be summarized by the equation

$$\phi_t + u_o\phi_x + A\phi\phi_x + B\phi_{xxx} = 0 \tag{2}$$

where ϕ is vertical displacement, u_o is the velocity given in (1), $A \equiv 3u_o/2d$ and $B \equiv u_o[d^2/b - T/2\rho g]$, T and ρ are the surface tension and the density of water, and subscripts denote partial derivatives. They found traveling wave solutions

$$\phi(x,t) = \tilde{\phi}(x - ut) \tag{3}$$

to be a family of periodic elliptic functions, called *cnoidal waves*, that could be interpreted as Russell's solitary wave in the infinite wavelength limit. From a physical perspective this solitary wave maintains a dynamic balance between the *ying* of nonlinearity (expressed by the term $A\phi\phi_x$) and the *yang* of dispersion (expressed by the term $B\phi_{xxx}$).

Solitary waves can also be found that maintain a dynamic balance between nonlinearity and diffusion; a simple example is the ordinary candle. The heat from the flame diffuses into the wax, vaporizing it at the rate required to provide fuel for the flame. If the energy stored in the wax is E(joules/cm) and the power input required by the flame is P(joules/sec), the flame will propagate at the velocity u for which

$$P = uE . \tag{4}$$

The flame digests energy at the same rate it is eaten. The *ying* of nonlinear energy release is balanced by the *yang* of energetic diffusion. The reader is cautioned not to dismiss this as a trivial example; Michael Faraday didn't [8].

Perhaps the most important class of systems to display nonlinear diffusion is found in the nerve fibers of animal organisms, but during the first half of the nineteenth century physiologists assumed nervous activity to be propagated with the velocity of light. Contrary to the arguments of Luigi Galvani, there was nothing special about "animal electricity." Young Herman Helmholtz arrived on the scene with a calling to become a physicist but, through financial necessity, with training as a physician; and he chose the direct measurement of nerve signal propagation speed as one of his first experimental adventures. Using an apparatus (of which that shown in Fig. 4 is a refined version) he reported on January 15, 1850 a velocity of some 32 meters per second along the frog's sciatic nerve [9]. He missed, however, the essential explanation of nonlinear diffusion by supposing that the relatively small velocity implied that signals were carried by the motion of material particles; a notion that was to confuse neurophysiologists until well into the twentieth century. In retrospect it seems difficult to understand the failure of nineteenth century mathematical science to appreciate the descriptive power of nonlinear diffusion. It is certainly not a question of skill when one considers the names of Boussinesq and Rayleigh and Helmholtz. Indeed Helmholtz's analytical study of hydrodynamic vortex motion [10] was an early and prophetic contribution to ultimate developments in solitary wave theory.

It must have been that linear diffusion is so "unwavelike." Any pulse-like solution of $\phi_t = \phi_{xx}$ will fall where it has down-

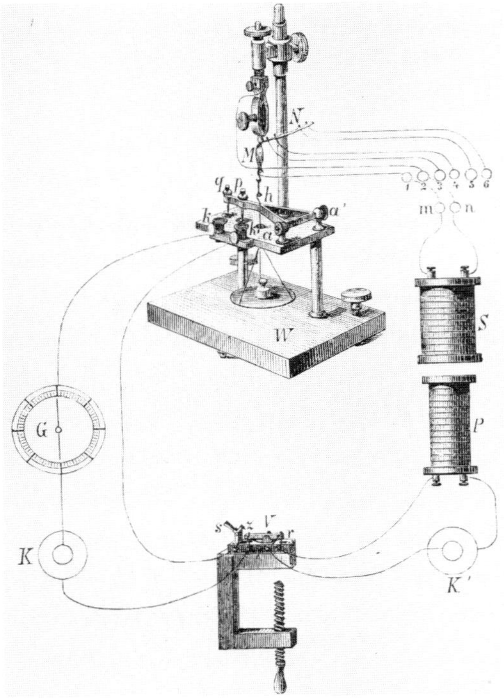

FIGURE 4. Apparatus used by Helmholtz to measure the speed of propagation on a frog's nerve-muscle (NM) preparation. Closure of switch (V) induces a pulse on the nerve and starts a time measurement on a ballistic galvanometer (G). When the muscle twitches, a mercury contact (h') is broken and the time measurement stops From L. Hermann, Handbuch der Physiologie (1879) .

ward curvature and rise where it curves upward, and therefore spread itself out. How could this qualitative effect be altered by the presence of nonlinearity? Consider augmenting the linear diffusion equation to

$$\phi_{xx} - \phi_t = F(\phi) \tag{5}$$

where $F(\phi)$ is chosen to be the simple cubic $\phi(\phi - a)(\phi - 1)$. A traveling wave solution, as is indicated in (3), was found by Huxley to be the solitary wave [11]

$$\tilde{\phi} = \left[1 + exp\left(\frac{x - ut}{\sqrt{2}}\right)\right]^{-1}$$

which propagates at the fixed velocity

$$u = (1 - 2a)/\sqrt{2} \ .$$

Simple and important as this solution is, it was not published until 1965!

Nonetheless if we had a time machine, it would be interesting to return to Paris (say) at the turn of the century for a few years of study. The literature of this golden age of applied mathematics is difficult to come by nowadays, but some important aspects have been reviewed recently by George Lamb [12]. Of particular interest is the "Bäcklund transform" which can be viewed as a technique for constructing solutions to a partial differential equation (pde) via the Pfaffian form: $d\phi = Pdx + Qdt$. Clearly one must require

$$\phi_x = P; \ \phi_t = Q$$

and the condition of integrability

$$P_t = Q_x \ .$$

If P and Q can be found as functions of ϕ and a known solution, ϕ_0, then a new known solution (ϕ_1) can be generated by integrating the first order pair

$$\phi_{1,x} = P(\phi_1, \phi_0)$$

$$\phi_{1,t} = Q(\phi_1, \phi_0) \ .$$

The generation process is therefore

$$\text{"known" solution } (\phi_0) \xrightarrow{\text{Bäcklund T.}} \text{"new" solution } (\phi_1)$$

after which, of course, the "new" solution is "known" and can be used to generate another new solution in an hierarchical order. It's easy to find a Bäcklund transform (BT) for *any* linear pde for which each "turn of the crank" introduces a new eigenfunction into the total solution. Only certain nonlinear pde's are found to have BT's, but Bäcklund showed that these include

$$\phi_{\xi\tau} = sin\phi \tag{6}$$

which arose in connection with research on surfaces of constant negative curvature. The BT for (6) is

$$\begin{aligned} \phi_{1,\xi} &= 2a \, sin\left(\frac{\phi_1 + \phi_0}{2}\right) + \phi_{0,\xi} \\ \phi_{1,\tau} &= \frac{2}{a} \, sin\left(\frac{\phi_1 - \phi_0}{2}\right) - \phi_{0,\tau} \end{aligned} \tag{7}$$

where a is an arbitrary constant. (To check this the reader needs merely to demonstrate that $\phi_{0,\xi\tau} = sin\phi_0$ implies that $\phi_{1,\xi\tau} = sin\phi_1$.) A known solution of (6) is clearly the "vacuum" $\phi_0 = 0$. Integration of (7) then gives

$$\phi_1 = 4 \, tan^{-1}[exp(a\xi + \tau/a)] \tag{8}$$

a function that was familiar enough in the nineteenth century to carry a special name: the gudermannian. Generation of (8) from the vacuum can be represented by the diagram

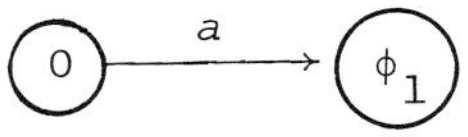

and subsequent development of a hierarchy of solutions by

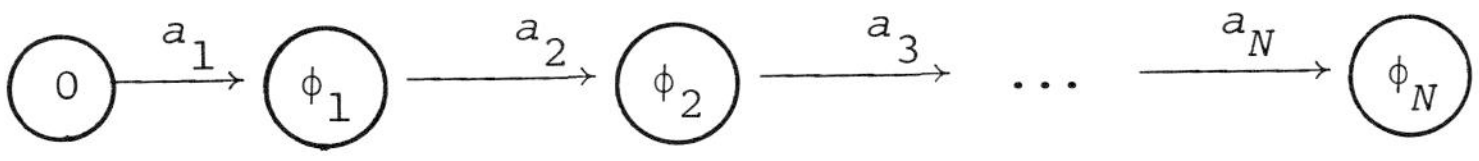

Each hierarchical level in such a diagram includes an additional component (or nonlinear eigenfunction) with the basic form indicated in (8). Direct integration to obtain higher level solutions becomes increasingly tedious, but Bianchi showed how these could be found by simple algebraic techniques obviating the need for integration beyond the first level [12]. The use of such analytical power to find solitary wave solutions for the nonlinear diffusion equation (5) would have been like cracking an almond with John Scott Russell's cannon.

Another thread of our story begins in 1912 when Gustav Mie began to publish his "theory of matter" [13]. In this brilliant series of papers, Mie suggested a nonlinear augmentation of the Maxwell equations from which the elementary particles (e.g., the electron) would arise in a natural way. To this end he defined a "world function" (Φ) as an energy functional depending upon electric field intensity ($\bar{E}$) magnetic flux density ($\bar{B}$) and the four components of electromagnetic potential ($\bar{A}$, $i\phi$). Requiring Φ to be a function of the parameters $\eta \equiv \sqrt{(\bar{E}^2 - \bar{B}^2)}$ and $\chi \equiv \sqrt{(\phi^2 - \bar{A}^2)}$ insured Lorentz invariance; and the specific choice

$$\Phi = -\frac{1}{2}\eta^2 + \frac{1}{6}a\chi^6 \tag{9}$$

led to a static, spherically symmetric electric potential (ϕ) satisfying $r\phi'' + 2\phi' + ar\phi^5 = 0$ with a solution

$$\phi \doteq \frac{[3r_o^2/a]^{\frac{1}{4}}}{\sqrt{r^2 + r_o^2}} \quad .$$

Setting $4\pi[3r_o^2/a]^{\frac{1}{4}} = e$ (the electronic charge) yields a spherically symmetric model for the electron with "radius" r_o and electric potential

$$\phi \rightarrow \frac{e}{4\pi r}$$

for $r >> r_o$. The Lorentz invariance that is built into (9) permits this solution to travel with any speed up to the limiting velocity of light exhibiting appropriate Lorentz contraction.

This brief sketch can hardly do justice to the depth and scope of Mie's ideas. Especially for discussions of the relationship of nonlinear electrodynamics with quantum theory and gravity the reader is urged to consult the original work.

II. BETWEEN THE WARS

Bäcklund transformations ceased to be of research interest after World War I. George Lamb believes that this was due, at least in part, to the untimely deaths of many young scientists active in the field. Be that as it may, all solitary wave research seemed to sleep until 1934 when Max Born began to reconsider Mie's nonlinear electromagnetics. Born was particularly concerned with establishing a "gauge invariant theory" (for which solutions would be independent of the electromagnetic potential) that would be compatible with the requirements of quantum theory. To this end he eliminated the χ dependence in Mie's functional ansatz [see (9)] and took instead a Lagrangian density of the form

$$L = E_o^2 \sqrt{1 + (\bar{H}^2 - \bar{E}^2)/E_o^2} - 1 \qquad (10)$$

where E_o is a nonlinear limit to the magnitude of field intensities, and $\bar{H}$, of course, is magnetic field intensity [14]. In the limit of low field amplitudes ($|\bar{E}|^2$ and $|\bar{H}|^2 \ll E_o^2$) this clearly reduces to the classical $L \doteq \frac{1}{2}(\bar{H}^2 - \bar{E}^2)$. Born and Infeld found a spherically symmetric model electron for which the electric field ($\bar{E}$) was everywhere finite although electric displacement ($\bar{D}$) exhibited a singularity at the origin. For plane wave solutions the Lagrangian density (10) is of the general class

$$L = L(\phi_t^2 - \phi_x^2) \qquad (11)$$

which implies a solitary wave solution of arbitrary shape but with a speed equal to the velocity of light.

An entirely unrelated event of the mid-thirties was the first mathematical study of the nonlinear diffusion equation (5) by Kolmogoroff, Petrovsky and Piscounoff in the Soviet Union [15]. Since this study was motivated by the problem of genetic diffusion, rather than nerve propagation, the nonlinear function $F(\phi)$ was chosen to be of the form $\phi(1 - \phi)$ which does not lead to a

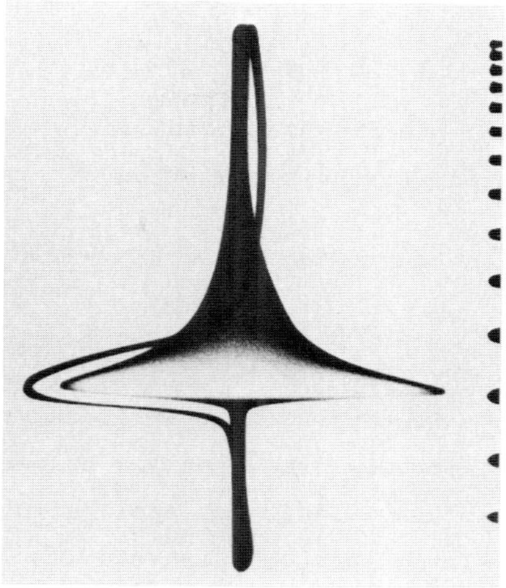

FIGURE 5. Direct measurement of the increase in membrane conductance (band) during the action potential (line) on the squid giant axon. Time marks are 1 millisecond K.S. Cole and H.J. Curtis, Nature 142, 209 (1938).

solitary wave solution. How much more rapidly might neurophysiology have advanced if these researchers had been aware of the nerve problem? It was in 1936 that discovery of the giant axon of the squid was announced by J.Z. Young and in 1938 that Cole and Curtis published the classic cathode ray recording of an impulse on the squid nerve which is shown in Fig. 5.

Also from the Soviet Union there emerged during this period an equally unrelated line of research motivated by a fundamental problem in solid state physics: the relation between dislocation dynamics and plastic deformation of crystalline material. In this study Frenkel and Kontorova showed that a basic equation to describe dislocation motion takes the form [16]

$$\phi_{xx} - \phi_{tt} = \sin\phi. \tag{12}$$

This can be viewed as a nonlinear augmentation of the linear wave equation ($\phi_{xx} - \phi_{tt} = 0$) in the same sense that (5) is a nonlinear augmentation of the linear diffusion equation ($\phi_{xx} - \phi_t = 0$). Furthermore (12) is identical to (6) under the independent variable transformation

$$\xi = \tfrac{1}{2}(x - t)$$

$$\tau = \tfrac{1}{2}(x + t)$$

The gudermannian solution (8) corresponds, under this transformation, to the propagation of a single dislocation described by

$$\phi = 4\, tan^{-1}\left[exp\left(\frac{x - ut}{\sqrt{1 - u^2}}\right)\right] \tag{13}$$

where the velocity $u = \sqrt{(1 - a^2)/(1 + a^2)}$.

To the extent that it existed at all between the wars, nonlinear wave research was fragmented beyond belief. Old knowledge was unavailable for application to new problems. The fundamental relationships between studies in nonlinear electrodynamics, solid state physics, hydrodynamics, neurodynamics, genetic diffusion and applied mathematics were entirely overlooked. The prophetic insights of John Scott Russell were completely forgotten.

III. SOLITON RESEARCH FROM 1945 TO 1974

If scientific research had suffered from the spilling of its young blood in the first world war, this mistake was not to be repeated during the second. Poets, perhaps, could still be driven into the guns but not engineers and scientists; they were drafted for more important tasks. U.S. science, in particular, was greatly strengthened by the war. Not only did we receive several dozen of the very best European scientists as permanent citizens, researchers and teachers; but the forced development of our technological power (in such diverse areas as electronics, microwaves, communications and control, and the manipulation of elementary particles) left us in a state of high morale. We could do anything.

One thing we did was to develop the digital computer, and Enrico Fermi suggested one of the very first scientific problems to which it was addressed: the dynamics of energy equipartition in a slightly nonlinear multimode mechanical system. This system consisted of 64 equal mass particles connected by slightly nonlinear springs. It was expected that if all the initial energy were put into a single mode, the slight nonlinearity would cause a gradual progress toward equipartition of this energy among all 64 modes in what could be considered a thermalized (or ergodic) state. The computer results obtained by Pasta and Ulam were surprising; no tendency toward thermalization was observed. If the energy was originally put into the lowest frequency mode, it returned almost entirely into that mode after a period of interaction with a few other low frequency modes [17]. John Pasta has commented that, shortly before his untimely death, Fermi felt this to be one of the most important problems he (Fermi) had studied [18]. It was certainly one of the most mysterious. Pursuit of this mystery led Zabusky and Kruskal to approximate the nonlinear spring-mass system by a rediscovered Korteweg-deVries (KdV) equation [see (2)] (which they found to apply also to wave motion in a collisionless plasma) and to observe numerically that KdV solitary waves pass through each other with no change in shape or speed [19]. It should be noted that Zabusky and Kruskal were not the first to observe nondestructive interactions of solitary waves. Apart from the above mentioned tank measurements by Scott Russell, Perring and Skyrme had published results of a numerical study of (12) in which two solitary waves of the form (13) underwent a collision. They observed perfect recovery of wave shapes and speeds after the collision and were led to the analytic description

$$\phi = 4 \tan^{-1}\left[\frac{u \sinh (x/\sqrt{1-u^2})}{\cosh (ut/\sqrt{1-u^2})}\right] . \tag{14}$$

This result would have been no surprise to Bäcklund or to Bianchi; it is merely the second (i.e., ϕ_2) in the hierarchy of solutions generated from the vacuum by the BT given above in (7). Nor would it have been a surprise to Seeger, Donth and Kochendörfer who had noted in 1953 the connection between this early work and the paper by Frenkel and Kontorova [16]. But Perring and Skyrme were interested in (12) as a one-dimensional model for an elementary particle, and, in this context, one supposes that the complete absence of scattering might have been a bit disappointing. Throughout the 1960's (12) arose in a wide variety of problems in applied science (including the propagation of ferromagnetic domain walls, the "self-induced transparency" effect in nonlinear optics, and propagation of magnetic flux quanta on Josephson transmission lines) and eventually became known as the "sine-Gordon" equation [22].

The important contribution by Zabusky and Kruskal in their 1965 publication [19] was to recognize the relation between nondestructive solitary wave collisions and the mystery of Fermi-Pasta-Ulma (FPU) recurrence effect. The solitary wave solutions of the KdV equation were viewed as independent and localized dynamic entities (called *solitons*) out of which more complex behavior could be constructed. By 1967 this insight had led Gardner, Greene, Kruskal and Miura to a truly brilliant scheme for developing a general solution to the KdV equation through a series of linear calculations [23]. This method was soon expressed in the following elegant and general form by Lax [24].

We are interested in a general nonlinear wave equation, $\phi_t = N(\phi)$, where $N(\cdot)$ denotes a nonlinear operator on some suitable space of functions. suppose we can find two linear operators, L and B, which depend upon $\phi(x,t)$ (a solution of the nonlinear pde) and which satisfy the operator equation

$$iL_t = BL - LB. \tag{15}$$

If L is viewed as a scattering operator with potential $\phi(x,t)$, its eigenvalues (λ) are found from study of

$$L\psi = \lambda\psi. \tag{16}$$

Now if the time dependence of the scattered wave is taken to be

$$i\psi_t = B\psi, \tag{17}$$

(15) implies that the eigenvalues in (16) are independent of time. Thus the computation

$$\phi(x,0) \longrightarrow \phi(x,t)$$

can be effected through the following three steps.

1) *Direct problem.* Calculate scattering parameters (such as the reflection and transmission coeffieincts of L) for ψ at $|x| = \infty$ and $t = 0$ from a knowledge of $\phi(x,0)$.

2) *Time evolution of the scattering data.* Use (17), together with the asymptotic form of B at $|x| = \infty$, to calculate the time evolution of the scattering data.

3) *Inverse problem.* From a knowledge of the scattering data of L as a function of time, construct $\phi(x,t)$.

Each bound state eigenfunction of (16) corresponds to a particular soliton component in the general solution. In the small amplitude (linear) limit there are no solitons present in the solution and the above described procedure degenerates into the usual Fourier transform method for linear pde's. Thus is has become known as the "inverse scattering transform method" (ISTM) wherein we see soliton components of the solution acting as generalized Fourier components.

Another extremely important development of the 1960's was the discovery by Morikazu Toda of *exact* two soliton interactions on a nonlinear-spring mass system in which the spring potential took the form [25]

$$\text{potential} = \frac{a}{b}\,[\exp(-br_n) - 1] + ar_n \;.$$

This model is of great flexibility because it can be varied between the harmonic limit ($a \to \infty$ and $b \to 0$ with ab finite) and the hard sphere limit ($a \to 0$ and $b \to \infty$ with ab finite).

Up to this point we have been concentrating our attention on the dynamics of solitary waves. In 1965, however, Whitham began a series of papers that investigated the dynamics of *periodic* traveling waves [26]. These take the general form

$$\phi(x,t) = \tilde{\phi}(kx + \omega t) \equiv \tilde{\phi}\,(\theta)$$

and include the cnoidal waves obtained by Korteweg and deVries for the equation that carries their names. Whitham supposed the solution to be locally periodic (though *not* sinusoidal) and assumed that the underlying pde could be derived from a Lagrangian density function, L. From this he obtained an averaged Lagrangian over a period of the wave by

$$<L> \equiv \frac{1}{2\pi}\int_0^{2\pi} L\, d\,\theta$$

as a function of the local values of k, ω, and an amplitude parameter, A. Euler variation of the averaged Lagrangian with respect to A and θ then gives two equations

$$\frac{\delta <L>}{\delta A} = 0 \Longrightarrow \frac{\partial <L>}{\partial A} = 0 \tag{18}$$

$$\frac{\delta <L>}{\delta \theta} = 0 \Longrightarrow \frac{\partial}{\partial x}\frac{\partial <L>}{\partial k} + \frac{\partial}{\partial t}\frac{\partial <L>}{\partial \omega} = 0\,. \tag{19}$$

These together with the pulse conservation equation

$$\frac{\partial \omega}{\partial x} = \frac{\partial k}{\partial t} \tag{20}$$

suffice to fix the slow x and t variation of k, ω and A. Eq. (18) can be viewed in this context as a *nonlinear* dispersion equation

$$D(k,\ \omega,\ A) = 0 \tag{18'}$$

which depends upon the local value of the wave amplitude, A.

The above listed items are only the most important features in a growing panorama of nonlinear wave activities which became increasingly less parochial throughout the nineteen-sixties. Solid state physicists began to see some relationship between their solitary waves (domain walls, self-shaping light pulses, magnetic flux quanta) and those from classical hydrodynamics, and applied mathematicians began to suspect that the ISTM might apply to a broader set of nonlinear wave equations. It was in the context of this growing excitement and self-awareness that Alan Newell organized a research conference for three and a half weeks during the summer of seventy-two in which the participants "ranged over a wide spectrum of ages (from graduate students to senior scientists), background interests (biology, electrical engineering, geology, geophysics, mathematics, physics) and countries of origin (United States, Canada, Great Britain, Australia)" [27]. It is difficult to overemphasize the importance of this conference to solitary wave research in the english-speaking world. Countless cross-disciplinary bonds of collaborative interaction and friendship were formed, and a sense of the existence of nonlinear wave study as a broad and vigorous activity was established. Solitary wave research came out of the closet.

One of the most significant inputs to Newell's conference was from the Soviet Union. In a paper first published in 1971 [28], Zakharov and Shabat showed that the "Lax-operators" L and B [see (15) - (17)] could be found for the "nonlinear Schrödinger" (NLS) equation

$$i\phi_t - \phi_{xx} = k\, |\phi|^2 \phi \tag{21}$$

and proceeded in an elegant and systematic way to develop an ISTM and derive an infinite set of conservation laws corresponding to those previously found for the KdV equation by Robert Miura [29]. It should perhaps be emphasized that (21) is not a precious object invented by Zakharov and Shabat to display their analytical wizardry; it had arisen in practice to describe envelope waves in hydrodynamics, nonlinear optics, nonlinear acoustics and plasma waves [30]. The Zakharov and Shabat paper was widely circulated and read and discussed (formally and in the evenings over Newell's endless supplies of beer), and everyone left the conference realizing that four of the most fundamental nonlinear wave systems [KdV(2), the "sine-Gordon" equation (SGE)(12), NLS (21), and the nonlinear spring mass system described by Toda]displayed solitary waves with the special behavior that had led Zabusky and Kruskal to coin the term *soliton*.

Within the next two years the basic ingredients (i.e., Lax operators) for the ISTM had been constructed for the SGE [31] and the Toda lattice [32]. Following the discovery, through a very general approach based on differential geometry, of a Bäcklund transform for the KdV equation [33], Newell demonstrated the equivalence of the BT and ISTM for a rather wide class of nonlinear wave systems [34].

IV. PARTICLE PHYSICS TO THE PRESENT

Einstein's conviction that a consistent theory for particle physics must be based on singularity free solutions of partial differential equations [35] was shared by some of the most distinguished of his colleagues. In addition to the above mentioned interest of Born and Infeld in this question, both Heisenberg [36] and deBroglie [37] have described nonlinear field theories which, in their simplest representations, can be viewed as the augmentation of classical e.m. field equations by a nonlinear term of the form $|\phi|^2\phi$ as in (21). The ideas of deBroglie bear an interesting relationship to the ISTM. In his "theory of the double solution" the real particle is a singularity free solution of a nonlinear wave equation $\phi = \Phi\, exp(i\ \theta')$, but associated with it is a corresponding solution of a linear wave equation $\psi = \Psi\, exp(i\ \theta)$ for which

$$\theta = \theta'$$

except inside a small sphere surrounding the particle. The function ψ is taken to be a solution of Schrödinger's equation and the above phase condition allows the particle to be "guided" by ψ. In the same sense the nonlinear solution of (15) can be viewed as being guided through space-time by the linear asymptotic solution of (17).

Several investigators during the 1960's gave consideration to the sine-Gordon equation (12) as a (1 + 1) dimensional field theory for elementary particles [20, 22, 38] but this work was of little general interest until the early 1970's when it became clear that the special properties of the SGE allowed it to be completely quantized [39]. This showed that classical solutions [e.g., (13)] "survive quantization." The classical field energy was found to be a useful first approximation for the solition mass with quantum effects introducing a second order correction. Following this line there has been a dramatic rise in research related to the solitary wave picture of elementary particles as shown in Fig. 6. Several useful reviews are available for readers interested in following recent developments [40]. It is

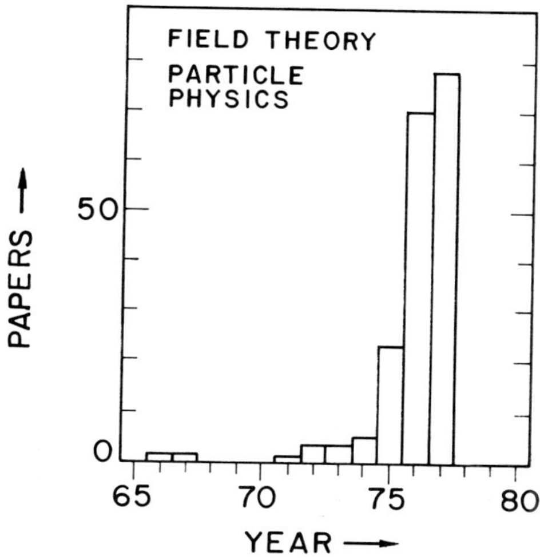

FIGURE 6. Annaul number of publications on solitary waves in elementary particle physics (Y.H. Ichikawa, private communication).

interesting to observe that the "topological solitons" currently being studied are three dimensional analogs of the vortex which was originally described by Helmholtz [10].

V. NEURODYNAMICS TO THE PRESENT

Post World War II research in electrophysiology was galvanized by the substantial increase in electronic measurement technology developed in support of military communications and RADAR. Peacetime dividends were not long in coming; a theoretical basis for pulse propagation on the giant axon of the squid was soon developed. The general outlines for a theory of nerve fiber dynamics had been known since the mid-thirties and can be sketched as follows [see Fig. 7]. If $v(x,t)$ is transmembrane voltage and $i\,(x,t)$ is axial current, then

$$v_x = -ri$$
$$i_x = -cv_t - j_i \quad . \qquad (22a,b)$$

In (22a) the parameter r is the series resistance per unit length of the ionic core (axoplasm); thus (22a) is merely a statement of Ohm's law. Eq. (22b), on the other hand, is a conservation law

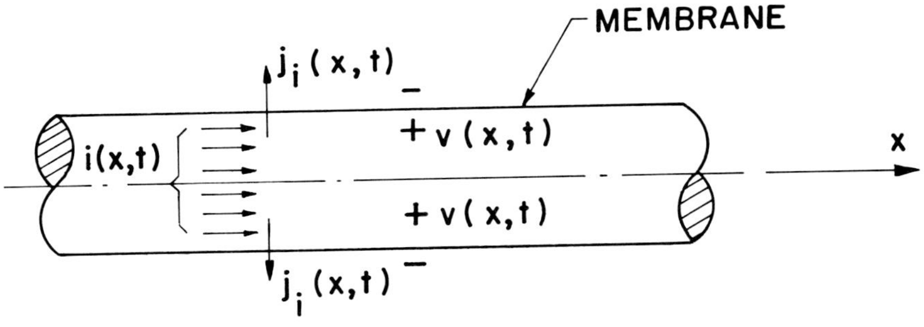

FIGURE 7. Currents and voltage in a nerve fiber.

for electrical charge inside the fiber for which c is the capacitance per unit length of the membrane and j_i is the ionic current per unit length crossing the membrane. If r is independent of x, (22) takes the form

$$v_{xx} - rc\, v_t = rj_i \tag{23}$$

which is a linear diffusion operator augmented by the nonlinear effects of transmembrane ionic current (j_i). The main problem facing post-war researchers was to represent the dynamics of j_i, and a rather definitive answer to this question was provided in 1952 by Hodgkin and Huxley [41]. Their theoretical picture introduced three additional phenomenological variables to represent the opening and closing of various ionic channels across the membrane, but they managed to obtain an excellent prediction of both shape and speed of the solitary wave shown in Fig. 5.

A simpler dynamical description for j_i was introduced in 1962 by Nagumo, Arimoto and Yoshizawa and was based upon previous work by FitzHugh [42]. In this picture [called the FitzHugh-Nagumo equation (FNE)],

$$\begin{aligned} j_i &= F(v) + R \\ R_t &= \varepsilon(V + a - bR) \end{aligned} \tag{24a,b}$$

where R is a "recovery variable" and ε, a and b are adjustable constants. In the limiting case $\varepsilon = 0$ it is clear from (24b) that R remains constant; thus substitution of (24a) into (23) gives a result that is identical to (5). In fact R does remain relatively constant during the leading edge of a typical pulse so analysis of (5) yields useful first approximations to both the shape of the leading edge and the speed of the entire pulse.

It is interesting to emphasize at this point how our basic picture of the nerve impulse emerges as a singularity-free solution of a nonlinear pde (23) just as Einstein had prescribed for the elementary particles of matter. In this context we might refer to the nerve impulse as an "elementary particle of thought."

The dynamic behavior of nerve impulses as they collide with others, propagate past irregularities in the fiber structure, interact with others at points where a fiber branches, etc. is currently a subject of considerable analytical and experimental interest. For a survey of this work in the Soviet Union the reader should consult the fine book by Khodorov [43]. Somewhat more recent surveys have been published by the present author [44] in which it has been suggested that physical scientists (physicists and electrical engineers in particular) should become seriously interested in neurodynamics. An excellent recent discussion of mathematical results in the study of "reaction-diffusion systems" (of which the nerve fiber system is a special case) has been provided by Paul Fife [45].

VI. CURRENT SOLITARY WAVE RESEARCH

After a gestation period of almost a century and a half, the solitary wave concept has finally been established. It was a difficult birth, but a new paradigm is now a part of our collective scientific thought. The "thingness" of solitary wave in general (and solitons in particular) is widely accepted as a structural basis for viewing and understanding the dynamic behavior of complex nonlinear systems.

But the sheer bulk of current solitary wave research prohibits a survey with any degree of bibliographic detail. Fortunately some books are becoming available that should help the interested reader find a way to activities and results [46-56]. Also a new review and research journal has recently been established that will draw together nonlinear wave research from the widely scattered scientific workshops [57].

From Fig. 1 we see that the solitary wave paradigm is being exercised in an impressive number of studies throughout science. One of the most fundamental areas of application is in statistical mechanics where soliton (or solitary wave) modes are found, from both experimental and theoretical investigations, to participate in energy equipartition. Such studies are related to soliton perturbation theory wherein the dynamic behavior of solitary waves that are "almost solitons" is calculated as a slow variation in soliton parameters (speeds and location) induced by the difference between the actual pde and some approximation that is integrable via the ISTM. The discovery of solitary wave solutions to pde's of physical interest is now being vigorously pursued using the tools of "modern" differential geometry which in turn is directly in the nineteenth century tradition that died (or was badly wounded) in the first world war. Finally there is a vast array of specific solitary wave applications: from hydrodynamics to meteorology, from computer technology to shock wave dynamics, from nonlinear filter theory to the propagation of crystal dislocations and domain walls, from the elementary particles of matter to the elementary particles of thought.

With our present perspective John Scott Russell's last book [6], far from being a catalog of "extraordinary and groundless speculations," seems rather conservative.

REFERENCES

1. Russell, J.S., "Report on waves," 14th Meeting of the British Assoc. Adv. Sci. (1844). See also J. Robison and J.S. Russell, 7th Meeting (1838).
2. Readers wishing to learn more of efforts to protect this historic canal can contact the Linlithgow Union Canal Society c/o Bruce Jamieson, 121 Baronshill Ave., Linlithgow EH 49 7JQ, Scotland.
3. See *John Scott Russell - A Great Victorian Engineer and Naval Architect* by George S. Emmerson (John Murray, London, 1977) for a timely and sympathetic biography of this "zealous educator, idealistic social reformer, and would-be peacemaker between nations, as well as the undisputed and respected leader of his profession." Emmerson's account is particularly valuable since L.T.C. Rolt's *Isambard Kingdom Brunel* (London, 1957) ascribes the difficulties between these two

brilliant engineers to base elements in Russell's character. Rolt's view of Russell is not expressed by Brunel's son [see I. Brunel, *The Life of Isambard Kingdom Brunel*, London (1870)] and it has been rejected by both A.M. Robb [see "John Scott Russell, the 'Great Eastern' and some other matters," *College Courant*, Glasgow, Martinmas 1958, Whitsun 1959] and by G.P. Mabon [J. Roy. Soc. Arts, pp. 204-8 and 299-302 (1967)]. The reader is left to judge whether Rolt's taste in human conduct is of service even to the memory of Brunel.

4. See J.W. Miles, Annual Reviews of Fluid Mechanics, *12* for a very careful survey of these discussions.
5. Lamb, H., *Hydrodynamics*, sixth ed. (1932) Dover.
6. Russell, J.S., *The Wave of Translation in the Oceans of Water, Air and Ether*, London (1885).
7. Korteweg, D.J. and DeVries, G., *Phil. Mag. 39*, 422-443 (1895).
8. Faraday, M., The Harvard Classics *30*, P.F. Collier & Sons, New York (1910).
9. Helmholtz, H., *Arch. Anat. Physiol.*, 276-364 (1850).
10. Helmholtz, H., *Ostwald's Klassiker 79* (1896).
11. By Nagumo, J., *et al.*, in *Trans. IEEE CT-12*, pp. 400-412 (1965) and attributed to A.F. Huxley as a private communication. Analysis of a nerve fiber model equivalent to (5) with $F(\phi)$ a piecewise linear function was first published by F.F. Offner *et al.*, in *Bull. Math. Biophys. 2*, pp. 89-103 (1940); unfortunately an algebraic error renders their final results incorrect. The first correct solitary wave solution to (5) was published by the present author in *Trans. IRE Ct-9*, pp. 192-195 (1962) without knowledge of Offner's work and only a vague notion of the relation between (5) and nerve conduction.
12. Lamb, G.L., Jr., "Bäcklund transformations at the turn of the century," in *Bäcklund Transformations* (R.M. Miura, ed.) Springer Math. Series #515 (1976). [A substantial portion of the French literature can be found in the library of Clark University, Worcester, Mass.]
13. Mie, G., *Ann. d. Physik 37*, 511-534 (1912); *39*, 1-40 (1912); *40*, 1-66 (1913).
14. Born, M., *Proc. Roy. Sco. A 143*, 410 (1934). For additional discussions see M. Born and L. Infeld, ibid. *144*, 425 (1934); *147*, 522 (1934); and *150*, 141 (1935); Also J. Frenkel, ibid. *146*, 930 (1930); E. Feenberg, *Phys. Rev. 47*, 148 (1935); B.M. Barbashov and N.A. Chernikov, *Sov. Phys. - JETP 24*, 437-442 (1967).
15. Kolmogoroff, A., Petrovsky, I. and Piscounoff, N., *Bull. Univ. Moscow, Série Int., A1*, 1-25 (1937). See also R.A. Fisher, *Ann. Eugen.* (now *Ann. Human Genet.*) *7*, 355-369 (1937).
16. Frenkel, J. and Kontorova, T., *J. Phys. (USSR) 1*, 137-149 (1939).
17. Fermi, E., Pasta, J.R. and Ulam, S.M., "Studies of nonlinear problems," LASL Rept. No. LA-1940 (1955); also in *Collected Works of Enrico Fermi Vol. II*, U. of Chicago Press, 978

(1965), and in [27].

18. During anecdotal reminiscences presented at the Como Conference on Nonlinear Stochastic Problems organized by J. Ford and G. Casati in June, 1977. It is probably correct to consider Fermi a casualty of the war on the scientific front.
19. Zabusky, N.J. and Kruskal, M.D., *Phys. Rev. Lett. 15*, 240-243 (1965).
20. Perring, J.K. and Skyrme, T.H.R., *Nucl. Phys. 31*, 550-555 (1962).
21. Seeger, A., Donth, H. and Kochendörfer, A., *Z. Phys. 130,* 321-336 (1951); *134*, 173-193 (1953).
22. This catchy name first appeared in the paper: J. Rubinstein, *J. Math. Phys. 11*, 258-266 (1970) although rumor has it that the coinage was actually by Kruskal. Some physicists have rather immoderately objected to the term, which seems strange because in general physicists lead (the academic world at least) in the invention of whimsical jargon. For additional reviews of sine-Gordon applications during the 1960's see: Lamb, G.L., Jr., *Rev. Mod. Phys. 43,* 99-124 (1971); and also Barone, A., *et al., Riv. Nuovo Cimento, 1,* 227-267 (1971).
23. Gardner, C.S., Green, J.M., Kruskal, M.D. and Miura, R.M., *Phys. Rev. Lett. 19*, 1095-1097 (1967).
24. Lax, P.D., *Commun. Pure Appl. Math. 21*, 467-490 (1968).
25. Toda, M., *J. Phys. Soc. Japan, 22*, 431-436 (1967). See also ibid. *23*, 501-506 (1967); *26* 235-237 (1969); and *34*, 18-25 (1973); *Phys. Reports 18*, 1 (1975).
26. Whitham, G.B., *J. Fluid Mech. 22*, 273-283 (1965); ibid. *27*, 399-412 (1967); ibid. *44*, 373-395 (1970). Also *Proc. Roy. Soc. A 283*, 238-261 (1965) and *299*, 6-25 (1967). This work is conveniently summarized in Whitham's recent book: *Linear and Nonlinear Waves*, Wiley (1975).
27. Newell, A.C. (ed.) *Nonlinear Wave Motion,* AMS Lectures in *Appl. Math. Vol. 15* (1974).
28. Zakharov, V.E. and Shabat, A.B., *Sov. Phys.-JETP 34*, 62-69 (1972).
29. Miura, R.M., *J. Math. Phys. 9*, 1202-1204 (1968). See also Miura *et al.*, ibid. 1204-1209.
30. Benney, D.J. and Newell, A.C., *J. Math. Phys. 46*, 133-139 (1967); Bespalov, V.I. and Talanov, V.I., *JETP Lett. 3*, 307-310 (1966); Kelley, P.L., *Phys. Rev. Lett. 15*, 1005-1008 (1965); Tappert, F. and Varma, C.M., *Phys. Rev. Lett. 25*, 1108-1111 (1970); and Ichikawa, Y.H., Imamura, T. and Taniuti, T., *J. Phys. Soc. Japan 33*, 189-197 (1972).
31. Ablowitz, J.J., Kaup, D.J., Newell, A.C. and Segur, H., *Phys. Rev. Lett. 30*, 1262-1264 (1973); Takhtadzhyan, L.A. and Faddeev, L.D., *Theor. and Math. Phys. 21*, 1046-1057 (1974).
32. Hénon, M., *Phys. Rev. B 9*, 1921-1923 (1974); Flaschka, H., ibid. 1924-1925.
33. Wahlquist, H.D. and Estabrook, F.B., *Phys. Rev. Lett. 31*, 1386-1390 (1973).

34. Ablowitz, M.J., Kaup, D.J., Newell, A.C. and Segur, H., *Studies in Appl. Math. 53*, 249-315 (1974); see also Chu, F.Y.F. and Scott, A.C., *Phys. Lett. A 47* 303-304 (1974).
35. Einstein, A., *Ideas and Opinions*, Crown, New York (1954), pp. 306-307.
36. Heisenberg, W., *Introduction to the Unified Field Theory of Elementary Particles*, Interscience (1966).
37. DeBroglie, L., *Nonlinear Wave Mechanics*, Elsevier (1960); and *Introduction to the Vigier Theory of Elementary Particles*, Elsevier (1963).
38. Enz, U., *Phys. Rev. 131*, 1392 (1963); Hobart, R.H., *Proc. Phys. Soc. 82*, 201 (1963); Derrick, G.H., *J. Math. Phys. 5*, 1252 (1964); Rosen, G., ibid. *6*, 1269 (1965); Skyrme, T.H.R., ibid. *12*, 1735 (1971).
39. Dashen, R.F., Hasslacher, B. and Neveu, A., *Phys. Rev. D 10*, 4114-4142 (1974); *11*, 3424-3450 (1975); Goldstone, J. and Jakiw, R., ibid. *11*, 1486 (1975); Faddeev, L.D., *JETP Lett. 21*, 64-65 (1975).
40. Rebbi, C., Scientific American, Feb. 1979, p. 92; *Z. Parsa, Am. J. Phys. 47*, 56-62 (1979); Jakiw, R., *Rev. Mod. Phys. 49*, 681 (1977); Makhankov, V.G., *Phys. Repts. 35*, No. 1, 1-128 (1978).
41. Hodgkin, A.L. and Huxley, A.F., *J. Physiol. 117*, 500-544 (1952).
42. Nagumo, J., Arimoto, S. and Yoshizawa, S., *Proc. IRE 50*, 2061-2070 (1962).
43. Khodorov, B.I., *The Problem of Excitability*, Plenum, New York (1974).
44. Scott, A.C., *Rev. Mod. Phys. 47*, 487-533 (1975); and *Neurophysics*, Wiley-Interscience (1977).
45. Fife, P.C., *Mathematical Aspects of Reacting and Diffusing Systems*, Springer Lecture Notes in Biomathematics (1979).
46. Calogero, F. (ed.) *Nonlinear Evolution Equations Solvable by the Spectral Transform*, Pitman, London (1978).
47. Miura, R.M., *SIAM Rev. 18*, 412-459 (1976).
48. Lonngren, K. and Scott, A. (eds.) *Solitons in Action*, Academic Press (1978).
49. Caudrey, P.J. and Bullough, R.K. (eds.) *Solitons*, Springer (to appear).
50. Hermann, R. (ed.) *The Geometric Theory of Non-Linear Waves*, Math Sci. Press, Brookline (1977).
51. Hermann, R., *Toda Lattices Cosymplectic Manifolds, Bäcklund Transformations and Solitons*, Math Sci Press, Brookline (1977).
52. Hermann, R., *Geometric Theory of Non-Linear Differential Equations, Bäcklund Transformations and Solitons*, Math Sci Press, Brookline (1977).
53. Barut, A.O. (ed.) *Nonlinear Equations in Physics and Mathematics*, Reidel (1978).

54. Bishop, A.R. and Schneider, T. (eds.) *Solitons and Condensed Matter Physics*, Springer (1979).
55. Lamb, G.L., Jr., *Elements of Soliton Theory*, Wiley-Interscience (to appear).
56. Wilhelmsson, H. (ed.) *Solitons in Physics*, Physica Scripta. 20, 289 (1979).
57. *Physica D - Nonlinear Phenomena*. It is to be edited by H. Flaschka, J. Ford, A.C. Newell and A.C. Scott and published by the North-Holland Publishing Co. The first issue is expected to appear in March of 1980.

A PHYSICAL DESCRIPTION OF THE SPECTRAL TRANSFORM[1]

David W. McLaughlin

Department of Mathematics and
Program in Applied Mathematics
University of Arizona
Tucson, Arizona

In this chapter the inverse spectral representation of certain nonlinear dispersive waves is described through a physical model of the resonant interaction of a laser pulse with a two level medium. The particle-wave duality of the nonlinear wave is emphasized. References to appropriate review articles are included.

I. INTRODUCTION

When one examines one dimensional, nonlinear, dispersive waves experimentally (either by numerical simulation or by laboratory experiments on water waves, waves in plasmas, condensed matter excitations, pulses on nonlinear transmission lines, laser pulses, and so forth), one observes that such waves quickly evolve into an asymptotic or "far-field" state which is comprised of a few localized pulses, together with dispersive radiation [6]. The localized pulses are the most striking component of the "far-field" state. These pulses are very localized in space and translate at constant speed. They are very stable; indeed, they

[1] *Supported in part by N.S.F. Grant No. MSC-7903533 and by D.O.E. (Los Alamos Scientific Laboratory, T-7 Group).*

ISBN 0-12-709660-4

emerge from collisions with the identical amplitudes, velocities, and shapes with which they approached the collision. The only effect of collisions is a phase shift in the location of their space-time trajectories. Two types of localized pulses exist -- those with and those without internal degrees of freedom. (The localized waves with internal degrees of freedom pulsate as they translate.) These localized pulses in the "far-field" state are certainly not anticipated from our experiences with linear dispersive waves. On the other hand, the radiative component of the "far-field" state behaves similarly to linear, dispersive waves.

Upon examining experimental results on nonlinear dispersive waves, any scientist will certainly notice that the localized components ("solitons") behave as particles. This wave-particle duality in the "far-field" asymptotic state of one-dimensional, nonlinear, dispersive waves is extremely intriguing.

Given our vast experience with the analysis of linear waves by transform methods, it seems natural to inquire about the possibiligy of a transform which would permit the analysis and synthesis of a general nonlinear dispersive wave in terms of more basic components. To be useful, this collection of basic components must include the stable, localized pulses, i.e., the "solitons." The collection of basic components should also include the radiation in the "far-field" state. Since the wave under consideration is *non*linear, such a transform could not synthesize a general wave by *linear* superposition of the basic components. The synthesis procedure must be nonlinear. With such a transform, one would hope (i) to predict from properties of the initial data the number and type of solitons which eventually comprise the "far-field" state, as well as their velocities and pulsation frequencies; (ii) to predict the radiation density of the asymptotic state from properties of the initial data; (iii) to extend concepts such as *group velocity* to nonlinear waves; (iv) to understand more completely the wave-particle duality of nonlinear dispersive waves; (v) to investigate the response of solitons to external perturbations, inhomogeneities, and exterior forces; (vi) to investigate the statistical behavior of collections of solitons; (vii) to study quantum effects on nonlinear waves; ....

During the past decade, a new transform has been developed which accomplishes these goals. It is called the "Inverse Spectral Transform," and amounts to a mathematical tool for the description of one-dimensional, nonlinear, dispersive waves in general and solitons in particular. My goal in this chapter is merely to provide a little intuition about this transform. I will try to be descriptive, with very few mathematical formulas. For more details, one can refer to the references. There is a vast literature on this subject, with several very good general surveys. For my references I have chosen, for the most part, reviews rather than the original technical articles. Many scientists have contributed to this field. A list of credits would be too long to include here. However, two scientists must be mentioned -- M. Kruskal and V. Zakharov. Without the contributions of these two persons, our understanding of nonlinear dispersive waves would be reduced immensely.

Most of this chapter is a description of the inverse spectral transform in the context of a specific physical example -- the passage of a laser pulse through a medium which resonates with the electromagnetic pulse. This physical model provides intuitive insight about the spectral transform. For definiteness, explicit formulas for solitons of the sine-Gordon equation are included in Section IV. This chapter concludes with some general comments about applications of the transform, together with references to the literature.

II. A PHYSICAL MODEL OF THE INVERSE SPECTRAL TRANSFORM

Consider a laser pulse propagating through a medium which consists of two-level atoms. In an idealized model, the medium is described by a collection of identical oscillators, each with a ground state and exactly one excited state. I emphasize that this is a coupled system. The electromagnetic field of the laser pulse will polarize the medium which, in turn, will modify the field. This coupling is a nonlinear effect; it becomes important only for large amplitude electromagnetic fields. This coupling becomes particularly strong when the medium resonates with the carrier wave of the laser pulse.

Under such large amplitude and resonant conditions, the following result can be observed [1,2]. On a "macroscopic scale," the envelope of the laser wave shapes into localized soliton pulses which propagate through the medium at constant speed without distortion. In laser physics this effect is called "self induced transparency." On this "macroscopic" scale, the medium appears transparent to these pulses. On the other hand, on a "microscopic" scale, substantial interaction is taking place. The leading edge of the laser pulse excites the individual atoms in the medium. A large number of these excited atoms then radiate collectively as they return to their ground state. This

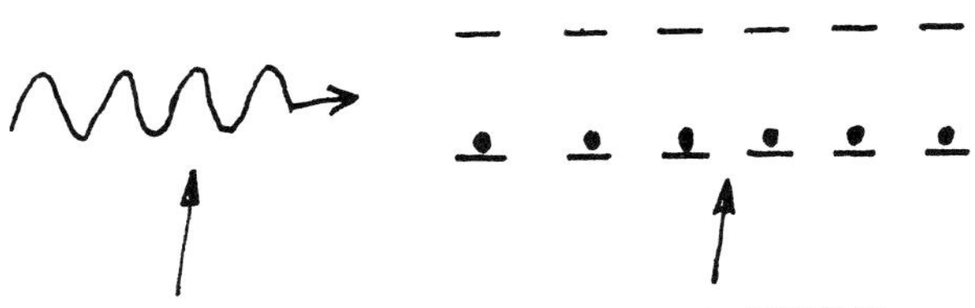

FIGURE 1. A laser pulse approaching a two-level medium.

radiation, which is resonant with the laser pulse, is then recaptured by the trailing edge of the pulse. Thus, the solitons in the laser pulse are actually shaped by the transferal of energy from the pulse to the medium, and back to the pulse again.

From one (macroscopic) point of view, solitons are *very localized* "particle-like" states in the field. From another (microscopic) viewpoint, they result from *collective, long range* excitations of a great many of the oscillators which comprise the medium. Both viewpoints are useful when studying the physics of solitons.

To model this situation mathematically, we use Maxwell's equations to describe the laser pulse and a simple form of Schroedinger's equation to describe each two level atom:

$$[\partial_{tt} - c^2\partial_{xx}]E = \partial_{tt} P$$

$$i\hbar\partial_t \begin{pmatrix} \psi_+ \\ \psi_- \end{pmatrix} = \frac{1}{2}\begin{bmatrix} \hbar\omega & 2E \\ 2E & -\hbar\omega \end{bmatrix}\begin{pmatrix} \psi_+ \\ \psi_- \end{pmatrix} \tag{1a,b}$$

Here E is the electric field of the laser pulse, ψ_+ denotes the excited state of the two-level atom, ψ_- is its ground state. For simplicity, most physical constants have been set at unity (except Planck's constant $\hbar$ and the speed of light c). P denotes the induced polarization and is given in terms of the two level states by

$$P \equiv \langle \psi_+ \psi_-^* + \psi_- \psi_+^* \rangle , \tag{2}$$

where <•> denotes a micro to macro map which sums the polarizations of many microscopic oscillators into a single macroscopic polarization P. Before interaction, the laser pulse is a plane wave and the medium is in its ground state. Thus, the boundary conditions for system (1) are

$$\left.\begin{aligned} &E = 2\mathcal{E}\cos[k_c x - \omega_c t] \\ &\begin{pmatrix}\psi_+\\ \psi_-\end{pmatrix} = \begin{pmatrix}0\\ e^{i\frac{\omega}{2}t}\end{pmatrix}\end{aligned}\right\} \text{(before interaction)}, \tag{3}$$

where $\omega_c \equiv ck_c$. Notice before interaction the medium is not polarized because $P = \langle\psi_+ \psi_-^* + \psi_- \psi_+^*\rangle = 0$ (by boundary conditions (3)). The entire polarization P is induced by the action of the laser pulse on the medium. This interaction is described by the coupling term in the Schroedinger equation (1b):

$$\frac{1}{2}\begin{bmatrix}0 & 2E\\ 2E & 0\end{bmatrix}\begin{bmatrix}\psi_+\\ \psi_-\end{bmatrix}$$

Finally, the resonance criterion is imposed by setting the oscillation frequency ω of the medium equal to the carrier frequency ω_c of the laser pulse:

$$\omega = \omega_c \quad \text{(at resonance)}.$$

To analyse system (1) with boundary conditions (3) mathematically, one makes an ansatz of the form

$$\begin{aligned} &E \simeq \mathcal{E}(X,T)e^{-i\theta} + \mathcal{E}^*(X,T)e^{+i\theta} \\ &\psi_\pm \simeq \Psi_\pm(X,T)e^{\mp i\theta/2}\end{aligned} \tag{4}$$

Here X,T denote slow space and time scales; the phase $\theta \equiv k_c x - \omega_c t$. The assumption is that the envelopes ($\mathcal{E}$ and $\Psi_\pm$) vary slowly when compared to the rapid oscillations of the carrier $\cos\theta = \cos(k_c x - \omega_c t)$. Using multiscale perturbation theory, one averages out these rapid oscillations and reduces the system to one for the slowly varying envelope. The result of this reduction (see [3,4,25] for details) is

$$\partial_X \mathcal{E} = \langle\Psi_+ \Psi_-^*\rangle \tag{5a}$$

$$i\partial_T \begin{pmatrix} \Psi_+ \\ \Psi_- \end{pmatrix} = \begin{pmatrix} \zeta & E \\ E^* & -\zeta \end{pmatrix} \begin{pmatrix} \Psi_+ \\ \Psi_- \end{pmatrix}, \tag{5b}$$

with boundary conditions

$$E(X,T)\Big|_{X=0} = E_o(T) \to 0 \quad \text{as} \quad |T| \to \infty \tag{6}$$

$$\begin{pmatrix} \Psi_+ \\ \Psi_- \end{pmatrix} \simeq \begin{pmatrix} 0 \\ e^{i\zeta T} \end{pmatrix} \quad \text{(before interaction)}$$

Here $\zeta = (\omega - \omega_c)/2$ measures how far the given two level atom is from perfect resonance, and the envelope polarization $\Psi_+ \Psi_-^*$ is a measure of the excitation of this atom. These equations (5a,b), together with boundary conditions (6), are fundamental in our description of the transform.

The boundary condition on E, $E_o(T) \to 0$ as $|T| \to \infty$, guarantees that the laser field is a pulse which vanishes long before and long after the interaction has taken place. This situation is depicted in figure 2.

Consider a two level atom at X_o. Before the laser pulse arrives at X_o, and after it has completely passed X_o, the two level state $\Psi_+(X_o,T)$ can be written down explicitly because *at those times* $E(X_o,T) = 0$:

Before the Laser Pulse Arrives at X_o:

$$\vec{\Psi}(X_o,T) = \begin{pmatrix} 0 \\ e^{i\zeta T} \end{pmatrix} \tag{7a}$$

After the Laser Pulse Has Passed X_o:

$$\vec{\Psi}(X_o,T) = \begin{pmatrix} b(\zeta;X_o)e^{-i\zeta T} \\ a(\zeta;X_o)e^{i\zeta T} \end{pmatrix}. \tag{7b}$$

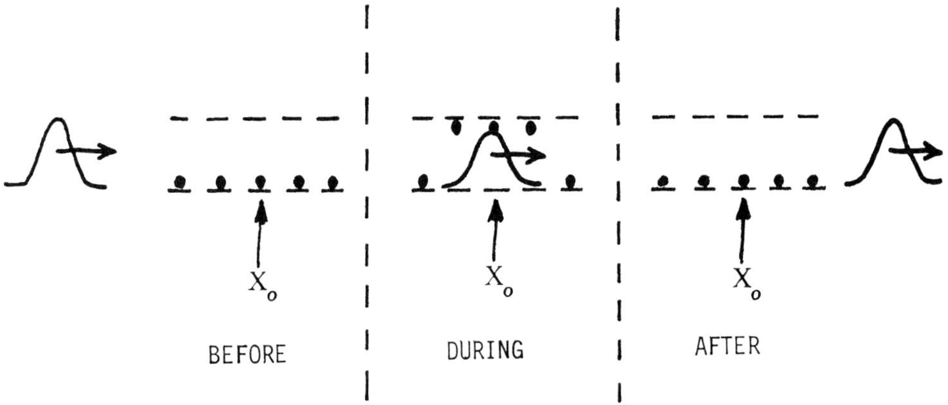

FIGURE 2. The medium at X_o (before, during, and after) the laser pulse passes.

Notice that (7a) just states that the medium is in its ground state before interaction, while (7b) shows that $b(\zeta;X_o)$ measures the excitation in the medium at X_o after the laser pulse has passed. (The laser pulse passes X_o, excites the medium at X_o, and leaves the medium in an excited state which is measured by $b(\zeta;X_o)$.)

In this context, the spectral transform may be described in the following manner: One uses the medium at X_o, after it has interacted with and shaped the laser pulse E, to measure the E pulse. Notice the intrinsically nonlinear nature of this measuring device. One half of the (nonlinear) interacting system (the medium), which has just participated in an essential manner in the shaping of the laser pulse, is then used to measure the pulse which it has just helped to shape.

Mathematically, this measuring device is described by the linear Schroedinger equation (5b) at $X = X_o$,

$$i\partial_T \begin{pmatrix} \Psi_+ \\ \Psi_- \end{pmatrix} = \begin{pmatrix} \zeta & E(X_o,T) \\ E^*(X_o,T) & -\zeta \end{pmatrix} \begin{pmatrix} \Psi_+ \\ \Psi_- \end{pmatrix} \tag{8}$$

which is considered as an eigenvalue problem with eigenvalue parameter ζ. The eigenfunction $\vec{\Psi}(X_o,T;\zeta)$ is normalized as $T \to -\infty$ by the condition

$$\begin{pmatrix} \Psi_+ \\ \Psi_- \end{pmatrix} \simeq \begin{pmatrix} 0 \\ e^{i\zeta T} \end{pmatrix} \quad \text{as } T \to -\infty \ , \tag{9}$$

and satisfies, as $T \to +\infty$

$$\begin{pmatrix} \Psi_+ \\ \Psi_- \end{pmatrix} \simeq \begin{pmatrix} b(\zeta;X_o)e^{-i\zeta T} \\ a(\zeta;X_o)e^{i\zeta T} \end{pmatrix} . \tag{10}$$

In fact, this asymptotic statement as $T \to +\infty$ defines the amplitudes $b(\zeta,X_o)$, $a(\zeta;X_o)$. The eigenvalues of this eigenvalue problem are defined as those values of ζ for which $\vec{\Psi}(X_o,T;\zeta) \to 0$ as

$T \to \pm \infty$. Clearly from the normalization condition (9),

$$\begin{pmatrix} \Psi_+ \\ \Psi_- \end{pmatrix} \simeq \begin{pmatrix} 0 \\ e^{i\zeta T} \end{pmatrix} = \begin{pmatrix} 0 \\ e^{i\zeta_r T} \; e^{-\zeta_I T} \end{pmatrix}$$

$$\text{as } T \to -\infty, \; \zeta = \zeta_r + i\zeta_I ,$$

the eigenvalues $\zeta = \zeta_r + i\zeta_I$ must lie in the lower half complex ζ plane if Ψ_- is to vanish as $T \to -\infty$. In addition, by asymptotic behavior (10),

$$\begin{pmatrix} \Psi_+ \\ \Psi_- \end{pmatrix} \simeq \begin{pmatrix} b(\zeta;X_0)e^{-i\zeta_r T} \; e^{+\zeta_I T} \\ a(\zeta,X_0)e^{i\zeta_r T} \; e^{-\zeta_I T} \end{pmatrix} \qquad \text{as } T \to +\infty,$$

these eigenvalues must be lower half plane zeros of the amplitude $a(\zeta;X_0)$ if Ψ_- is to vanish as $T \to +\infty$. We label these eigenvalues by $\{\zeta_1, \ldots, \zeta_N\}$.

In this manner, one defines the following *spectral data S:*

$$S = \{(\zeta_1, \ldots, \zeta_N); \; b(\zeta;X_0) \text{ for all real } \zeta \text{ and for } \zeta \in (\zeta_1, \ldots, \zeta_N)\} .$$

The spectral transform can be described in terms of this data S:

Theorem. Fix X_0 and consider $E(X_0,T)$ as a function of T which vanishes sufficiently rapidly as $T \to \pm \infty$. This function of T is equivalent to the spectral data $b(\zeta;X_0)$, considered as a function of ζ. That is,

$$\{E(X_0,T) \; \forall T \in (-\infty,\infty)\} \simeq \{b(\zeta;X_0) \; \forall \text{ real } \zeta \text{ and for } \zeta \in (\zeta_1, \ldots, \zeta_N)\}. \tag{11}$$

A proof of this theorem involves the Gelfand-Levitan-Marchenko integral equations of inverse spectral theory [5], and is too detailed to present here. Instead, in the next section, I will summarize and expand upon the interpretation of this theorem in the physical context of the resonant interaction of a laser pulse with a two level medium.

III. COMMENTS ABOUT THIS INTERPRETATION OF THE SPECTRAL TRANSFORM

In the physical context of this interacting system, the spectral transform may be interpreted as a change of variables from the laser pulse at X_o for all time T, $\{E(X_o,T)$ for all $T\}$, to the medium at X_o (after the pulse has passed) for all values of the spectral parameter ζ, $\{b(\zeta,X_o)$ for all $\zeta\}$. The transform is a change of variables at fixed X_o from functions of time T to functions of the spectral parameter ζ. In this manner, the spectral transform reminds us of the Fourier transform,

$$\beta(\zeta;X_o) = \int_{-\infty}^{\infty} e^{i\zeta T} E(X_o,T)\,dT ,$$

from functions of T to functions of ζ. Indeed, for small E, a "Born perturbation series" of $\vec{\Psi}$ [5] shows that the spectral transform reduces to the Fourier transform for small amplitude laser pulses,

$$b(\zeta;X_o) \simeq \beta(2\zeta;X_o) + O(E^2).$$

The Fourier transform is linear in the field amplitude E, while the spectral transform is nonlinear in E, as is clearly shown by this Born series.

Notice that the spectral data splits naturally into two components:

$$\{b(\zeta;X_o) \text{ for all real } \zeta\} \text{ and } \{\zeta_j, b(\zeta_j) \text{ for } j = 1, 2, \ldots, N\}.$$

These two components describe the radiation and soliton components of the nonlinear wave as discussed in the introduction. To see this division, recall that $b(\zeta;X_o)$ measures the excitation of the medium at X_o after the laser pulse has passed. This excitation of the medium can radiate into the E-field. On the other hand, the medium is transparent to solitons. If the laser pulse were a perfect soliton, the medium would be entirely in its ground state after the pulse has passed; that is, $b(\zeta,X_o) = 0$ for real ζ. Thus the map from $E(T) \to b(\zeta)$ would be of the form

$$E_{SOL}(T) \to b(\zeta) = 0 \text{ for real } \zeta,\ b(\zeta_j) \neq 0. \quad \text{(Perfect Soliton)}$$

We see that the discrete components of the spectral transform represent solitons, while its continuous components represent radiation.

The reason for the terminology "direct spectral transform" and "inverse spectral transform" is clear in this physical model. Given the laser pulse $E(X_o,T)$, the computation of the medium's response to this pulse $(b(\zeta,X_o))$ is the "direct problem;" conversely, given the medium's future state

$$\Psi_+(X_o,T;\zeta) = b(\zeta;X_o)e^{-i\zeta T}$$

after the pulse has passed, the reconstruction of that laser pulse $E(X_o,T)$ which placed the medium in this state is an "inverse problem." Given the continuous and discrete spectral data, *there is exactly one* laser pulse which produced this data. This existence and uniqueness statement is the content of the thoerem in the last section.

In general, a closed form expression which gives the laser pulse E in terms of the spectral data is not known. However, if all continuous components vanish, closed form expressions for the laser pulse E in terms of the spectral data $\{\zeta_j,b(\zeta_j)\}$ are known. These are called "N-soliton formulas" [6]. These formulas show that the number of solitons in the laser pulse is fixed by the number of discrete eigenvalues; the locations of the eigenvalues in the lower half complex plane fix the type of solitons (with or without an internal degree of freedom), as well as their speeds and pulsation frequencies. The other half of the discrete data

$\{b(\zeta_j)\}$ fixes the location of each soliton. These facts follow, for example, from the explicit solution of the Gelfand-Levitan integral equations of inverse spectral theory (when the continuous component $b(\zeta;X_o) = 0$ for all real ζ) [6].

This description of the spectral transform in terms of the interaction of a laser pulse with a two-level medium aids our understanding of the concept of this transform; however, this description sheds no light on another crucial property of soliton equations -- their *complete integrability* by the spectral transform. The complete integrability of soliton systems has been concretely described elsewhere [4,7]. Here, I just mention this property in the framework of this physical model.

So far our entire discussion has taken place at a fixed location X_o in the medium. For example, the scattering data $b(\zeta;X_o)$ and the laser pulse $E(X_o,T)$ have both been computed at that location X_o. As one moves around in the medium by changing X_o, one would expect the location of the eigenvalues $\zeta_j(X_o)$ and even their number N to change with X_o. Certainly the laser pulse $E(X_o,T)$ will change; in fact, its evolution in X_o is governed by equation (5a). Nevertheless, it turns out that as $E(X_o,T)$ evolves in X_o by equation (5a), the X_o evolution of the scattering data is trivial. *In particular, the number and location of the eigenvalues* ζ_j *do not change with* X_o! This invariance of the eigenvalues with X_o describes the remarkable stability of solitons. Since the speeds of the solitons are fixed by the eigenvalues, these speeds are not altered even by direct collisions with other solitons.

Finally, consider a different X_o evolution than (5a); for example, if the field E experiences some decay because of losses in the medium, equation (5a) might be replaced by

$$\partial_X E = \langle \Psi_+ \Psi_-^* \rangle - \sigma E, \quad \sigma > 0.$$

The spectral transform from $E(X_o,T) \to b(\zeta;X_o)$ is still valid;

however, the X evolution of the spectral data is no longer trivial. The eigenvalues are no longer constant in X; in general, they satisfy very complicated evolution equations. For some reason the spectral transform is naturally suited to the specific interacting system (5), just as the Fourier transform is naturally suited to linear wave equations with constant coefficients. Our physical model of the transform has provided no insight into the reasons behind this trivial X flow of the spectral data.

IV. APPLICATIONS OF THE SPECTRAL TRANSFORM

During the past few years the spectral transform has been used in many ways to clarify our understanding of the physics of nonlinear dispersive waves. In this final section, I will briefly mention several of these applications. For details, one can refer to the references. In particular, I want to emphasize the particle wave duality of a nonlinear dispersive wave.

Consider, once again, system (5),

$$\partial_X E = \langle \Psi_+ \Psi_-^* \rangle$$

$$i\partial_T \begin{pmatrix} \Psi_+ \\ \Psi_- \end{pmatrix} = \begin{pmatrix} \zeta & E \\ E^* & -\zeta \end{pmatrix} \begin{pmatrix} \Psi_+ \\ \Psi_- \end{pmatrix} .$$

For simplicity, we examine a special case: only that part of medium at *perfect resonance* with the carrier wave is coupled to the laser pulse. Since $\zeta = (\omega - \omega_c)/2$, this assumption essentially means that the micro to macro map $\langle \cdot \rangle$ is given by

$$\langle \Psi_+(X,T;\zeta)\ \Psi_-^*(X,T;\zeta) \rangle = \Psi_+(X,T;\zeta)\ \Psi_-(X,T;\zeta) \Big|_{\zeta=0} .$$

With this micro to macro map, we find a real solution (E,Ψ) which satisfies boundary conditions (6):

$$\Psi_- = \cos \int_{-\infty}^{T} E(X,T')dT'$$

$$\Psi_+ = \sin \int_{-\infty}^{T} E(X,T')dT'$$

$$E_X = \left(\sin \int_{-\infty}^{T} E(X,T')dT' \right) \left(\cos \int_{-\infty}^{T} E(X,T')dT' \right) .$$

In this case, the E field decouples from the medium at the expense of introducing a sinusoidal nonlinearity. Indeed, if we define

$$\sigma(X,T) \equiv 2 \int_{-\infty}^{T} E(X,T')dT' ,$$

we find that σ satisfies the nonlinear equation

$$\sigma_{XT} = \sin \sigma .$$

For convenience, we change space-time variables from $(X,T) \rightarrow (x,t)$,

$$x = X + T$$

$$t = X - T,$$

and find that σ satisfies the "sine-Gordon equation"

$$\sigma_{tt} - \sigma_{xx} = -\sin \sigma . \tag{12}$$

The spectral transform maps this nonlinear wave equation into a trivial (linear and decoupled) evolution equation for the spectral data, just as the Fourier transform maps its linear analogue, the Klein-Gordon equation

$$\tilde{\sigma}_{tt} - \tilde{\sigma}_{xx} = -\tilde{\sigma} , \tag{13}$$

to trivial form.

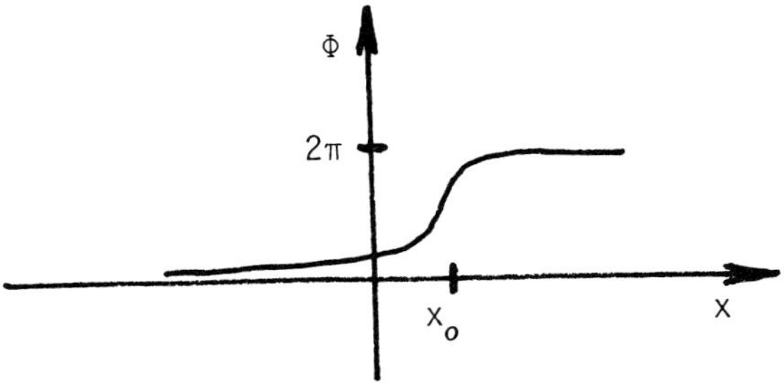

FIGURE 3. A single sine-Gordon "soliton"

Even though the linear Klein-Gordon equation (13) has no localized traveling wave solutions, the nonlinear sine-Gordon (12) has a two parameter family of localized traveling waves [8]:

$$\phi_{\pm}(x,t;u,x_o) = 4 \tan^{-1}\left[\exp\left(\pm \frac{x - x_o - ut}{1 - u^2}\right)\right] . \tag{14}$$

This solution is sketched in figure (3). It represents a soliton traveling at constant speed u, $-1 < u < 1$, located at $x = x_o + ut$. This soliton has no internal degree of freedom. Its spectral transform [9] has $b(\zeta) = 0$ for all real ζ and has a single eigenvalue ζ_1, which is given in terms of its velocity u by

$$\zeta_1 = -i\,\frac{1}{4}\sqrt{\frac{1-u}{1+u}} .$$

For the sine-Gordon equation, a soliton with one internal degree of freedom is called a "breather." In its rest frame, the breather solution is given by

$$\phi_B = 4\tan^{-1}\left[\frac{\tan v \,\sin[(\cos v)(t-t_o)]}{\cosh[(\sin v)(x-x_o)]}\right] .$$

This wave form is sketched in figure 4. Note that it pulsates with frequency $\cos v$ as it translates. The spectral transform of this soliton has vanishing $b(\zeta)$ for all real ζ, and two eigen-

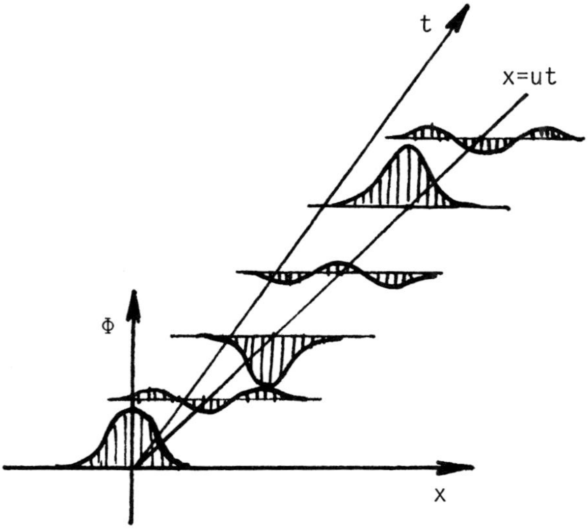

FIGURE 4. A sine-Gordon "breather"

values in the lower half ζ plane, ζ_1 and $\zeta_2 = -\zeta_1^*$. The magnitude of ζ_1 fixes the speed of translation, while its phase fixes the pulsation frequency.

A general solution of the sine-Gordon equation (which is localized in x) consists in N solitons, L breathers, and radiation. The solitons and breathers constitute the particle component of the wave. Their number and final speeds can be determined by a spectral transform of the initial data. The solitons behave as free particles which move at constant velocity, and may or may not possess an internal degree of freedom. Moreover, the analogy between the discrete soliton component of the nonlinear dispersive wave and particles is even stronger than just described.

Consider the soliton component of a nonlinear dispersive wave. If the dynamics of the wave is described by the sine-Gordon equation (12), the solitons have infinite lifetime, travel at constant speed, and generally behave as *free particles*. However, if their dynamics is described by a perturbation of the sine-Gordon equation,

$$\sigma_{tt} - \sigma_{xx} + \sin\sigma = \varepsilon f(x,t,\sigma) \ , \tag{15}$$

many things can happen to these particles. Depending upon the particular nature of the perturbation f, the solitons can slow down, accelerate, radiate, couple to each other; solitons with an internal degree of freedom can fission into two solitons without internal degrees of freedom; solitons can be created or annihilated; and so forth. To describe these effects, perturbation theories have been developed [10,11,12,13] which use the inverse spectral transform in much the same way as near-linear mode coupling calculations use the Fourier transform. A particularly beautiful result of these perturbation theories [10] describes the response of a solition ϕ_+ to an exterior force. Consider a single soliton with velocity u propagating under exterior force $f = f(x)$ (independent of σ). The velocity u is no longer constant since the soliton must respond to the exterior force; perturbation calcualtions [11] show the velocity of the soliton satisfies

$$\frac{du}{dt} = -\frac{\partial}{\partial X} U_{eff} \ , \tag{16}$$

where

$$U_{eff} \equiv \frac{\varepsilon}{8}(1-u^2)^{3/2}\int_{-\infty}^{\infty} f[(1-u^2)^{1/2}\,\theta - X]\phi_+(\theta)\,d\theta \ .$$

This equation (16) for the rate of change of the velocity u is similar to Newton's equation ($F = ma$) (with, of course, a relativistic correction). The effective potential to which the soliton responds is not $U = -\int^X f\,dx$, but $f(x)$ "averaged over the soliton." In any case, equation (16) is more evidence for the particle interpretation of the soliton.

The spectral transform can be described mathematically as a canonical transformation (in the sense of Hamiltonian mechanics) from the field variables E to the spectral data [14,7,4]. In this canonical description of the spectral data, the conjugate momentum $P_j(\zeta_j)$ is a function only of the eigenvalue ζ_j (which fixes the soliton's speed), while its canonically conjugate position variable fixes the soliton's location. Again, we note consistency with the particle interpretation.

Given the interpretation of the soliton component of a nonlinear wave as the particle component, it seems natural to inquire about the statistical mechanics of a "gas of solitons." This topic is particularly relevant in condensed matter physics, where the solitons are always connected to, and often driven by, a heat bath. Using techniques related to the spectral transform, considerable progress has been made in the statistical description of a sparse gas of solitons. A good survey may be found in [15].

Another natural question concerns the quantum mechanical behavior of these nonlinear waves. One approach involves a semiclassical or "WKB" quantization of the (quantum) fluctuations about the soliton state. This approach is described in [16]. A second approach adapts the inverse spectral transform to the fully quantized nonlinear field [17,18]. I believe neither approach has been applied to the interaction of a laser pulse with a two level medium.

In short, the inverse spectral transform has provided a great deal of concrete information about solitons in isolation and about sparse configurations of solitons. To date, it has not provided much concrete physical information about the radiation component of a nonlinear dispersive wave, nor about dense configurations of solitons. In this context, the concept of nonlinear group velocity has been introduced [19,20,21] and is beginning to be investigated by inverse spectral methods [22,23]. These studies show that a connection exists between dense configurations of solitons and the radiation component [22,23], but the application of the inverse spectral transform to these problems is in its infancy.

V. CONCLUSION

In this chapter I have introduced the inverse spectral transform in the physical context of the resonant interaction of a laser pulse with a two level medium. I must remark that this description has severe limitations. (For example, the inverse spectral transform has wide applicability to many equations which do not seem to arise from the interaction of a wave with a medium.) Never-the-less, I believe that this physical context helps our intuitive understanding of the transform mechanism.

Secondly, I have emphasized that the appropriate mathematical tool for studying the physics of one dimensional, nonlinear dispersive waves (the physics of solitons) is the inverse spectral transform. I would recommend that those desiring more information about this transform begin with the excellent survey by R. Miura [24], then study the more mathematical survey of H. Flaschka and A. Newell [5], and then proceed into literature such as referenced in this chapter.

1. McCall, S.L. and Hahn, E.L., "Self induced transparency," *Phys. Rev. 183*, 457-485 (1969).
2. Lamb, G.L., Jr., "Analytical descriptions of ultrashort optical pulse propagation in a resonant medium," *Rev. Mod. Physics 43*, 99-124 (1971).
3. McLaughlin, D.W. and Corones, J., "Semiclassical radiation theory and the inverse method," *Phys. Rev. A 10*, 2051-2062 (1974).
4. Lamb, G.L., Jr. and McLaughlin, D.W., "Interacting systems and the inverse method," to appear in a volume on Solitons, (R. Bullough, ed.), Springer-Verlag.
5. Flaschka, H. and Newell, A., "Integrable systems of nonlinear evolution equations," *in* "Dynamical Systems Theory and Applications" (J. Moser, ed.), pp. 355-440. Springer-Verlag,

New York, (1975).
6. Scott, A.C., Chu, F.Y.F. and McLaughlin, D.W., "The soliton: A new concept in applied science," *Proc. IEEE 61*, 1443-1483 (1973).
7. Flaschka, H. and McLaughlin, D.W., "Some comments on Bäcklund transformations, canonical transformations, and the inverse scattering method," *in* "Bäcklund Transformations" (R. Miura, ed.), pp. 253-295, Springer-Verlag, New York, (1976).
8. McLaughlin, D.W. and Scott, A.C., "Perturbation analysis of fluxon dynamics," *Phys. Rev. A 18*, 1652-1680 (1978).
9. Takhtadjian, L.A. and Faddeev, L.D., "Essentially nonlinear one-dimensional model of classical field theory," *Theor. Math. Phys. 21*, 1046 (1974).
10. Kaup, D.J. and Newell, A.C., "Solitons as particles and oscillators," *Proc. R. Soc. Lond. A 361*, 413-446 (1978).
11. McLaughlin, D.W. and Scott, A.C., "Soliton perturbation theory," *in* "Nonlinear Evolution Equations Solvable by the Spectral Transform" (F. Calagero, ed.), pp. 225-243. Putman, London, (1978).
12. McLaughlin, D.W. and Scott, A.C., "A multisoliton perturbation theory," *in* "Solitons in Action" (K. Lonngren and A.C. Scott, eds.), pp. 201-256. Academic Press, New York, (1978).
13. Karpman, V.I., "Soliton evolution in the presence of perturbation," Physica Scripta *20*, 462-475 (1979). This issue is devoted to solitons in physics.
14. Zakharov, V.E. and Faddeev, L.D., "Korteweg de Vries equation: a completely integrable Hamiltonian system," translated in *Funct. Anal. Appl. 5*, 280-287 (1972).
15. Bishop, A., "Statistical mechanics of nonlinear dispersive systems," *in* "Solitons in Action" (K. Lonngren and A.C. Scott, eds.), Academic Press, New York, (1978).
16. Hasslacher, B. and Newell, A., "Nonlinear quantum field theory," *Rocky Mountain J. of Math. 8*, 341-369 (1978). This entire issue is devoted to general articles on solitons and their applications.
17. Takhtadjian, L.A. and Faddeev, L.D., Uspekhi Mathematischeskikk Nauk *34*, 13-63 (1979), (to be translated in *Russian Math. Surveys*).
18. Thacker, H.B. and Wilkinson, D., "The inverse scattering transform as an operator method in quantum field theory," Preprint, *Fermi National Acc. Lab.* (1979).
19. Whitham, G.B., "Linear and Nonlinear Waves" Ch. 15. Wiley-Interscience, New York, (1974).
20. Zakharov, V.E. and Manakov, S.V., "Asymptotic behavior of the non-linear wave systems integrated by the inverse scattering method," *Sov. Phys. JETP 44*, 106-112 (1976).
21. Segur, H. and Ablowitz, M.J., "Asymptotic solutions and conservation laws for the nonlinear Schroedinger equation. I," *J. Math. Phys. 17*, 710-713 (1976).

22. Flaschka, H., Forest, G. and McLaughlin, D.W., "Multiphase averaging and the inverse spectral solution of K.dV." to appear in: *Comm. Pure Appl. Math.* (1980).
23. Lax, P. and Levermore, C.D., "The zero dispersion limit of K.dV.," *Proc. U.S. National Acad. Sciences, 76*, 3602 (1979).
24. Miura, R., "The Korteweg de Vries equation: a survey of results," *SIAM Review 18*, 412-459 (1976).
25. Haus, H.A., "Physical interpretation of the inverse scattering formalism applied to self induced transparency," *Rev. Mod. Phys. 51*, 331-341 (1979). In the same issue, there are two reviews about the inverse spectral solution of the equations of the three wave interaction by D.J. Kaup, A. Reiman, and A. Bers and by A. Reiman.

SOLITONS IN RANDOMLY INHOMOGENEOUS MEDIA

Ioannis M. Besieris

Department of Electrical Engineering
Virginia Polytechnic Institute and State University
Blacksburg, Virginia

I. INTRODUCTION

One of the most significant accomplishments in the area of mathematical physics during the past twelve years has been the development of the *inverse scattering method* [1-3], whereby initial and boundary value problems for nonlinear wave systems are solved exactly by the successive application of linear techniques. Such solutions consist of *solitons* [4] -- nonlinear wave packets emerging from collisions with similar packets with unaltered shape and speed by virtue of the balance of various physical mechanisms (e.g., nonlinearity and dispersion).

An important nonlinear partial differential equation which arises in electromagnetic propagation and is amenable to the "inverse method" is the nonlinear complex parabolic equation, wherein nonlinear corrections to the index of refraction are assumed to exhibit a pronounced Kerr effect, i.e., they are proportional to the field intensity (irradiance). Within the framework of the quasi-optical description, it models exceedingly well the self-interaction of light beams, which may lead to self-focusing, self-modulation, etc.

There exist physical situations (e.g., graded lightguides, laser-induced plasmas, optical channels arising from laser

ISBN 0-12-709660-4

radiation-induced evaporation of water drops in dense clouds and mist, etc.) which require that one take into account the presence of statistical fluctuations in addition to nonlinearities and dispersion. A model stochastic nonlinear complex parabolic equation incorporating all these effects is studied in this exposition.

The statistical technique expounded here is based on a fundamental ansatz [5]. This ansatz constitutes essentially an embedding process: through a transformation, the nonlinear stochastic parabolic equation is brought into a one-to-one correspondence with the nonlinear parabolic equation characterizing the unperturbed (deterministic) medium, together with a set of stochastic nonlinear ordinary differential equations. The basic statistical analysis of the original stochastic nonlinear partial differential equation is thus simplified considerably since one can now combine the results of the inverse scattering technique with the already highly developed theory of stochastic ordinary differential equations. Many physically interesting averaged observables can be calculated by this approach. Specifically, we shall report on the behavior of single-envelope solitons in the presence of statistical fluctuations.

II. THE NONLINEAR STOCHASTIC COMPLEX PARABOLIC EQUATION

Ignoring depolarization effects, the transverse, complex, electric field of radiation in a nondissipative, nonmagnetic, random, nonlinear medium is governed by the nonlinear stochastic Helmholtz equation [6]

$$\nabla^2 E(\underline{r},\omega;\alpha) + (\omega/c)^2 \varepsilon_r[\underline{r},\ \omega, |E(\underline{r},\omega;\alpha)|^2;\alpha]E(\underline{r},\omega;\alpha) = 0,\quad \underline{r} \in R^3. \tag{1}$$

Here, ω is the angular frequency, c is the speed of light *in vacuo*, and $\varepsilon_r[\underline{r},\ \omega, |E(\underline{r},\omega;\alpha)|^2]$ is the relative permittivity which is assumed to be a real function having the following form:

$$\varepsilon_r[\underline{r},\ \omega, |E(\underline{r},\omega;\alpha)|;\alpha] = \varepsilon_{rr}(\underline{r},\omega;\alpha) + \varepsilon_{rn}[\omega,\ |E(\underline{r},\omega;\alpha)|^2]. \tag{2}$$

In this expression, $\varepsilon_{rr}(\underline{r},\omega;\alpha)$ accounts for possible regular refraction, as well as random inhomogeneities in the channel [$\alpha \in A$, (A, F, P) being an underlying probability measure space]; on the other hand, $\varepsilon_{rn}[\omega,\ |E(\underline{r},\omega;\alpha)|^2]$, a functional of the irradiance $|E(\underline{r},\omega;\alpha)|^2$, is due to the existing nonlinearities. For the sake of convenience, we shall assume in the sequel that the medium exhibits a pronounced Kerr effect, i.e., $\varepsilon_{rn}[\omega,\ |E(\underline{r},\omega;\alpha)|^2]$ is proportional to the intensity of $E(\underline{r},\omega;\alpha)$, vis., $\varepsilon_{rn}[\omega,\ |E(\underline{r},\omega;\alpha)|^2] = \tilde{\varepsilon}_3(\omega)|E(\underline{r},\omega;\alpha)|^2$.

Let $E\{\varepsilon_{rr}(\underline{r},\omega;\alpha)\}$ and $\delta\varepsilon_{rr}(\underline{r},\omega;\alpha)$ denote respectively the average and fluctuating parts of $\varepsilon_{rr}(\underline{r},\omega;\alpha)$. Let, furthermore, $\varepsilon_0(\omega)$ be a convenient "reference" quantity. It may, for example, coincide with $E\{\varepsilon_{rr}(\underline{r},\omega;\alpha)\}$ if the latter is independent of the position variable $\underline{r}$. We introduce, next, four new quantities as follows:

$$K^2(\omega) = (\omega/c)^2\varepsilon_0(\omega), \tag{3}$$

$$\varepsilon_1(\underline{r},\omega;\alpha) = \delta\varepsilon_{rr}(\underline{r},\omega;\alpha)/\varepsilon_0(\omega), \tag{4}$$

$$\varepsilon_2(\underline{r},\omega) = [E\{\varepsilon_{rr}(\underline{r},\omega;\alpha)\} - \varepsilon_0(\omega)]/\varepsilon_0(\omega), \tag{5}$$

$$\varepsilon_3(\omega) = \tilde{\varepsilon}_3(\omega)/\varepsilon_0(\omega). \tag{6}$$

With these definitions, (1) assumes the form

$$\nabla^2 E(\underline{r},\omega;\alpha) + K^2(\omega)\varepsilon_3(\omega)|E(\underline{r},\omega;\alpha)|^2 E(\underline{r},\omega;\alpha) + K^2(\omega)[1 + \varepsilon_2(\underline{r},\omega) + \varepsilon_1(\underline{r},\omega;\alpha)]E(\underline{r},\omega;\alpha) = 0. \tag{7}$$

It is clear that $\varepsilon_2(\underline{r},\omega)$ accounts for the deterministic inhomogeneities in the medium and, in light of a statement made earlier, it vanishes in the absence of such background profiles.

On the other hand, $\varepsilon_1(\underline{r},\omega;\alpha)$, a zero-mean random function, is directly associated with the superimposed random effects. Of course, the term in (7) proportional to $\varepsilon_3(\omega)$ arises from the presence of nonlinearities.

For plane or beam propagation in the z-direction, it is convenient to resort to the transformation

$$E(\underline{r},\omega;\alpha) = \psi(\underline{x},z,\omega;\alpha)\,exp(iKz); \quad r = (\underline{x},z),\ K = K(\omega). \tag{8}$$

In the quasi-optical description, the slowly varying, complex, random, amplitude function $\psi(\underline{x},z,\omega;\alpha)$ is described exceedingly well by the nonlinear stochastic complex parabolic equation [7,8]

$$\frac{i}{K}\frac{\partial}{\partial z}\psi(\underline{x},z,\omega;\alpha) = -\frac{1}{2K^2}\nabla_{\underline{x}}^2\,\psi(\underline{x},z,\omega;\alpha)$$

$$+\ V(\underline{x},z,\omega;\alpha)\psi(\underline{x},z,\omega;\alpha),\ z > 0, \tag{9a}$$

$$V(\underline{x},z,\omega;\alpha) = -\frac{1}{2}\,[\varepsilon_3(\omega)\,|\psi(\underline{x},z,\omega;\alpha)|^2$$

$$+\ \varepsilon_2(\underline{x},\omega) + \varepsilon_1(\underline{x},z,\omega;\alpha)]. \tag{9b}$$

In the presence of a deterministic profile ($\varepsilon_2 \neq 0$) and nonlinearity ($\varepsilon_3 \neq 0$), the parabolic equation (9) is a valid approximation to (7) if the normals to the wavefronts in the unperturbed problem, where $\varepsilon_1 = 0$, remain close to the z-axis.

Corresponding to the boundary condition $E(\underline{x},0,\omega;\alpha) \equiv E_0(\underline{x},\omega;\alpha)$ for (7), one has the "initial" condition

$$\psi(x,0,\omega;\alpha) \equiv \psi_0(\underline{x},\omega;\alpha) = E_0(\underline{x},\omega;\alpha), \tag{9c}$$

which incorporates all the information concerning the temporal frequency spectrum and the spatial distribution of the source at the initial plane $z = 0$.

A statistical analysis of (9) has already been considered by various workers under the following restrictive assumptions:

Case (i): ε_2 = constant, $\varepsilon_3 = 0$ (linear)

This problem has been studied by Besieris and Tappert [9] in connection with the quantized motion of a particle in a randomly perturbed potential field. Extensive accounts of this problem, with application to optical propagation in atmospheric turbulence, have been given by Tatarskii and Klyatskin [10] on the basis of the pure Markovian approximation. More recent contributions by Furutsu [11], Dashen [12], and Rose and Besieris [13] deal primarily with the computation of second- and higher-order statistics for CW and pulsed signals.

Case (ii): $\varepsilon_3 = 0$ (linear)

The interaction of random inhomogeneities with a deterministic inhomogeneous background has been examined in connection with the quantum mechanical harmonic oscillator [14], laser propagation in lens-like media [5; 15-18] and underwater sound random wave propagation [19, 20].

Case (iii): $\varepsilon_1 = 0$, $\varepsilon_2 =$ constant (nonlinear)

Assuming that $\psi_0(\underline{x},\omega;\alpha)$ [cf. Eq. (9c)] is a complex random function specified at $z = 0$, this problem is of interest in connection with the self-interaction of incoherent light beams in a cubic medium [21]. Its statistical analysis so far is based on the random phase approximation.

Case (iv): $\varepsilon_2 =$ constant (nonlinear)

This case has been considered by Jokipii and Marburger [22] using the Rytov approximation for wave propagation in random media generalized to include self-focusing effects, and by Vorob'ev [23] within the framework of the pure Markovian random approximation -- a technique which allows large-scale fluctuations and is valid for long ranges.

In the work of Vorob'ev the effect of the nonlinearity in the medium on scattering is taken into consideration by neglecting the fluctuations of the nonlinear part of the relative permittivity on the assumption that they are small. This leads to the approximation

$$V(\underline{x},z,\omega;\alpha) = -\frac{1}{2}\left[\varepsilon_3(\omega)E\{|\psi(\underline{x},z,\omega;\alpha)|^2\} + \varepsilon_2(\underline{x},\omega) + \varepsilon_1(\underline{x},z,\omega;\alpha)\right] \tag{10}$$

in (9b). Using the resulting simplified parabolic equation, Vorob'ev has derived an exact equation for the single-frequency mutual coherence function $E\{\psi^*(\underline{x}_2,z,\omega;\alpha)\psi(\underline{x}_1,z,\omega;\alpha)\}$ in the pure Markovian approximation. (In this approximation, $\varepsilon_1(\underline{x},z,\omega;\alpha)$ is taken to be a homogeneous, wide-sense stationary, Gaussian random process, δ-correlated in z.) An approximate (within the confines of a quadratic Kolmogorov transverse spectrum) solution of this transport equation has revealed that "the presence of nonlinearity in the medium may substantially reduce the random broadening of a beam if the power in the beam is higher than a critical one, while the presence of strong fluctuations may impede the collapse of the beam."

III. SOLITONS IN A NONLINEAR MEDIUM WITH ADDITIVE STATISTICAL FLUCTUATIONS

It is our intent to consider the general problem (9), without imposing the random phase approximation [cf. Sec. II, Case (iii)], or the serious restrictions introduced in Vorob'ev's work [cf. Sec. II, Case (iv)]. More specifically, the fluctuations of the nonlinear part of the relative permittivity will not be neglected.

In principle, one may lift the aforementioned restrictions by solving (9) in variational derivatives for the characteristic functional (Hopf technique) in order to derive information about the statistical moments of the wave function $\psi(\underline{x},z,\omega;\alpha)$. Although the resulting functional equation turns out to be linear, its solution lies beyond the scope of presently available mathematical techniques.

A more fruitful approach is to make use of asymptotic statistical methods (e.g., the direct-interaction approximation) which have been proven useful in the study of hydrodynamic and plasma turbulence [24-26]. A characteristic feature shared by these techniques is that an equation, say for the nth-order moment of the wave function $\psi(\underline{x},z,\omega;\alpha)$, must be derived first (usually by a truncation of an infinite hierarchy), and then solved in order to obtain specific information about nth-order or lower-order averaged observables. (Because of the nonlinearity, the equation for the nth-order moment is coupled to averaged equations for the lower-order moments.) The derivation of such averaged equations is very difficult if realistic information is incorporated concerning the deterministic and statistical fluctuations of the medium parameters. In addition, the problem of integrating these equations is challenging, even with the availability of high-speed computers.

It is our specific intent in this exposition to present a novel statistical technique which allows the evaluation of second- and higher-order moments related to the nonlinear stochastic comples parabolic equation (9), without having to derive first equations for the second- and higher-order coherence functions. As outlined in the Introduction, this technique is based on a fundamental ansatz which, in turn, requires the integration of the deterministic nonlinear complex parabolic equation by means of the inverse scattering method.

In the absence of statistical fluctuations ($\varepsilon_1 = 0$) and a deterministic profile ($\varepsilon_2 = 0$), the inverse scattering method is fully developed to-date for the one-dimensional ($\underline{x} \rightarrow x \in R^1$) nonlinear complex parabolic equation. For this reason, our discussion in the sequel will be restricted to a planar (x,z) geometry. It is well known [27] in this case that the inverse scattering method can accommodate a deterministic profile: exactly in the case $\varepsilon_2(x,\omega) \propto \varepsilon_{22}(\omega)x$, or asymptotically (adiabatically) in general. For the sake of convenience, we shall eliminate the deterministic profile altogether in the following discussion. We shall limit the scope of our investigation in one more respect: only single-frequency coherence functions will be computed. This is, again, a matter of convenience, and not a limitation of the proposed statistical technique.

A model nonlinear stochastic complex parabolic equation incorporating the aforementioned restrictions is the following:

$$\frac{i}{K}\frac{\partial}{\partial z}\psi(x,z;\alpha) = -\frac{1}{2K^2}\frac{\partial^2}{\partial x^2}\psi(x,z;\alpha)$$
$$-\frac{1}{2}\varepsilon_3|\psi(x,z;\alpha)|^2\psi(x,z;\alpha)$$
$$-\frac{1}{2}\varepsilon_1(x,z;\alpha)\psi(x,z;\alpha), \quad z > 0, \tag{11a}$$

$$\psi(x,0;\alpha) = \psi_0(x). \tag{11b}$$

It should be noted that the dependence on the radian frequency ω is suppressed since our discussion will be confined to cw propagation.

A basic question that we shall endeavor to answer in this chapter is the following: Given that a single-soliton solution to the deterministic ($\varepsilon_1 = 0$) equation corresponding to (11) enters the channel at some range $z = z_0$, how will it be affected by the presence of the additive statistical fluctuations ($\varepsilon_1 \neq 0$) for ranges $z \geq z_0$?

The structure of the rest of the chapter is as follows: In Sections 4-6 the Donsker-Furutsu-Novikov functional formalism together with the assumption of a simplified (quadratic) Kolmogorov spectrum for $\varepsilon_1(x,z;\alpha)$ result in an effective stochastic nonlinear complex parabolic equation which is equivalent (in a sense to be made specific) to (11) and is amenable to the fundamental ansatz. The fundamental ansatz itself is discussed briefly in Section 5, and its implementation, which yields second- and higher-order averaged observables, is carried out in Sections 6-11.

IV. THE DONSKER-FURUTSU-NOVIKOV FUNCTIONAL FORMALISM

Our ultimate goal in connection with (11) is to compute the nth-order moments $E\{\psi^{(n)}(\underline{X},z;\alpha)\}$, where

$$\psi^{(n)}(\underline{X},z;\alpha) = \prod_{p=1}^{n/2} \psi^*(x_{2p},z;\alpha)\psi(x_{2p-1},z;\alpha), \tag{12a}$$

$$\underline{X} = (x_1, x_2, \ldots, x_n) \in R^n \quad , \tag{12b}$$

and n is assumed to be an even integer.

Remark: The choice of n even is made on the strength of physical evidence -- at least in the linear case -- that moments $E\{\psi^{(n)}(\underline{X},z;\alpha)\}$ with unequal number of conjugated and unconjugated terms in (12a) decay relatively fast with increasing range z.

An equation of evolution for $\psi^{(n)}(\underline{X},z;\alpha)$ follows readily from the definition (12a) and the basic equation for $\psi(x,z;\alpha)$ [cf. Eq. 11]:

$$\begin{aligned}\frac{i}{K}\frac{\partial}{\partial z}\psi^{(n)}(\underline{X},z;\alpha) = {} & \frac{1}{2K^2}\sum_{p=1}^{n}\xi_p\frac{\partial^2}{\partial x_p^2}\psi^{(n)}(\underline{X},z;\alpha) \\ & - \frac{1}{2}\varepsilon_3\sum_{p=1}^{n}\xi_p|\psi(x_p,z;\alpha)|^2\psi^{(n)}(\underline{X},z;\alpha) \\ & - \frac{1}{2}\sum_{p=1}^{n}\xi_p\varepsilon_1(x_p,z;\alpha)\psi^{(n)}(\underline{X},z;\alpha), \\ & z > 0,\end{aligned} \tag{13a}$$

$$\psi^{(n)}(\underline{X},0;\alpha) = \psi_0^{(n)}(\underline{X}). \tag{13b}$$

The quantity ξ_p entering into (13a) assumes the value 1 if p is odd, and becomes -1 if p is even.

We ensemble-average next both sides of (13) over the realizations $\alpha \in A$:

$$\frac{i}{K}\frac{\partial}{\partial z}E\{\psi^{(n)}(\underline{X},z;\alpha)\} = \frac{1}{2K^2}\sum_{p=1}^{n}\xi_p\frac{\partial^2}{\partial x_p^2}E\{\psi^{(n)}(\underline{X},z;\alpha)\}$$

$$-\frac{1}{2}\varepsilon_3\sum_{p=1}^{n}\xi_p E\{|\psi(x_p,z;\alpha)|^2$$

$$\times\ \psi^{(n)}(\underline{X},z;\alpha)\}$$

$$-\frac{1}{2}\sum_{p=1}^{n}\xi_p E\{\varepsilon_1(x_p,z;\alpha)\psi^{(n)}(\underline{X},z;\alpha)\},$$

$$z > 0, \tag{14a}$$

$$E\{\psi^{(n)}(\underline{X},0;\alpha)\} = \psi_0^{(n)}(\underline{X}). \tag{14b}$$

This, of course, is not a closed equation for $E\{\psi^{(n)}(\underline{X},z;\alpha)\}$ by virtue of the second and third terms on the right-hand side of (14a). A partial closure, however, can be effected for a special class of random processes $\varepsilon_1(x,z;\alpha)$.

Let $\varepsilon_1(x,z;\alpha)$ be a zero-mean, homogeneous, wide-sense stationary, Gaussian random process, δ-correlated in range, viz.,

$$E\{\varepsilon_1(x,z;\alpha)\varepsilon_1(x',z';\alpha)\} = A(x-x')\delta(z-z'). \tag{15}$$

It follows, then, from the Donsker-Furutsu-Novikov functional formalism [28-30] that

$$E\{\varepsilon_1(x_p,z;\alpha)\psi^{(n)}(\underline{X},z;\alpha)\} = \int_{-\infty}^{\infty} dx_p' A(x_p - x_p')$$

$$\times\ E\{\delta\psi^{(n)}(\underline{X},z;\alpha)/\delta\varepsilon_1(x_p',z;\alpha)\}. \tag{16}$$

The functional derivative $\delta\psi^{(n)}(\underline{X},z;\alpha)/\delta\varepsilon_1(x_p',z;\alpha)$ required in (16) can be found from the original stochastic nonlinear complex parabolic equation (11). Omitting intermediate steps, we present here the final result:

$$\frac{\delta\psi^{(n)}(X,z;\alpha)}{\delta\varepsilon_1(x_p',z;\alpha)} = i\,\frac{K}{4}\sum_{q=1}^{n}\xi_q\,\delta(x_q - x_q')\psi^{(n)}(\underline{X},z;\alpha). \tag{17}$$

Introducing (17) into (16), we obtain

$$E\{\varepsilon_1(x_p,z;\alpha)\psi^{(n)}(\underline{X},z;\alpha)\}$$

$$= i\,\frac{K}{4}\sum_{q=1}^{n}\xi_q\,A(x_p-x_q)\,E\{\psi^{(n)}(\underline{X},z;\alpha)\}. \tag{18}$$

Finally, (14) assumes the form

$$\frac{i}{K}\,\frac{\partial}{\partial z}E\{\psi^{(n)}(\underline{X},z;\alpha)\} = \frac{1}{2K^2}\sum_{p=1}^{n}\xi_p\,\frac{\partial^2}{\partial x_p^2}E\{\psi^{(n)}(\underline{X},z;\alpha)\}$$

$$-\frac{1}{2}\,\varepsilon_3\sum_{p=1}^{n}\xi_p\,E\{|\psi(x_p,z;\alpha)|^2$$

$$\times\,\psi^{(n)}(\underline{X},z;\alpha)\}$$

$$-\,i\,\frac{K}{8}\sum_{p=1}^{n}\sum_{q=1}^{n}\xi_p\xi_q\,A(x_p-x_q)$$

$$\times\,E\{\psi^{(n)}(\underline{X},z;\alpha)\},\quad z>0 \tag{19a}$$

$$E\{\psi^{(n)}(\underline{X},0;\alpha)\} = \psi^{(n)}(\underline{X}). \tag{19b}$$

Remark: Under the restrictions imposed so far on the process $\varepsilon_1(x,z;\alpha)$, (19) is an exact equation. It is clear, however, that it is not a closed transport equation for $E\{\psi^{(n)}(\underline{X},z;\alpha)\}$ due to the presence of nonlinearities. An attempt to effect a closure of the second term on the right-hand side of (19) will result in the infinite hierarchy of coupled moment equations mentioned in the previous section.

V. SIMPLIFIED (QUADRATIC) KOLMOGOROV SPECTRUM

In many physical situations, the transverse correlation $A(x_p-x_q)$ [cf. Eq. (19a)] is isotropic and of power-law type, viz.,

$$A(x_p-x_q) = A(0)\left\{1-\frac{1}{2}\left[\frac{1}{L_0}\,|x_p-x_q|\,\right]^{\beta}\right\}\ , \tag{20}$$

where L_0 is a characteristic length, and the power parameter β is usually within the range $1 < \beta < 4$.

Even in the absence of nonlinearity ($\varepsilon_3 = 0$), it is impossible to evaluate $E\{\psi^{(n)}(\underline{X},z;\alpha)\}$, with $n > 2$, exactly unless the parameter β introduced in (20) is equal to 2. For $n = 2$ and $\varepsilon_3 = 0$, (19) can be integrated exactly for arbitrary values of β.

In the sequel we shall restrict the discussion to the case $\beta = 2$ (simplified or quadratic Kolmogorov spectrum). In this case, (19) changes to

$$\frac{i}{K}\frac{\partial}{\partial z}E\{\psi^{(n)}(\underline{X},z;\alpha)\} = \frac{1}{2K^2}\sum_{p=1}^{n}\xi_p\frac{\partial^2}{\partial x_p^2}E\{\psi^{(n)}(\underline{X},z;\alpha)\}$$
$$-\frac{1}{2}\varepsilon_3\sum_{p=1}^{n}\xi_p E\{|\psi(x_p,z;\alpha)|^2 \times \psi^{(n)}(\underline{X},z;\alpha)\}$$
$$+\frac{i}{4}KD\sum_{\substack{p=1\\p<q}}^{n}\sum_{q=1}^{n}\xi_p\xi_q(x_p - x_q)^2 \times E\{\psi^{(n)}(\underline{X},z;\alpha)\},\quad z > 0 \tag{21a}$$

$$E\{\psi^{(n)}(\underline{X},0;\alpha)\} = \psi^{(n)}(\underline{X}), \tag{21b}$$

where $D = A(0)/2L_0^2$.

VI. EFFECTIVE NONLINEAR STOCHASTIC COMPLEX PARABOLIC EQUATION

Consider the auxiliary equation

$$\frac{i}{K}\frac{\partial}{\partial z}\psi_e(x,z;\alpha) = -\frac{1}{2K^2}\frac{\partial^2}{\partial x^2}\psi_e(x,z;\alpha)$$
$$-\frac{1}{2}\varepsilon_3|\psi_e(x,z;\alpha)|^2\psi_e(x,z;\alpha)$$
$$-xH(z;\alpha)\psi_e(x,z;\alpha),\quad z > 0, \tag{22a}$$

$$\psi_e(x,0;\alpha) = \psi_0(x). \tag{22b}$$

Let, furthermore, $H(z;\alpha)$ be a zero-mean, wide-sense stationary, δ-correlated, real Gaussian random process characterized by the autocovariance $E\{H(z;\alpha)H(z';\alpha)\} = (D/2)\delta(z-z')$, D being the quantity defined in the previous section. Then, the original nonlinear stochastic complex parabolic (11) is equivalent to the "effective" equation (22) in the sense that

$$E\{\psi^{(n)}(\underline{X},z;\alpha)\} = E\{\psi_e^{(n)}(\underline{X},z;\alpha)\} \tag{23}$$

for all even values of n.

Remark: It must be emphasized that the aforementioned equivalence is valid under the assumption of the quadratic Kolmogorov spectrum for the random process $\varepsilon_1(x,z;\alpha)$.

In order to prove the conjecture embodied in (23), we begin with the equation of evolution for $\psi_e^{(n)}(\underline{X},z;\alpha)$:

$$\frac{i}{K}\frac{\partial}{\partial z}\psi_e^{(n)}(\underline{X},z;\alpha) = -\frac{1}{2K^2}\sum_{p=1}^{n}\xi_p\frac{\partial^2}{\partial x_p^2}\psi_e^{(n)}(\underline{X},z;\alpha)$$

$$-\frac{1}{2}\varepsilon_3\sum_{p=1}^{n}\xi_p|\psi_e(x_p,z;\alpha)|^2\psi_e^{(n)}(\underline{X},z;\alpha)$$

$$-\sum_{p=1}^{n}\xi_p x_p H(z;\alpha)\psi_e^{(n)}(\underline{X},z;\alpha),$$

$$z > 0, \tag{24a}$$

$$\psi_e^{(n)}(\underline{X},0;\alpha) = \psi_0^{(n)}(\underline{X}). \tag{24b}$$

Ensemble-averaging both sides of (24) over the realizations of $\alpha \in A$ results in the expression

$$\frac{i}{K}\frac{\partial}{\partial z}E\{\psi_e^{(n)}(\underline{X},z;\alpha)\} = -\frac{1}{2K^2}\sum_{p=1}^{n}\xi_p\frac{\partial^2}{\partial x_p^2}E\{\psi_e^{(n)}(\underline{X},z;\alpha)\}$$

$$-\frac{1}{2}\varepsilon_3\sum_{p=1}^{n}\xi_p E\{|\psi_e(x_p,z;\alpha)|^2$$

$$\times \psi_e^{(n)}(\underline{X},z;\alpha)\}$$

$$- \sum_{p=1}^{n} \xi_p x_p E\{H(z;\alpha)\psi_e^{(n)}(\underline{X},z;\alpha)\},$$

$$z > 0, \tag{25a}$$

$$E\{\psi_e^{(n)}(\underline{X},0;\alpha) = \psi_0^{(n)}(\underline{X}). \tag{25b}$$

The statistical averaging in the third term on the right-hand side of (25a) can be handled by the Donsker-Furutsu-Novikov formulá; specifically,

$$E\{H(z;\alpha)\psi_e^{(n)}(\underline{X},z;\alpha)\} = \frac{D}{2} E\{\delta\psi_e^{(n)}(\underline{X},z;\alpha)/\delta H(z;\alpha)\}. \tag{26}$$

The functional derivative appearing in (26) can be computed from the effective nonlinear stochastic parabolic equation (22). The final result is

$$\delta\psi_e^{(n)}(\underline{X},z;\alpha)/\delta H(z;\alpha) = i\,\frac{K}{4} \sum_{q=1}^{n} \xi_q x_q \psi_e^{(n)}(\underline{X},z;\alpha). \tag{27}$$

Substituting (27) into (26), we have

$$E\{H(z;\alpha)\psi_e^{(n)}(\underline{X},z;\alpha)\} = \frac{i}{4} KD \sum_{q=1}^{n} \xi_q x_q E\{\psi_e^{(n)}(\underline{X},z;\alpha)\}, \tag{28}$$

and (25) is recast into the form

$$\frac{i}{K}\frac{\partial}{\partial z} E\{\psi_e^{(n)}(\underline{X},z;\alpha)\} = -\frac{1}{2K^2} \sum_{p=1}^{n} \xi_p \frac{\partial^2}{\partial x_p^2} E\{\psi_e^{(n)}(\underline{X},z;\alpha)\}$$

$$- \frac{1}{2}\varepsilon_3 \sum_{p=1}^{n} \xi_p E\{|\psi_e(x_p,z;\alpha)|^2$$

$$\times \psi_e^{(n)}(\underline{X},z;\alpha)\}$$

$$- \frac{i}{4} KD \sum_{p=1}^{n} \sum_{q=1}^{n} \xi_p \xi_q x_p x_q$$

$$\times \; E\{\psi_e^{(n)}(\underline{X},z;\alpha)\}, \quad z > 0, \tag{29a}$$

$$E\{\psi_e^{(n)}(\underline{X},0;\alpha)\} = \psi_0^{(n)}(\underline{X}). \tag{29b}$$

We observe, next, that

$$-\sum_{\substack{p=1 \\ p<q}}^{n} \sum_{q=1}^{n} \xi_p \xi_q (x_p - x_q)^2 = \sum_{p=1}^{n} \sum_{q=1}^{n} \xi_p \xi_q x_p x_q \, . \tag{30}$$

Bearing this relationship in mind, a direct comparison of (29) with (21) shows that the "equations" for $E\{\psi^{(n)}(\underline{X},z;\alpha)\}$ and $E\{\psi_e^{(n)}(\underline{X},z;\alpha)\}$ are identical. In this sense, the nonlinear stochastic parabolic equations (11) and (22) are equivalent.

Remark: Equation (29), being the same with (21), is not a closed transport equation for $E\{\psi_e^{(n)}(\underline{X},z;\alpha)\}$. In this respect, no real progress has been made, at least at the level of *local* transport or moment equations. Beginning at the next section, however, it will be shown that, in contradistinction to (11), the effective nonlinear stochastic complex parabolic equation (22) lends itself to a *global* statistical analysis.

VII. THE FUNDAMENTAL ANSATZ

In the effective nonlinear stochastic parabolic equation (22) we make a change of the transverse (with respect to z) spatial variable corresponding to a transition to a "moving" coordinate system,

$$y(z;\alpha) = x - u(z;\alpha), \tag{31}$$

and represent the wave function $\psi_e(x,z;\alpha)$ in the form

$$\psi_e(x,z;\alpha) = \phi[y(z;\alpha),z]\exp\{iK[\dot{u}(z;\alpha)y(z;\alpha) + \gamma(z;\alpha)]\}. \tag{32}$$

The new wave function $\phi(y,z)$ and the quantities $u(z;\alpha)$ and $\gamma(z;\alpha)$ can be shown [5] to satisfy the following relationships:

$$\text{(i)}\quad \frac{i}{K}\frac{\partial}{\partial z}\phi(y,z) = -\frac{1}{2K^2}\frac{\partial^2}{\partial y^2}\phi(y,z) - \frac{1}{2}\varepsilon_3|\phi(y,z)|^2\phi(y,z), \quad z > 0, \tag{33a}$$

$$\phi(y,0) = \psi_0(y); \tag{33b}$$

$$\text{(ii)}\quad \ddot{u}(z;\alpha) = H(z;\alpha), \quad z > 0, \tag{34a}$$

$$u(0;\alpha) = \dot{u}(0;\alpha) = 0; \tag{34b}$$

$$\text{(iii)}\quad \dot{\gamma}(z;\alpha) = \frac{1}{2}\dot{u}^2(z;\alpha) + u(z;\alpha) + u(z;\alpha)H(z;\alpha), \quad z > 0, \tag{35a}$$

$$\gamma(0;\alpha) = 0. \tag{35b}$$

It is seen that within the framework of this formulation, $\phi(y,z)$ is governed by the nonlinear complex parabolic equation characterizing the unperturbed medium. It should be noted, however, that ϕ is a random function by virtue of its implicit dependence on $u(z;\alpha)$, viz., $\phi = \phi[x - u(z;\alpha), z]$.

Our procedure in the sequel can be outlined as follows: The deterministic nonlinear parabolic equation (33) will be solved first for the wave function ϕ. The latter is a functional of the random function $u(z;\alpha)$ as mentioned earlier. This solution for ϕ will then be used in the expression (32) for the wave function $\psi_e(x,z;\alpha)$. This wave function, in turn, is a functional of $u(z;\alpha)$, $\dot{u}(z;\alpha)$ and $\gamma(z;\alpha)$ because of its dependence on ϕ and the presence of the exponential factor in (32).

In general, the computation of statistical moments of $\psi_e(x,z;\alpha)$ require knowledge of the joint probability density functional of the random processes $u(z;\alpha)$, $\dot{u}(z;\alpha)$ and $\gamma(z;\alpha)$. In this exposition, however, we are only interested in the special class of nth-order moments $E\{\psi^{(n)}(\underline{X},z;\alpha)\} = E\{\psi_e^{(n)}(\underline{X},z;\alpha)\}$. This introduces a significant simplification since $\psi_e^{(n)}(\underline{X},z;\alpha)$ is independent of $\gamma(z;\alpha)$. Thus, our statistical analysis need only be confined to finding the joint density functional of $u(z;\alpha)$ and $\dot{u}(z;\alpha)$.

VIII. ANALYSIS OF THE DETERMINISTIC BACKGROUND PROBLEM; INVERSE SCATTERING METHOD; SOLITONS

Consider the deterministic nonlinear complex parabolic equation (33), viz.,

$$\frac{i}{K}\frac{\partial}{\partial z}\phi(y,z) = -\frac{1}{2K^2}\frac{\partial^2}{\partial y^2}\phi(y,z) - \frac{1}{2}\varepsilon_3|\phi(y,z)|^2\phi(y,z). \tag{36}$$

For $\varepsilon_3 > 0$, (36) describes stationary two-dimensional self-focusing.

Equation (36) can be integrated exactly by the inverse scattering method [3]. A particular solution is the following:

$$\phi(y,z) = \sqrt{\varepsilon_3}\,\eta\, sech[2\sqrt{2}\,\eta\, K(y-y_0) - 8\eta\xi Kz] \times exp[-i4(\xi^2 - \eta^2)Kz + i2\sqrt{2}\xi Ky + i\theta_0]. \tag{37}$$

Such a solution is referred to as a soliton. The soliton is characterized by four constants: η, ξ, y_0, and θ_0. The parameter η specifies the amplitude of the soliton, y_0 its initial transverse (with respect to z) position, ξ its "velocity," and θ_0 its phase. The constants η and ξ are independent and can be chosen arbitrarily.

Remark: In physical situations where (36) describes self-modulation rather than self-focusing, the independent variable z has the meaning of time, and the solution (37) corresponds to a wave packet propagating without distortion of its envelope with speed equal to $(4/\sqrt{2})\xi$. On the other hand, in the self-focusing problem under consideration here, the soliton has the meaning of a structure with intensity $\varepsilon_3\eta^2 sech^2[2\sqrt{2}\eta K(y-y_0) - 8\eta\xi Kz]$ across a waveguide-like channel induced in the medium by the intense field making an angle $tan^{-1}[(4/\sqrt{2})\xi]$ to the z-axis [3,31].

We use, next, the exact soliton solution (37) (with $y_0 = 0$ and $\theta_0 = 0$ for the sake of simplicity) in conjunction with the fundamental ansatz (32):

$$\psi_e(x,z;\alpha) = \sqrt{\varepsilon_3}\,\eta\, sech\{2\sqrt{2}\,\eta K[x - u(z;\alpha)] - 8\eta\xi Kz\}$$
$$\times\, exp\{i2\sqrt{2}\,\xi K[x - u(z;\alpha)] - i4(\xi^2 - \eta^2)Kz$$
$$\times\, exp\{iK[[\dot{u}[x - u(z;\alpha)] + \gamma(z;\alpha)]]\}\,. \tag{38}$$

Remark: The expression given in (38) for $\psi_e(x,z;\alpha)$ constitutes a complete integration of the effective nonlinear stochastic parabolic equation (22), valid for each realization $\alpha \in A$.

It is clear from (38) that $\psi_e^{(n)}(\underline{X},z;\alpha)$ is a function only of the random processes $u(z;\alpha)$ and $\dot{u}(z;\alpha)$. Therefore, in order to compute the desired moments $E\{\psi^{(n)}(\underline{X},z;\alpha)\} = E\{\psi_e^{(n)}(\underline{X},z;\alpha)\}$, we shall have to examine first the stochastic differential equation (34).

IX. ANALYSIS OF THE BASIC STATISTICAL PROBLEM

The second-order, Langevin-type, stochastic ordinary differential equation (34) may be recast into the form

$$\frac{d}{dz}\,v(z;\alpha) = H(z;\alpha), \tag{39a}$$

$$\frac{d}{dz}\,u(z;\alpha) = v(z;\alpha),\ z > 0, \tag{39b}$$

$$u(0;\alpha) = v(0;\alpha) = 0. \tag{39c}$$

The "fine-grained" density, or "phase-space" distribution function, associated with (39) is introduced next as follows:

$$f(u,v,z;\alpha) = \delta[u - u(z;\alpha)]\delta[v - v(z;\alpha)], \tag{40a}$$

$$f(u,v,0;\alpha) = \delta(u)\delta(v). \tag{40b}$$

It is governed by the continuity, or Liouville, equation

$$\frac{\partial}{\partial z} f(u,v,z;\alpha) + v\frac{\partial}{\partial u} f(u,v,z;\alpha)$$
$$+ H(z;\alpha)\frac{\partial}{\partial v} f(u,v,z;\alpha) = 0. \tag{41}$$

Under the asusmptions made in Sec. 6 about the process $H(z;\alpha)$, it follows from the Donsker-Furutsu-Novikov formula that the average phase-space distribution function obeys the Fokker-Planck equation

$$\frac{\partial}{\partial z} E\{f(u,v,z;\alpha)\} + v \frac{\partial}{\partial u} E\{f(u,v,z;\alpha)\} - \frac{1}{2} D \frac{\partial^2}{\partial v^2} E\{f(u,v,z;\alpha)\} = 0, \tag{42a}$$

$$E\{f(u,v,0;\alpha)\} = \delta(u)\delta(v). \tag{42b}$$

The transport equation (42) can be solved easily using the method of characteristics. Its exact solution is a two-dimensional Gaussian distribution in u and v.

Let

$$\Sigma = E\{\underline{w}(z;\alpha)\underline{w}^T(z;\alpha)\} \tag{43}$$

denote the auto-covariance matrix of the zero-mean vector-valued process $\underline{w}(z;\alpha) = [u(z;\alpha), v(z;\alpha)]$. It is given explicitly as follows:

$$\Sigma = \{\sigma_{ij}^2\}, \quad i,j = 1,2; \tag{44a}$$

$$\sigma_{11}^2(z) = (D/3)z^3, \tag{44b}$$

$$\sigma_{12}^2(z) = \sigma_{21}^2(z) = (D/2)z^2, \tag{44c}$$

$$\sigma_{22}^2(z) = Dz. \tag{44d}$$

The desired normal distribution density function is

$$E\{f(u,v,z;\alpha)\} \equiv F(u,v,z) = F(\underline{w},z) = (2\pi)^{-1}(\det \Sigma)^{-\frac{1}{2}} \exp(-\frac{1}{2}\underline{w}^T \Sigma^{-1}\underline{w}), \tag{45}$$

where Σ^{-1} is the inverse of the covariance matrix. More explicitly,

$$F(u,v,z) = (2\pi)^{-1}(det\ \Sigma)^{-\frac{1}{2}}$$

$$exp\left[-\frac{1}{2}\frac{1}{det\ \Sigma}\left(\sigma_{22}^2 u^2 + \sigma_{11}^2 v^2 - 2\sigma_{12}^2 uv\right)\right], \qquad (46)$$

where the σ_{ij}^2, $i,j = 1,2$, are the entries of the covariance matrix [cf. Eqs. (44b) - (44d)] and $det\ \Sigma = (D/12)z^4$.

The quantity $F(u,v,z)$ given in (46) is nonnegative and normalized to unity. It can be considered as the joint probability density function of u and v.

The "marginal" probability density functions of the random functions $u(z;\alpha)$ and $v(z;\alpha)$ can be readily found from (46):

$$F(u,z) = \int_{-\infty}^{\infty} dv\ F(u,v,z) = (2\pi)^{-\frac{1}{2}}\frac{1}{\sigma_{11}}\ exp\left(-\frac{1}{2}\frac{1}{\sigma_{11}^2}u^2\right), \qquad (47a)$$

$$F(v,z) = \int_{-\infty}^{\infty} du\ F(u,v,z) = (2\pi)^{-\frac{1}{2}}\frac{1}{\sigma_{22}}\ exp\left(-\frac{1}{2}\frac{1}{\sigma_{22}^2}v^2\right). \qquad (47b)$$

X. SECOND-ORDER AVERAGED OVSERVABLES

We set as our first task the computation of the *mean intensity* $E\{|\psi(x,z;\alpha)|^2\} = E\{|\psi_e(x,z;\alpha)|^2\}$. Toward this end, we note from (38) that

$$i(x,z;\alpha) = (2\varepsilon_3\eta^2)^{-1}|\psi_e(x,z;\alpha)|^2$$

$$= \frac{1}{2}\ sech^2\{2\sqrt{2}\ \eta K[x - u(z;\alpha)] - 8\eta\xi Kz\}. \qquad (48)$$

This quantity is normalized to unity with respect to x for all $z \geq 0$.

The irradiance function $i(x,z;\alpha)$ defined above is a function only of the random process $u(z;\alpha)$. Its ensemble average, therefore, is given by

$$E\{i(x,z;\alpha)\} = \int_{-\infty}^{\infty} du\ F(u,z)i(x,z;\alpha), \qquad (49)$$

where $F(u,z)$ is the marginal density function found in the previous section.

It will prove convenient at this stage to introduce dimensionless quantities; specifically,

$$\sqrt{2}\, Kx \to \bar{x}, \tag{50a}$$

$$\sqrt{2}\, Ku \to \bar{u}, \tag{50b}$$

$$Kz \to \bar{z}, \tag{50c}$$

$$2D/3K \to c. \tag{50d}$$

(The last parameter, c, is a measure of the strength of the statistical fluctuations in the medium.) Without loss in generality, we assume, also, that the soliton parameters η and ξ are equal to 1/2 and 1/4, respectively.

In this case, the normalized irradiance given in (48) becomes

$$i(\bar{x},\bar{z};\alpha) = \frac{1}{2}\, sech^2(\bar{x} - \bar{u} - \bar{z}). \tag{51}$$

In the absence of random fluctuations in the medium ($\bar{u} \to 0$), the quantity $(1/2)sech^2(\bar{x} - \bar{z})$ is the intensity across a waveguide-like channel inclined with respect to the z-axis at an angle of 45°.

The normalized mean intensity [cf. Eq. (49)] can be written out explicitly as follows:

$$E\{i(\bar{x},\bar{z};\alpha)\} \equiv I(\bar{x},\bar{z};c)$$

$$= \frac{1}{2\sqrt{2\pi}} \int_{-\infty}^{\infty} d\bar{u}\, \frac{1}{\sqrt{c\bar{z}^3}}\, exp\left(-\frac{1}{2}\,\frac{\bar{u}^2}{c\bar{z}^3}\right)$$

$$\times\, sech^2(\bar{x} - \bar{u} - \bar{z}). \tag{52}$$

Let the limit $c \to 0$ (no statistical fluctuations), one has the relationship

$$\lim_{c \to 0} \frac{1}{\sqrt{2\pi}} \frac{1}{\sqrt{c\bar{z}^3}} exp\left(- \frac{1}{2} \frac{\bar{u}^2}{c\bar{z}^3} \right) = \delta(\bar{u}) . \tag{53}$$

Performing, then, the integration over $\bar{u}$ in (52) results in the expression

$$I(\bar{x},\bar{z};0) = \frac{1}{2} sech^2(\bar{x} - \bar{z}) \tag{54}$$

for the soliton intensity in the absence of random disturbances in the channel. The "evolution" of $I(\bar{x},\bar{z};0)$ is shown in Fig. 1.

The integration in (52) has been carried out numerically. A graphical representation of $I(\bar{x},\bar{z};c)$ for two values of c, $c = 5 \times 10^{-4}$ and $c = 5 \times 10^{-2}$, is shown in Figs. 2a and 2b, respectively. It is clear from these figures that the mean intensity spreads monotonically due to the irreversible effects of randomness.

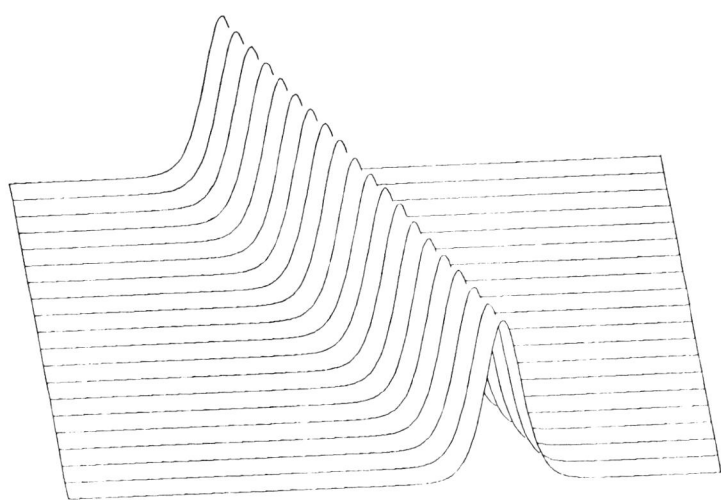

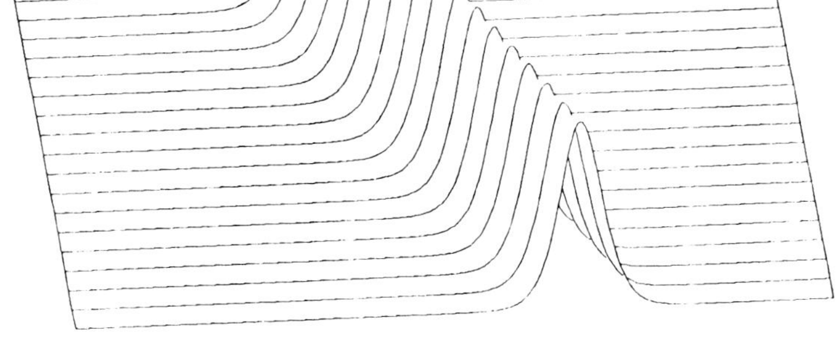

FIGURE 1. Evolution of the soliton intensity $I(\bar{x},\bar{z};0)$ in the absence of random inhomogeneities.

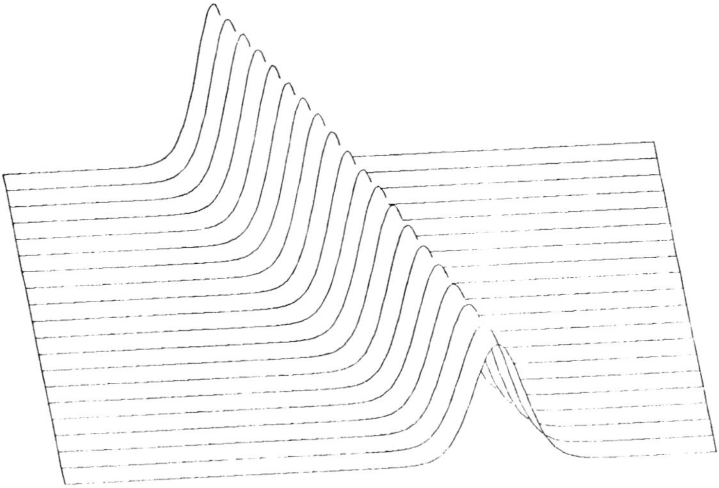

FIGURE 2a. Evolution of the mean soliton intensity in the presence of random fluctuations ($c = 5 \times 10^{-4}$).

An important question is whether the soliton channel axis, which is inclined with respect to the $\bar{z}$-axis at an angle of 45° in the case $c = 0$, will be affected by the random fluctuations. To answer this question, we introduce the *mean centroid*

$$X(\bar{z}) \equiv \int_{-\infty}^{\infty} d\bar{x}\ \bar{x}\ I(\bar{x},\bar{z};c)$$

$$= \frac{1}{2\sqrt{2\pi}} \int_{-\infty}^{\infty} d\bar{x} \int_{-\infty}^{\infty} d\bar{u}\ \frac{\bar{x}}{\sqrt{c\bar{z}^3}}\ exp\left(-\frac{1}{2}\ \frac{\bar{u}^2}{c\bar{z}^3}\right)$$

$$\times\ sech^2(\bar{x} - \bar{u} - \bar{z}). \qquad (55)$$

Performing the integration over $\bar{x}$ first and afterwards the one over $\bar{u}$ yields the answer $X(\bar{z}) = \bar{z}$. This means that, on the average, the soliton channel axis direction is an invariant.

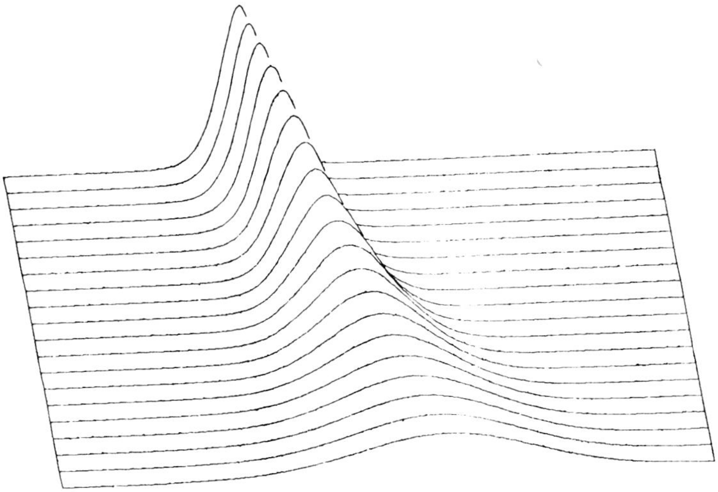

FIGURE 2b. Evolution of the mean soliton intensity in the presence of random fluctuations ($c = 5 \times 10^{-2}$).

(In the case of self-modulation, this statement is equivalent to the fact that the soliton average speed is not altered by the statistical fluctuations.)

It is seen from (52) that the mean intensity along the soliton channel axis is given by

$$I(\bar{x},\bar{z};c)\Big|_{\bar{x}=\bar{z}} = \frac{1}{2\sqrt{2\pi}} \int_{-\infty}^{\infty} d\bar{u}\, \frac{1}{\sqrt{c\bar{z}^3}} \times \exp\left(-\frac{1}{2}\,\frac{\bar{u}^2}{c\bar{z}^3}\right) \operatorname{sech}^2\bar{u} \ . \tag{56}$$

A plot of this quantity for $c = 5 \times 10^{-4}$ and $c = 5 \times 10^{-2}$ is shown in Figs. 3a and 3b.

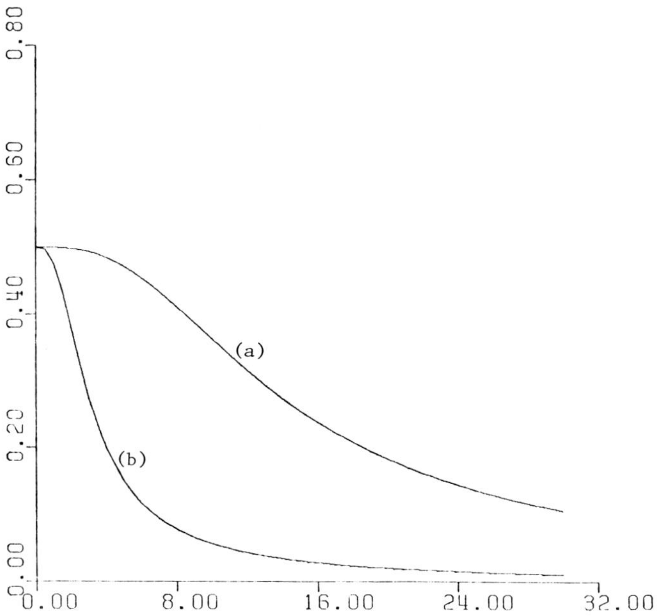

FIGURE 3. Mean intensity along the soliton channel axis. (a) $c = 5 \times 10^{-4}$; *(b)* $c = 5 \times 10^{-2}$.

A measure of the *spreading of the soliton* across its channel axis can be found by using the definition

$$\frac{1}{2}\sigma^2(\bar{z}) = \int_{-\infty}^{\infty} d\bar{x}[\bar{x} - X(\bar{z})]^2 I(x,z;c)$$

$$= \frac{1}{2\sqrt{2\pi}} \int_{-\infty}^{\infty} d\bar{x} \int_{-\infty}^{\infty} d\bar{u}(\bar{x} - \bar{z})^2 \frac{1}{\sqrt{c\bar{z}^3}}$$

$$\times \exp\left(-\frac{1}{2}\frac{\bar{u}^2}{c\bar{z}^3}\right) \operatorname{sech}^2(\bar{x} - \bar{u} - \bar{z}) . \quad (57)$$

Performing the $\bar{x}$ integration first and then the one over $\bar{u}$ we find that the *standard deviation* evolves in range as follows:

$$\sigma(z) = \left[2\left(\frac{\pi^2}{12} + c\bar{z}^3\right)\right]^{\frac{1}{2}} . \tag{58}$$

We close this section with one more second-order moment: the *spatial mutual coherence function*

$$\begin{aligned} I(\bar{x}_2,\bar{x}_1,\bar{z};c) &= (\varepsilon_3/2)^{-1} E\{\psi^{(2)}(\underline{X},z;\alpha)\} = (\varepsilon_3/2)^{-1} \\ &\quad \times E\{\psi_e^{(2)}(\underline{X},z;\alpha)\} \\ &= \frac{\sqrt{2}}{8\pi}\int_{-\infty}^{\infty} d\bar{u} \int_{-\infty}^{\infty} d\bar{v}\ \frac{1}{\sqrt{c\bar{z}^3}}\ \frac{1}{\sqrt{3c\bar{z}}} \\ &\quad \times exp\left(-\frac{1}{2}\ \frac{\bar{u}^2}{c\bar{z}^3}\right) exp\left(-\frac{\bar{v}^2}{3c\bar{z}}\right) \\ &\quad \times sech(\bar{x}_2 - \bar{u} - \bar{z})\, sech(\bar{x}_1 - \bar{u} - \bar{z}) \\ &\quad \times exp[-i\bar{v}(\bar{x}_2 - \bar{x}_1)]\, exp[-i(\bar{x}_2 - \bar{x}_1)], \end{aligned} \tag{59}$$

where $\bar{v} = v/\sqrt{2}$. This quantity is complex. Its amplitude and phase can be computed only numerically.

XI. HIGHER-ORDER AVERAGED OBSERVABLES

As an illustration of a physical situation which requires a fourth-order moment, consider the problem of "soliton wandering" across its mean channel axis.

We define the "centroid" of the soliton by

$$\bar{x}(z;\alpha) = \int_{-\infty}^{\infty} d\bar{x}\ \bar{x}\ i(\bar{x},\bar{z};a),\ \forall\, \alpha \in A \quad \text{[cf. Eq. (51)]}. \tag{60}$$

The mean centroid, $E\{\bar{x}(\bar{z};\alpha)\} = X(\bar{z})$, of the soliton has already been defined [cf. Eq. (55)] and found to be equal to $\bar{z}$. It is interesting, however, to consider also the variance of the random function $\bar{x}(z;\alpha)$, viz.,

$$
\begin{aligned}
var\{\bar{x}(\bar{z};\alpha)\} &= E\{[\bar{x}(z;\alpha) - X(z)]^2\} \\
&= \int_{-\infty}^{\infty} d\bar{x}_2 \int_{-\infty}^{\infty} d\bar{x}_1 \; (\bar{x}_2 - \bar{z})(\bar{x}_1 - \bar{z}) \\
&\qquad \times E\{i(\bar{x}_2,\bar{z};\alpha)\,i(\bar{x}_1,\bar{z};\alpha)\} \\
&= \frac{1}{4}\,\frac{1}{\sqrt{2\pi}} \int_{-\infty}^{\infty} d\bar{x}_2 \int_{-\infty}^{\infty} d\bar{x}_1 \int_{-\infty}^{\infty} d\bar{u}\; (\bar{x}_2 - \bar{z})(\bar{x}_1 - \bar{z}) \\
&\qquad \times \frac{1}{\sqrt{c\bar{z}^3}}\, exp\left(-\frac{1}{2}\,\frac{\bar{u}^2}{c\bar{z}^3}\right) \\
&\qquad \times sech^2(\bar{x}_2 - \bar{u} - \bar{z})\, sech^2(\bar{x}_1 - \bar{u} - \bar{z}) . \qquad (61)
\end{aligned}
$$

The integrations in (61) can be performed without difficulty. The final result is $var\{\bar{x}(\bar{z};\alpha)\} = \sigma_{11}^2(\bar{z}) = c\bar{z}^3$, an expression which vanishes at $\bar{z} = 0$ and becomes unbounded as $\bar{z} \to \infty$. This behavior is a manifestation of the instability of the soliton due to the presence of random fluctuations.

Let $i(\bar{x} = \bar{z},\bar{z};\alpha)$ denote the irradiance [cf. Eq. (51)] along the soliton axis for each realization $\alpha \in A$. A quantity of physical interest is its variance, viz.,

$$
\begin{aligned}
var\{i(\bar{x} = \bar{z},\bar{z};\alpha)\} &= E\{[i(\bar{x} = \bar{z},\bar{z};\alpha) - E\{i(\bar{x} = \bar{z},\bar{z};\alpha)\}]^2\} \\
&= E\{i^2(\bar{x} = \bar{z},\bar{z};\alpha)\} - [E\{i(\bar{x} = \bar{z},\bar{z};\alpha)\}]^2 \\
&= \frac{1}{4}\,\frac{1}{\sqrt{2\pi}} \int_{-\infty}^{\infty} d\bar{u}\; \frac{1}{\sqrt{c\bar{z}^3}}\, exp\left(-\frac{1}{2}\,\frac{\bar{u}^2}{c\bar{z}^3}\right) sech^4\bar{u} \\
&\qquad - \left[\frac{1}{2}\,\frac{1}{\sqrt{2\pi}} \int_{-\infty}^{\infty} d\bar{u}\; \frac{1}{\sqrt{c\bar{z}^3}} \right. \\
&\qquad \left. \times\, exp\left(-\frac{1}{2}\,\frac{\bar{u}^2}{c\bar{z}^3}\right) sech^2\bar{u} \right]^2 . \qquad (62)
\end{aligned}
$$

Figures 4a and 4b show the behavior of the on-axis fluctuations of the soliton intensity for two values of the parameter c,

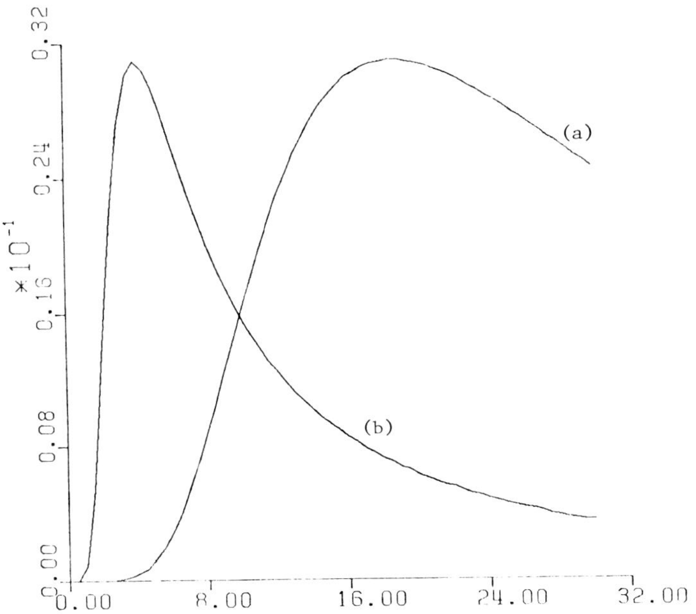

FIGURE 4. On-axis fluctuations of the soliton intensity (a) $c = 5 \times 10^{-4}$; (b) $c = 5 \times 10^{-2}$.

$c = 5 \times 10^{-4}$ and $c = 5 \times 10^{-2}$. As expected from physical considerations, $var\{i(\bar{x} = \bar{z}, \bar{z}; \alpha)\}$ vanishes at the plane of the source ($\bar{z} = 0$). It is also seen, however, that $var\{i(\bar{x} = \bar{z}, \bar{z}; \alpha)\}$ decays monotonically to zero as $\bar{z} \to \infty$, after it rises to a maximum at an intermediate range.

ACKNOWLEDGMENTS

The author wishes to thank George C. Papanicolaou and Conrad M. Rose for several informative discussions related to the subject of this chapter. He is also indebted to Anastasios I. Tsolakis for carrying out the numerical calculations leading to Figures 1-4 incorporated in Sections 10 and 11.

REFERENCES

1. Gardner, C.S., Greene, J.M., Kruskal, M.D., and Miura, R.M., *Phys. Rev. Lett. 19*, 1095 (1967).
2. Lax. P.D., *Comm. Pure Appl. Math. 21*, 467 (1968).
3. Zakharov, V.E. and Shabat, A.B., *Zh. Eksp. Teor. Fiz. 61*, 118 (1971).
4. Scott, A.C., Chu, F.Y.F., and McLaughlin, D.W., *Proc. IEEE 61*, 1443 (1973).
5. Besieris, I.M., *J. Math. Phys. 19*, 2533 (1978).
6. Bloembergen, N. "Nonlinear Optics," Benjamin, New York, (1965).
7. Bespalov, V.I., Litvak, A.G., and Talanov, V.I., in "Nonlinear Optics," (Transactions of the Second All-Union Symposium on Nonlinear Optics, Izd. Nauka, Novosibirsk, 1968).
8. Karpman, V.I., "Nonlinear Waves in Dispersive Media," Pergamon Press, New York, (1975).
9. Besieris, I.M. and Tappert, F.D., *J. Math. Phys. 14*, 1829 (1973).
10. Tatarskii, V.I., "Propagation of Waves in a Turbulent Atmosphere," (Nauka, Moskow, 1967); *Soviet Phys. - JETP 29*, 1133 (1969); Klyatskin, V.I. and Tatarskii, V.I., *Izv. VUZ., Radiofiz. 13*, 1061 (1970); *14*, 1400 (1971).
11. Furutsu, K., *J. Math. Phys. 20*, 617 (1979).
12. Dashen, R., *J. Math. Phys. 20*, 894 (1979).
13. Rose, C.M. and Besieris, I.M., *J. Math. Phys. 20*, 1530 (1979).
14. Besieris, I.M., Stasiak, W.B., and Tappert, F.D., *J. Math. Phys. 19*, 359 (1978).
15. Papanicolaou, G.C., McLaughlin, D., and Burridge, R., *J. Math. Phys. 14*, 84 (1973).
16. McLaughlin, D.W., *J. Math. Phys. 16*, 100 (1975).
17. Beran, M.J. and Whitman, A.M., *J. Math. Phys. 16*, 214 (1975).
18. Molodtsov, S.N. and Saichev, A.I., *Izv. VUZ., Radiofiz. 21*, 1785 (1978).
19. Besieris, I.M. and Kohler, W.E., *SIAM J. Appl. Math. 34*, 423 (1978).
20. Kohler, W.E. and Papanicolaou, G.C., "Wave Propagation in a Randomly Inhomogeneous Ocean," in Lecture Notes in Physics No. 70, "Wave Propagation and Underwater Acoustics," (J.B. Keller and J. Papadakis, ed.) Springer-Verlag, Berlin (1977).
21. Pasmarik, G.A., *Soviet Phys. - JETP 39*, 234 (1974).
22. Jokipii, J.R. and Marburger, J., *Appl. Phys. Lett. 23*, 696 (1973).
23. Vorob'ev, V.V., *Izv. VUZ., Radiofiz. 13*, 1053 (1970).
24. Kraichman, R.H., *J. Fluid Mech. 5*, 497 (1959); *J. Math. Phys. 2*, 124 (1961); Weinstock, J., *J. Phys. Fluids 12*, 1045 (1969); *13*, 2308 (1970); *15*, 454 (1972).

25. Balescu, R. and Misguich, J.H., *J. Plasma Phys. 13*, 33 (1975); Misguich, J.H. and Balescu, R., *J. Plasma Phys. 13*, 385, 419, 429 (1975).
26. Besieris, I.M. and Stasiak, W.B., *J. Math. Phys. 17*, 1711 (1976).
27. Chen, H.H. and Liu, C.S., *Phys. Rev. Lett. 37*, 693 (1977); *Phys. Fluids 21*, 377 (1978).
28. Donsker, M.D., "On Function Space Integrals," in "Analysis in Function Space," (W.T. Martin and I. Segal, ed.), M.I.T. Press, Cambridge, MA (1964).
29. Furutsu, K., *J. Res. Natl. Bur. Stand. Ser. D. 67*, 303 (1963).
30. Novikov, E.A., *Zh. Eksp. Teor. Fiz. 47*, 191 (1964).
31. Bullough, R.K., "Solitons in Physics," in "Nonlinear Equations in Physics and Mathematics," (A.O. Barut, ed.) D. Reidel Publishing Co., Dordrecht, Holland, (1978).

QUASIPARTICLE VIEW OF BEAM PROPAGATION IN NONLINEAR MEDIA[1]

Nathan Marcuvitz

Polytechnic Institute of New York
New York, New York

I. INTRODUCTION

The propagation of high power laser or microwave beams through material media frequently modifies the properties of the material in a manner dependent on the beam amplitude. One is thereby led to the problem of beam propagation in an inhomogeneous, nonlinear, dispersive medium. Purely analytic methods of solution are either nonexistent or of limited applicability, and hence one is forced to seek a mix of analytical and numerical techniques that in some sense provides an optimum method of solution.

In discussing nonlinear wave propagation, one distinguishes between the relatively simple "kinematic" problem of finding source-free wave solutions and the more difficult "dynamic" excitation problem. In linear waveguide media the kinematic or eigenvalue problem yields a spectrum of eigenvalues and eigenmodes in terms of which a solution of the excitation problem can be represented. In nonlinear media the kinematic wave solutions, defined by ordinary differential equations, may be discrete

[1]*The research reported herein has been supported by the Office of Naval Research under contract no. N 00014-77-c-0169.*

ISBN 0-12-709660-4

"stationary" wave solutions of the "soliton" type or a continuous spectrum of "cnoidal" periodic stationary wave solutions. However, the absence of orthogonality properties in the nonlinear case precludes a simple representation of the solution of the dynamic excitation problem in terms of stationary waves. In special cases, if a change of variables to an equivalent linear problem can be found, the method of inverse scattering [1] does provide a complete analytical procedure for solution of the excitation problem. But this method is often too difficult and too restricted for many practical nonlinear problems and one often resorts to averaging methods of solution.

The method of averaging is applicable to wave problems wherein there exist relatively "fast" carrier and "slow" modulational scales and the primary concern is with the slow average evolution of beam amplitude, wavenumbers, etc. In such cases averaging methods, of which Whitham's variational procedure [2] is an elegant example, lead to dispersion and transport equations for the desired average quantities; these latter equations are also nonlinear and usually require numerical techniques for their solution. An alternative method, likewise based on averaging out the fast wave behavior, associates a system of "quasiparticles" with the wave beam. The dynamics of the quasiparticle system is described kinetically in terms of a distribution function whose moments yield the averaged wave properties of interest. This quasiparticle view of wave propagation is indicative of a wave-particle duality that is familiar in quantum mechanics and is not unfamiliar in classical wave dynamics where at least at linear levels particle interpretations may be expressed in ray-optic terms. Although conventional ray treatments of wave propagation suffer from singularity difficulties, which may be partially removed by complex ray procedures [3], the quasiparticle method is not constrained by such difficulties. In the following we derive the quasiparticle description from the equations for the wave field and give an example of its use in nonlinear beam propagation. Since the quasiparticle description of beam propagation along a spatial direction is isomorphic to that of wave-packet evolution in time [4], much of the following terminology is based on quasiparticle dynamics that evolves in time rather than in distance.

II. THE QUASIPARTICLE APPROXIMATION

A beam wave formed of guided modes of a single type will be viewed at any plane z as a system of point quasiparticles, each described by its transverse position $x(z)$, transverse momentum $k(z)$, and longitudinal momentum $k_z(k,x,z)$. The individual quasiparticle motion is determined by the functional form of k_z as determined from the dispersion relation for the mode type. A kinetic description of the overall quasiparticle system in k, x space at plane z is given in terms of a distribution function $F(k,x,z)$ which provides a physical picture of how the beam evolves in z. A "fluid" description is obtained from the k moments of $F(k,x,z)$ and a "macro-quasiparticle" description is recovered from the k and x moments of $F(k,x,z)$. Our initial task is to derive a defining equation for $F(k,x,z)$ from the equations descriptive of the beam field.

Consider a single mode type of wave problem described by a complex wave function $\psi(x,z)$ satisfying

$$i\,\frac{\partial\psi}{\partial z} = P\psi \tag{1}$$

with the operator

$$P = P\left(\frac{1}{i}\,\frac{\partial}{\partial x},x,z,|\psi|\right)$$

and with ψ subject to a boundary condition $\psi(x,0)$. Such problems arise for example when one considers the "forward scattered" wave descriptive of wave propagation in a nonuniform, nonlinear, dispersive medium. Although the analysis is applicable to three dimensions, for simplicity of presentation only one transverse spatial dimension x will be considered. To identify the quasi-particle descriptors as well as the kinetic distribution function F from the defining equation for ψ, one introduces the double Fourier representation

$$\psi_1\psi_2{}^* = \iint_{-\infty}^{+\infty} \phi_1\phi_2{}^* \, exp[i(k_1x_1 - k_2x_2)]dk_1dk_2/(2\pi)^2 \tag{2}$$

where $\phi_1 = \phi(k_1,z)$ and $\phi_2 = \phi(k_2,z)$ are, respectively, the Fourier amplitudes of $\psi_1 = \psi(x_1,z)$ and $\psi_2 = \psi(x_2,z)$. The product $\psi_1\psi_2{}^*$ displays both a fast and slow spatial behavior and for the quasi-particle interpretation it is essential to average out the very fast behavior. An average over a suitably fast spatial period or phase is implicit in the following. After the averaging, "slow" (x or k) and "fast" (ξ or κ) spatial or wavenumber variables are defined by:

$$x = (x_1 + x_2)/2 \ , \ k = k_1 - k_2 \tag{3}$$

and

$$\xi = x_1 - x_2 \ , \ k = (k_1 + k_2)/2$$

whence since

$$k_1x_1 - k_2x_2 = k\xi + \kappa x \quad \text{and} \quad dk_1dk_2 = dkd\kappa,$$

equation (2) can be rewritten as

$$\psi_1\psi_2{}^* = \int_{-\infty}^{+\infty} F(k,x,z) \ exp(ik\xi)dk/2\pi \tag{3a}$$

where

$$F(k,x,z) = \int_{-\infty}^{+\infty} \phi_1\phi_2{}^* \ exp(i\kappa x)d\kappa/2\pi \tag{3b}$$

The reality property $F(k,x,z) = F^*(k,x,z)$ of the fast (correlation) spectrum of $\psi_1\psi_2{}^*$ follows readily from equation (3);

however, positive definiteness is not in general assured. Fourier transformations inverse to equation (3) are manifestly

$$F(k,x,z) = \int_{-\infty}^{+\infty} \psi_1\psi_2^* \, exp(ik\xi)d\xi \ , \tag{4a}$$

$$\phi_1\phi_2^* = \int_{-\infty}^{+\infty} F(k,x,z) \ exp(-i\kappa x)dx \ . \tag{4b}$$

For $\xi = 0$ and $k = 0$ equations (3a) and (4b) yield

$$|\psi(x,z)|^2 = \int_{-\infty}^{+\infty} F(k,x,z) \ dk/2\pi \ ,$$

$$|\phi(k,z)|^2 = \int_{-\infty}^{+\infty} F(k,x,z)dx \ ,$$

whence one infers that

$$\int_{-\infty}^{+\infty} |\psi(x,z)|^2 \ dx = \int_{-\infty}^{+\infty} |\phi(k,z)|^2 \ dk/2\pi$$

$$= \int_{-\infty}^{+\infty} F(k,x,z)dkdx/2\pi \ . \tag{5}$$

Even though only reality but not positive definiteness of $F(k,x,z)$ is assured, equation (5) suggests the identification of $F(k,x,z)$ as a number density of quasiparticles in k,x phase space. The consequent interpretation of $|\psi|^2$ as a quasiparticle number density in real space and of $|\phi|^2$ as a number density in momentum space are in conformity with a similar concept in quantum mechanics.

The defining equation for $F(k,x,z)$ can be deduced from equation (1) if one infers therefrom that

$$i \frac{\partial}{\partial z} (\psi_1\psi_2)^* = (P_1 - P_2^*)\psi_1\psi_2^* \tag{6}$$

where

$$P_1 = P\left(\frac{1}{i}\ \frac{\partial}{\partial x_1}, x_1, z\right) \qquad P_2{}^* = P\left(-\frac{1}{i}\ \frac{\partial}{\partial x_2}, x_2, z\right)$$

Noting the definitions (3) and forming the inverse transform (4a) of (6), one obtains

$$i\ \frac{\partial F}{\partial z} = \int_{-\infty}^{+\infty} exp(-ik\xi)\left[P\left(\frac{1}{i}\ \partial_\xi + \frac{1}{2i}\ \partial_x, x + \frac{\xi}{2}, z\right)\right.$$

$$\left. - P\left(\frac{1}{i}\ \partial_\xi - \frac{1}{2i}\ \partial_x, x - \frac{\xi}{2}, z\right)\right] \psi_1 \psi_2{}^*\ d\xi$$

After commuting the operator P so that it acts only on $exp(ik\xi)$ with a consequent change in sign of ∂_ξ, one finds as the formally exact defining equation for F:

$$i\ \frac{\partial F}{\partial z} = \left[P\left(k + \frac{1}{2i}\ \partial_k, x - \frac{1}{2i}\ \partial_k, z\right)\right.$$

$$\left. - P\left(k - \frac{1}{2i}\ \partial_x, x + \frac{1}{2i}\ \partial_k, z\right)\right] F\ . \tag{7}$$

Assuming a weak dependence of F and P on x, k, one expands P in first order to obtain as the "quasiparticle approximation" to (7):

$$\frac{\partial F}{\partial z} - \frac{\partial k_z}{\partial k}\ \frac{\partial F}{\partial x} + \frac{\partial k_z}{\partial x}\ \frac{\partial F}{\partial k} = 0 \tag{8}$$

where $k_z = P(k,x,z)$. Equation (8) has the form of a collisionless "kinetic" equation indicative of a distribution of "particles" in k, x phase space at plane z. The distribution function $F(k,x,z)$ defines the number density of such "particles" per unit volume of the phase space and the kinetic equation indicates the "characteristic" trajectories along which the "particles" move. The characteristic equations of the partial differential equation (8), namely:

$$\frac{dx}{dz} = -\frac{\partial k_z}{\partial k} \qquad \frac{dk}{kz} = \frac{\partial k_z}{\partial x} \tag{9}$$

identify the trajectories of quasiparticles of position $x(z)$, transverse momentum $k(z)$ and longitudinal momentum $k_z(k,x,z)$. On a quasiparticle trajectory, one observes in the quasiparticle approximation that for the mode type (1):

$$\frac{dF}{dz} = 0 \qquad \text{and} \qquad \frac{dk_z}{dz} = \frac{\partial k_z}{\partial z} \quad . \tag{10}$$

The implied constancy of F on a quasiparticle trajectory permits the z dependent solution of (8), which evolves from a given initial state $F(k,x,0)$, to be written conventionally as

$$F(k,x,z) = F(k_0(k,x,z), x_0(k,x,z), 0) \tag{11}$$

where $k_0(k,x,z)$, $x_0(k,x,z)$ define coordinates at plane $z = 0$ of a quasiparticle trajectory ending at k,x at plane z. However, for nonlinear potentials the solution (11) is difficult to employ in numerical computations because the indicated coordinates depend on $F(k,x,z)$. Note that in the quasiparticle approximation if $F(k,x,0)$ is positive definite, so also is $F(k,x,z)$.

"Fluid dynamic" properties of a quasiparticle system can be inferred from the k-moments of the kinetic distribution function F. Such properties are representative of macroscopic characteristics of beam solutions of (1). If a typical beam solution is represented as

$$\psi(x,z) = a(x,z)\ exp[i\theta(x,z)] \tag{12}$$

then as noted in equation (5) the local amplitude $a(x,z)$ follows from $F(k,x,z)$ via:

$$a^2(x,z) = \int_{-\infty}^{+\infty} F(k,x,z)\ dk/2\pi \ . \tag{13}$$

One observes that the amplitude squared $a^2(x,z)$ of the beam, or the zeroth moment of F, is indicative of the number density of quasiparticles. The local phase $\theta(x,z)$ of the beam can be determined from local transverse and longitudinal wavenumbers defined by:

$$K(x,z) = \frac{\partial\theta}{\partial x} \quad \text{and} \quad K_z(x,z) = \frac{\partial\theta}{\partial z}, \tag{14a}$$

by noting that

$$\theta\,(x,z) = \theta(x_0,0) + \int (K\,dx + K_z\,dz) \tag{14b}$$

where the integration in x,z space is over an appropriate path extending from $x_0,0$ to x,z. To find the local tranverse and propagation wavenumbers one first observes, in view of (14a), that

$$K(x,z) = -\,i\left(\psi_2{}^* \frac{\partial}{\partial x}\,\psi_1 - \psi_1 \frac{\partial}{\partial x}\,\psi_2{}^*\right)/2a^2\Big]_{\xi=0} \tag{14c}$$

$$-K_z(x,z) = i\left(\psi_2{}^* \frac{\partial}{\partial z}\,\psi_1 - \psi_1 \frac{\partial}{\partial z}\,\psi_2{}^*\right)/2a^2\Big]_{\xi=0}$$

whence, using (3a) and (1), one derives from (14c) in the limit $\xi = 0$ the relations

$$K(x,z) = \frac{1}{a^2}\int_{-\infty}^{+\infty} k\;F(k,x,z)\;dk/2\pi \tag{15a}$$

$$K_z(x,z) = \frac{1}{a^2}\int_{-\infty}^{+\infty}\left[k_z(k,x,z) - \frac{1}{8}\,\frac{\partial^2 k_z}{\partial k^2}\,\frac{\partial^2}{\partial x^2} - \frac{1}{8}\,\frac{\partial^2 k_z}{\partial x^2}\,\frac{\partial^2}{\partial k^2} + \cdots\right]F\;dk/2\pi \tag{15b}$$

which identify the local x,z dependent transverse wavenumber K and propagation wavenumber K_z as averages or moments of the quasiparticle distribution function F. Thus, via equations (3), (8), (13), and (14), the x,z dependent amplitude and phase characteristics of a propagating beam can be found directly from a kinetic description starting from initial values $F(k,x,0)$ and $\theta(x,0)$, which are derivable from $\psi(x,0)$. In particular, equation (15b) provides a x,z dependent "dispersion relation" relating $K_z(x,z)$ and $K(x,z)$.

III. EXAMPLE

To illustrate the applicability of the quasiparticle method we consider the problem of the propagation of a steady state beam in a nonlinear dispersive medium. The latter will be characterized by a field dependent index of refraction $n(x,z) = 1 + 2\delta n(x,z)$ that varies relatively weakly with x and z. The relevant wave equation is chosen as:

$$[\partial_z^2 + \partial_x^2 + (k_0 n)^2]\ \psi(x,z) = 0\ . \tag{1}$$

For a beam carrier having a fast wavenumber k_0 along z the wavefunction ψ can be expressed as:

$$\psi = exp[i(k_0 z)]\psi(x,z) \tag{2}$$

with the complex amplitude ψ being weakly dependent on x,z. On substitution of (2) into (1), assuming that $\partial_z\psi << ik_0\psi$, $\delta n << 1$ the equation defining $\psi(x,z)$ becomes:

$$i\partial_z\psi + \frac{1}{2}\partial_x^2\ \psi + k_0^2 d^2\delta n(x,z)\psi = 0 \tag{3a}$$

where d characterizes the scale of the field variations in the transverse x-direction. Equation (3a) is a single mode type ("forward scattering") equation. As an initial condition we choose a beam with the simple transverse form

$$\psi(x,0) = exp(-x^2/2), \tag{3b}$$

and observe that the x-dependence can be chosen quite arbitrarily. In numerical computations the quasiparticle technique becomes particularly useful when there is a relatively rapidly varying x-dependent phase. For nonlinear media, wherein the index depends nonlinearly on the amplitude $|\psi|$, equation (3a) is of the type (2.1) with the operator

$$P = -\frac{1}{2}\partial_x^2 + V(x,z,|\psi|) \quad \text{and} \quad V(x,z) = -k_0^2 d^2\delta n(x,z)\ . \tag{4}$$

One observes in (4) that the nonlinear potential $V(x,z)$ is a functional $V\{|\psi|\}$. More specifically the potential characterizing the medium will be assumed to have the form:

$$V = \frac{b^2}{2B}\left[exp(-B|\psi|^2)-1\right] \quad . \tag{5}$$

For the case $B = 0$, the above example has received considerable numerical [5] and analytical [6] attention in the literature under the heading of the "nonlinear parabolic or Schroedinger equation." The relevant potential is manifestly both inhomogeneous and nonstationary. In the $z = 0$ vicinity the potential follows from the entrance condition on $\psi(x,0)$ prescribed in (3b). From equation (2.3b) this condition implies that the quasiparticle phase space density at $z = 0$ is:

$$F(k,x,0) = \sqrt{4\pi}\ exp(-x^2 - k^2) \tag{6}$$

with the evolution in z being determined by the kinetic equation (2.8).

The longitudinal quasiparticle momentum, or dispersion relation, corresponding to the operator P of equation (4) is evidently:

$$-k_z = \frac{k^2}{2} + V(|\psi|) \tag{7}$$

with V given in (4). For quasiparticles with the indicated longitudinal momentum $k_z(x,z)$, the trajectory equations follow from equations (2.9) and (2.13) as:

$$\frac{dx}{dz} = k \ , \quad \frac{dk}{dz} = -\frac{b^2}{2B}\frac{\partial}{\partial x}\left[e^{-B\int_{-\infty}^{+\infty} F(k,x,z)\ dk/2\pi} - 1\right] \tag{8}$$

The dependence on $F(k,x,z)$ implies that these equations must be solved self consistently with the kinetic equation (2.8). The trajectory and kinetic equations may be solved by numerical computation starting from the entrance plane at $z = 0$. One notes that for nonlinear potentials the relation (2.11) does not

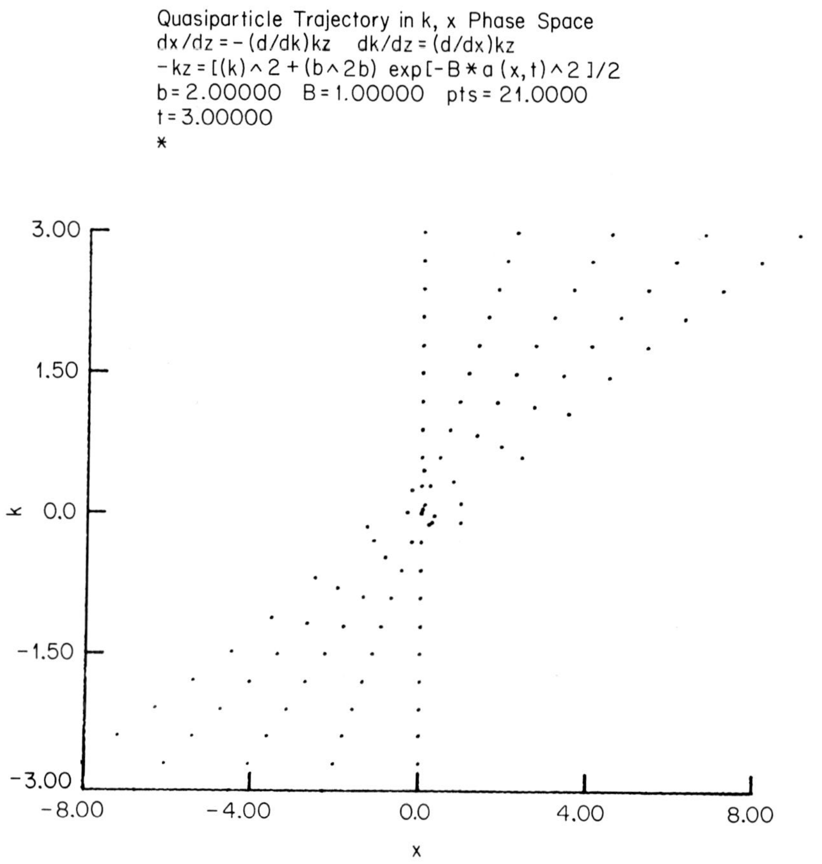

FIGURE 1. *Phase space locations of 21 quasiparticles originally in* $z = 0$ *plane at* $x = 0$ *with momenta* $-3 < k < 3$ *viewed at equi-spaced planes between* $z = 0$ *and* $z = 3.0$.

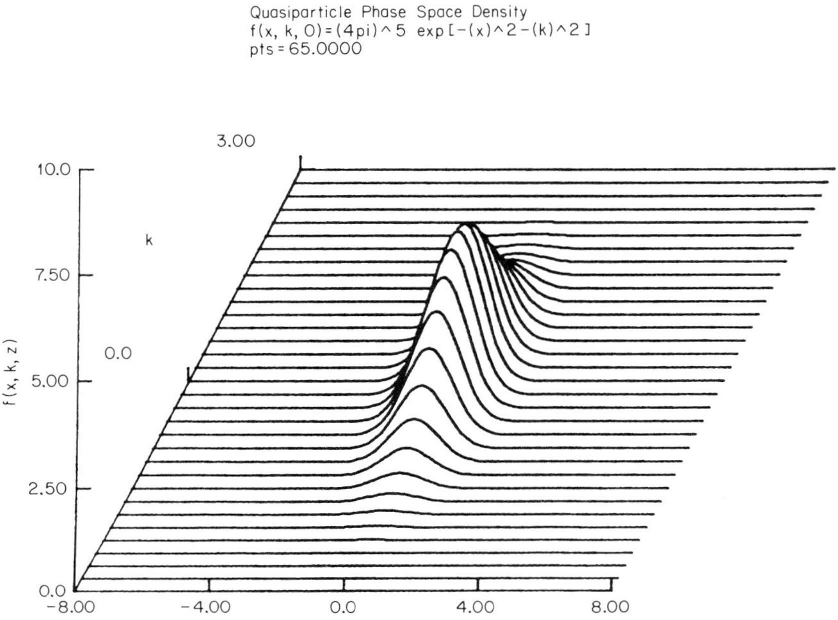

Figure 2a.

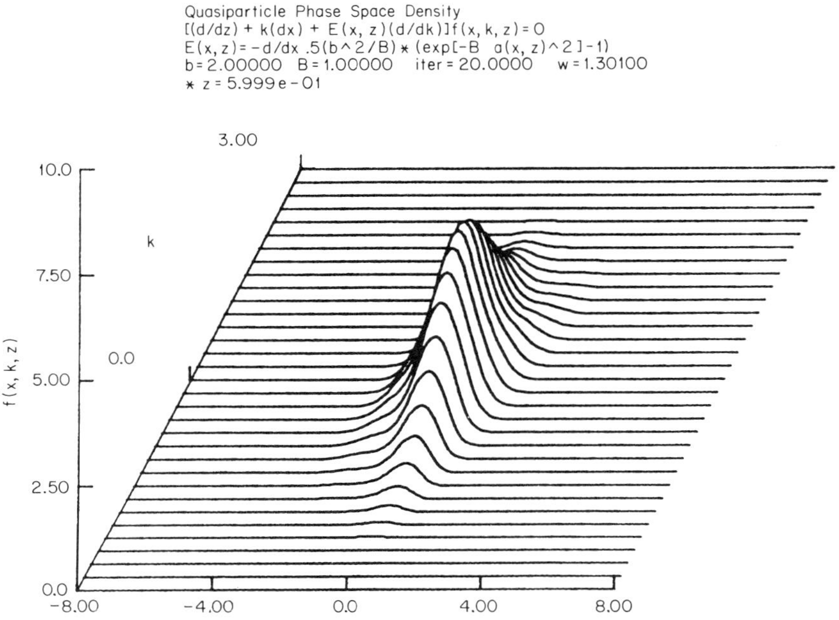

Figure 2b.

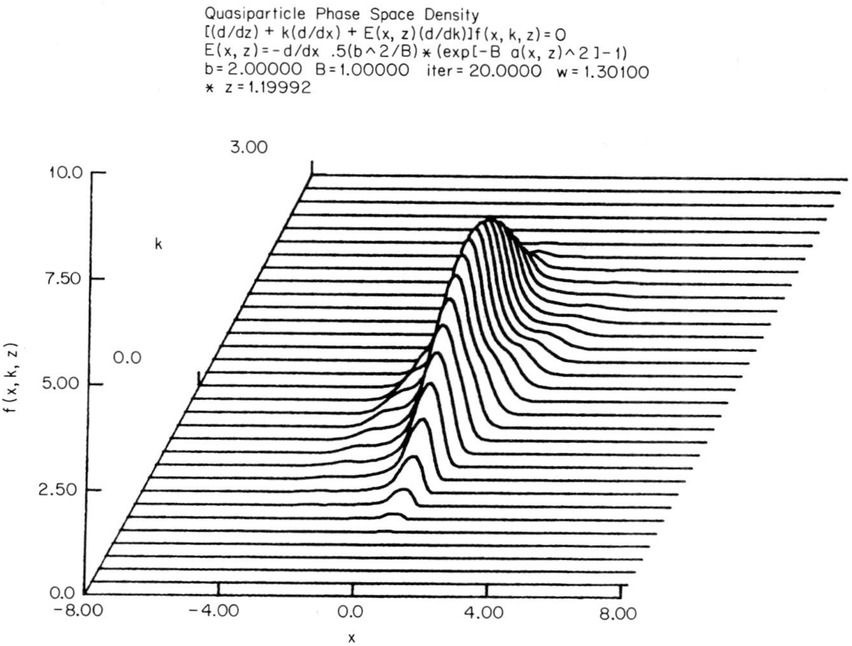

Figure 2c.

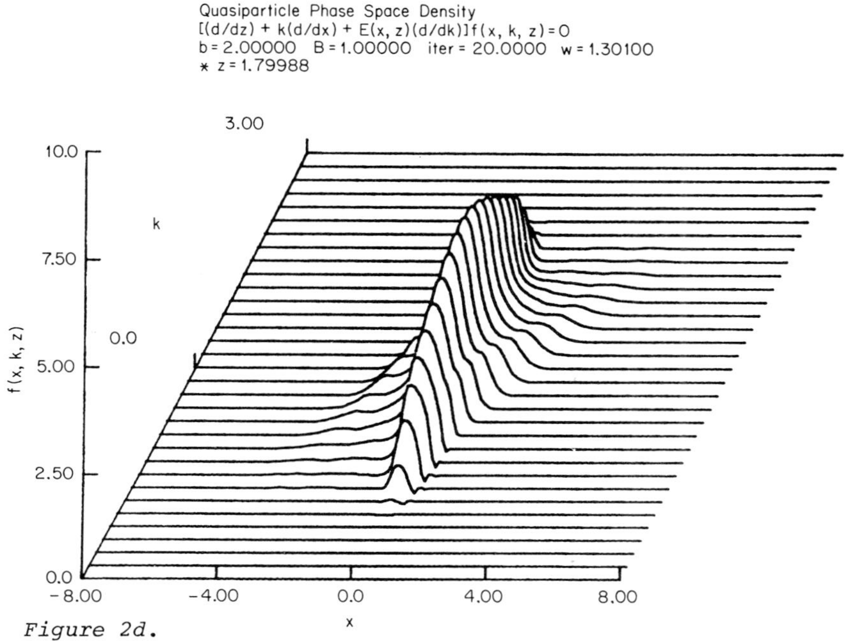

Figure 2d.

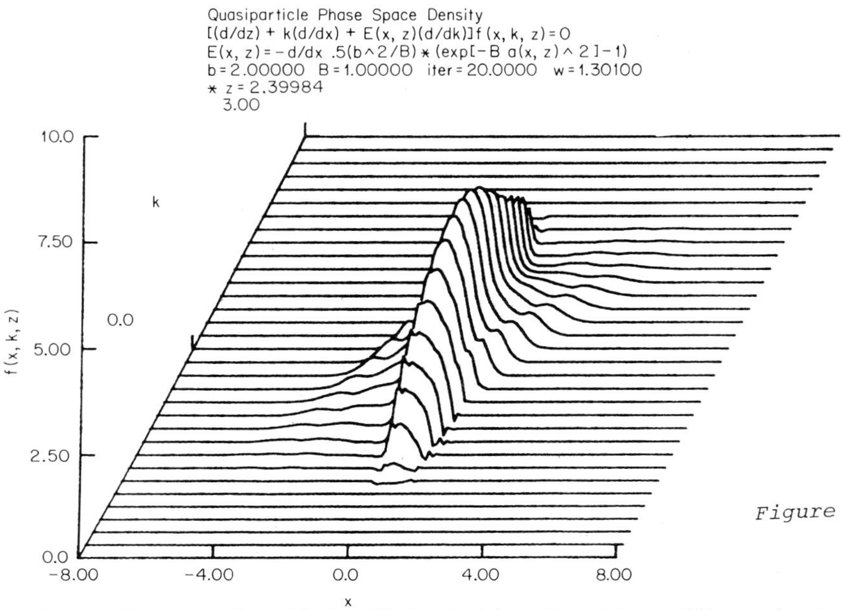

Figure 2e.

Figure 2. Quasiparticle distribution functions f(k,x,z) at equi-spaced planes between z = 0 and 3.0.

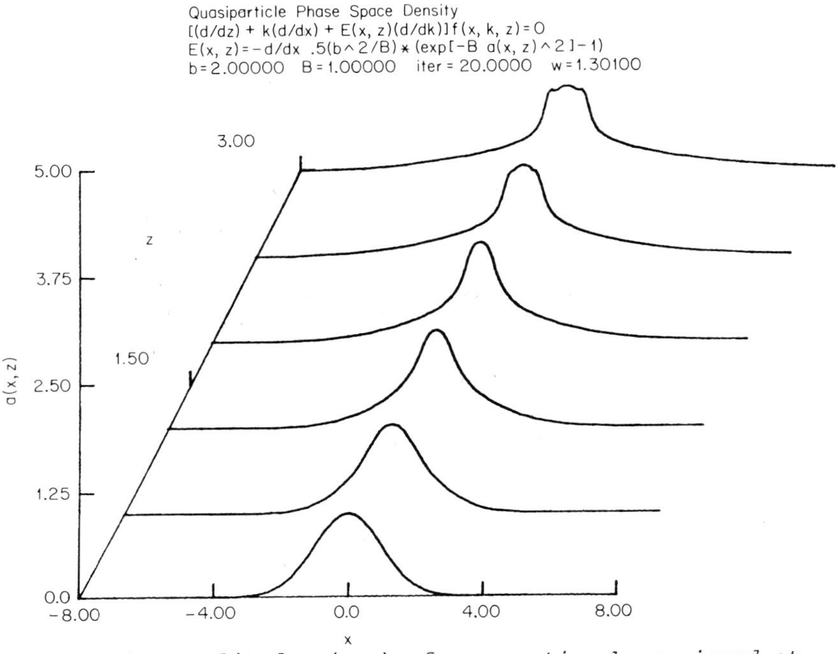

Figure 3. Amplitude a(x,z) of propagating beam viewed at equi-spaced planes between z = o and 3.0.

provide a simple numerical procedure for the determination of $F(k,x,z)$ from $F(k,x,0)$ since computational initialization is required on an exit plane z at which $F(k,x,z)$ is initially unknown. For the above nonlinear case a direct numerical solution of the kinetic equation (2.8) was employed using an alternating algorithm of second order accuracy in x and z.

We display in figure 1 for $b = 2$ and $B = 1$, at equal spaced planes ranging from $z = 0$ to $z = 3.0$, the successive phase space locations of quasiparticles initially at $x = 0$ with momenta evenly distributed within the interval $-3 < k < 3$. The z-dependent potential is seen to trap a number of quasiparticles in the $x = 0$ vicinity and permit others located at larger x to escape. Updating of the initial quasiparticle distribution (6) is effected by direct numerical solution of the kinetic equation (2.8). The resulting numerical calculations of $F(k,x,z)$ at equally spaced planes between $z = 0$ and $z = 3.0$ are displayed in figures 2a-d for the case $b = 2$ and $B = 1$. A quasi-singularity is apparent in the formation of cliff like structures in the quasiparticle distribution and leads to a numerical error which is presumed small. For $B = 0$ a singularity does become evident at finite z. This behavior arises because the nonlinear potential creates a negative central "pressure" that drives quasiparticles toward $x = 0$ with the result that negative momentum quasiparticles pile up on the positive side of $x = 0$ while positive momentum quasiparticles pile up on the negative $x = 0$ side. The different contributions of trapped and untrapped quasiparticles to this piling up process can also be inferred from careful following of individual quasiparticle paths in figure 1 plots. Numerical k-integration of $F(k,x,z)$ in accordance with equation (2.13) provides the data for the amplitude $a(x,z)$ plots in figure 3; the latter illustrate the effects of the piling up at $x = 0$ of the trapped quasiparticles and also the spreading effects due to the untrapped particles.

REFERENCES

1. Scott, A.C., Chu, F.Y.F., and McLaughlin, D.W., "The soliton a new concept in applied science," *Proc. IEEE 61*, 11, October (1973).
2. Whitham, G.B., "Linear and Nonlinear Waves" Wiley-Interscience, New York, (1974).
3. Choudhary, S. and Felsen, L.B., "Analysis of gaussian beam propagation," *Proc. IEEE 62*, 11, November (1974).
4. Marcuvitz, N., "Quasiparticle view of wave propagation," *PINY-MRI* Report 1402-79.
5. Zakharov, V.E. and Shabat, A.B., "Exact theory of nonlinear waves," *Soviet Physics JETP 34*, 62-9 (1972).
6. Zakharov, V.E. and Synakh, V.S., "The nature of the self-focusing singularity," *Soviet Physics JETP 41*, 3 (1975).

THE NONSTATIONARY EVOLUTION OF LOCALIZED WAVE FIELDS IN NONLINEAR DISPERSIVE MEDIA

A.B. Shvartsburg

IZMIRAN, Academgorodok
Moscow Region, USSR

INTRODUCTION

This chapter is devoted to the analytical theory of controlled nonlinear evolution of localized electromagnetic wave fields in dispersive media. Unlike the well-known soliton-like quasi-stationary wave distributions, in this review the attention is concentrated on the problems of nonstationary rebuilding of intense wave distributions inside nonlinear dispersive media. Such rebuilding possibilities are based on the competition between the nonlinear and dispersive phenomena which accompany the wave propagation, and on the sensitivity of the dynamics of such processes to the initial amplitude-phase profiles of the localized waves distribution. Recently, the analysis of wave groups envelopes has attracted attention only in gas and fluid mechanics, but lately, in connection with the appearance of power radiation sources in the radio and optical ranges, the whole complex of nonlinear optics and nonlinear electrodynamics problems has begun to extend. The investigation of nonlinear wave processes has become an important trend in radio physics, acoustics and plasma physics. Such investigations are presently of considerable theoretical and practical interest in connection with problems of high-frequency energy transfer, wave electronics, laboratory and

ISBN 0-12-709660-4

cosmic plasma processes, and physical phenomena peculiarities in high-power electromagnetic fields.

The analysis of wave processes shows a considerable analogy in the behavior of waves in different nonlinear media, regardless of the different physical nature of these waves. The wide range of interesting wave problems has stimulated the creation of one common approach and mathematical technique to analyse such phenomena. It is interesting to notice that approximately forty years ago a similar situation arose in connection with the analysis of nonlinear lumped-parameter systems oscillations in radio, mechanics and astronomy. Such a situation led to the elaboration of the common theory of nonlinear oscillations for lumped-parameter systems. During the last few years, the progress in intense waves dynamics pointed to the creation of the theory of waves in nonlinear distributive systems as a new branch of physics.

The analogy of nonlinear evolution tendencies among different wave fields permits the analysis of these tendencies in the framework of one individual group of wave processes without loss of generality. In this connection, it is worthwhile to concentrate attention on problems of nonlinear electrodynamics.

The perturbation of the medium's dispersive properties in the intense wave field changes the propagation conditions of the wave itself. The calculation of the influence of this perturbation on the wave field leads to the nonlinear equations of electromagnetic field evolution. The calculation of nonlinear perturbations in the real and imaginary parts of the refraction index corresponds to different processes. The imaginary part perturbations describe the effects of nonlinear absorption, harmonics generation, waves cross-modulation, and parametrical excitation of the oscillations [1]. These phenomena depend, as usual, upon the local field intensity only and are connected neither with the intensity distribution heterogeneity, nor with the field phase structure.

Unlike this, perturbations of the real part of the refractive index characterize the nonlinear corrections to the phase and group velocities. The role of such perturbations is essential in the evolution of the localized wave fields with heterogeneous phase and amplitude structure. These perturbations may stimulate the series of nonstationary wave processes: pulse self-constriction, spectrum broadening, self-stratification of the initially smooth wave distributions [2].

It is necessary to emphasize that the aforesaid division of nonlinear electrodynamics phenomena into two groups, connected with perturbations of real and imaginary parts of the refractive index, is to a marked degree conventional, because both types of perturbations are connected in many effects. Thus, the wave energy cumulation as a result of the refractive index real part perturbation, stimulates the amplification of the wave transformation phenomena near the intensity maximum; these transformations, in turn, are described with the help of the refractive index imaginary part perturbation. However, such division seems to be useful in this review, and the nonlinear phenomena connected with the imaginary part of the refractive index will not be discussed here. These effects were recently discussed in several books [3,4].

Unlike this, the dynamical picture of simultaneous rebuilding of amplitude and phase profiles of the localized wave distribution, connected with the real part of refractive index perturbation, is elaborated considerably less. The specific interest in this picture is stimulated by the problems of controlled evolution of intense wave pulses. The rate and the dynamics of such evolution display the essential dependence upon the initial pulse profiles. This dependence intensifies the interest in the exact analytical solutions of the wave evolution equations. One may hope that such solutions permit to predict the amplitude-phase profile of the pulse at some moment of the evolution, the initial wave distribution being known. Therefore, the development of the localized nonlinear waves nonstationary evolution theory seems to be very useful. The present review is devoted to the successes and the difficulties of this theory.

The method of nonlinear geometric optics in application to the aforesaid problems stands out through this chapter. It is well known that the geometric optics linear approximation is valid in the theory of optical instruments, in wave propagation problems, in the WKB method in quantum mechanics. Unlike this, the geometric optics nonlinear generalization, connected with the self-consistent picture of waves energy and spectrum deformation, is much more complicated. The wide classes of exact analytical solutions describing the localized electromagnetic wave pulse self-action will be constructed herein within the framework of nonlinear geometric optics. The useful analogy between nonlinear geometric optics and Eulerian hydrodynamical equations will be utilized in Section I. On the basis of these solutions the tendencies of controlled large scale self-constriction and self-stratification of the localized pulses will be illustrated. The nonlinear phase effects (spectrum broadening, nonlinear polarization) will be considered also. The same analogy is continued in application to the controlled pulses spreading-out effects in Section II. The peculiar properties of Riemann-like waves in a dispersive medium are considered. The limits of nonlinear geometric optics utilization are analysed in the framework of exact solutions for the nonlinear Schroedinger equation in Section III. The critical regimes, connected with nonstationary controlled evolution of the narrow high intensity region inside the medium, are illustrated here for the different solutions of these equations. The considerable enlargement of the limits of nonlinear geometric optics utilization in such critical regimes is shown. The approximate self-similar solutions of the two-dimensional nonlinear Schroedinger equation, connected with the arbitrary degree of the initial waves distribution anisotropy, are constructed in Section IV. The inverse scattering method, utilized in Section III to check the one-dimensional geometric optics solutions, is not valid here; therefore, the obtained two-dimensional solutions are verified in the framework of variational methods. Some actual modern problems of the geometric optics analysis of nonlinear wave processes are outlined in the Conclusion.

I. THE HYDRODYNAMICAL ANALOGY IN NONLINEAR ELECTROMAGNETIC WAVE THEORY

The localized wave pulse nonstationary rebuilding in the homogeneous transparent nonlinear medium is considered here. The nonlinearity is supposed to be connected with the dielectric permittivity dependence upon the electric field. In the isotropic medium, the nonlinearity may be described by a small additive term to the dielectric permittivity "linear" value, the nonlinear term being proportional to the wave intensity W:

$$\varepsilon = \varepsilon_0 + \Delta\varepsilon; \ \Delta\varepsilon = const \cdot W = \delta_1|\beta| \cdot |E|^2; \ \delta_1 = sgn\ \beta;$$

The constant β characterizes the nonlinear medium. It is worthwhile to notice that the pulse dynamics is very sensitive to the details of the dependence $\Delta\varepsilon(W)$. The aforesaid "quadratic" model represents the simplest case, connected formally with the first term in the function expansion in power series of W. However, even this simple form permits us to analyse many nonlinear phenomena. Other types of $\Delta\varepsilon(W)$ dependence, containing the saturation of nonlinearity and distorting the "quadratic" model results, were considered in [5].

The complex of phenomena of nonstationary amplitude-phase coupled evolution of intense electromagnetic pulses may be described in the framework of the nonlinear Schroedinger equation. In a simple one-dimensional problem one can write this equation in a dimensionless form:

$$i\frac{\partial\psi}{\partial\tau} + \frac{\partial^2\psi}{\partial q^2} + \chi|\psi|^2\psi = 0. \tag{1.1}$$

The meaning of the dimensionless variables τ and q depends upon the concrete type of wave process considered. In the nonstationary picture of evolution of high frequency electromagnetic wave pulses, the τ and q variables are connected with the coordinate Z and the time variable t:

$$\tau = \frac{Z}{L_S}\ ;\ L_S = \frac{2(V_0T_0)^2}{|V_\omega|}\ ;\ q = \frac{t - Z\cdot V_0^{-1}}{T_0}\ ;$$

$$V_0 = \frac{\partial\omega}{\partial k}\ ;\ V_\omega = \frac{\partial V_0}{\partial\omega} = \delta_2|V_\omega|\ . \tag{1.2}$$

The variable τ is connected with the "characteristic dispersion time" t_0 usually utilized for description of the dispersive spreading of short pulses: $\tau = (2\pi)^{-1}t_0^2 T_0^{-2}$; $t_0^2 = |-\pi Z V_\omega \cdot V_0^{-2}|$. Here ω is the wave angular frequency, V_0 is the wave group velocity, the dispersion of the medium is described by the parameter V_ω. The initial pulse duration is equal to $2T_0$. The parameter χ in eq. (1.1) depends upon the correlation between the dispersion and the nonlinearity of the medium:

$$\chi = \frac{L_S^2}{2L_0^2}\cdot\delta_1\delta_2;\ L_0^2 = \frac{2c(V_0\ T_0)^2}{\omega\cdot|V_\omega\cdot\beta|E_0|^2|}\ ;\ \delta_2 = sgn\ V_\omega; \tag{1.3}$$

In the following, the wave group velocity dependence upon the wave intensity is supposed to be negligible. The localized pulse evolution connected with the nonlinear additional term in the group velocity was considered in [6].

Eq. (1.1) describes the space-time wave pulse rebuilding, if the initial amplitude and phase profiles are known. However, one has to face the considerable mathematical difficulties connected with attempts to construct the solutions of this equation directly. Therefore, the role of results which show the general tendencies of nonlinear pulse evolution is very important. In order to reveal such tendencies, different methods were developed; it is worthwhile to point out the qualitative [7], the numerical [8], and the asymptotic [9] methods. The nonlinear character of intense pulses evolution restricts the possibilities of the numerical calculations and of the theory of perturbations, and stimulates the search of the exact solutions of self-action equations. The essential progress in the analysis of eq. (1.1) is associated with the results obtained in [10]. These results reduce the solution of eq. (1.1) to the inverse scattering problem. The procedure for solving eq. (1.1), proposed in [10], reduces this equation to a set of linear integral equations. However, new difficulties, connected with the solution of the above-mentioned set, arise in the framework of such an approach.

The pace of the evolution is determined by the competition between the processes of nonlinear distortion and the dispersive spreading. It is essential for intense pulses evolution, that the nonlinear distortions of the field structure due to the heterogeneity of the amplitude-phase distributions accumulate faster than the dispersive spreading of the initial amplitude profile occurs. Moreover, in a number of problems the size of the region occupied by the nonlinear medium is limited, so that the amplitude and the phase distributions at the boundary of the medium play an important role in the pulse self-action processes inside the medium. In such a situation, it is useful to consider the simplified problems which are connected with the application of the geometric optics method to the nonlinear Schroedinger equation.

I.1. *The Large Scale Nonstationary picture of Pulse Evolution in Nonlinear Geometric Optics*

The nonlinear geometric optics equations may be derived from the nonlinear Schroedinger equation (1.1). After the separation of the real and imaginary parts of the complex amplitude ψ in (1.1), one obtains the coupled equations for the wave intensity W and phase derivative U:

$$\frac{\partial W}{\partial \tau} + \frac{\partial (WU)}{\partial q} = 0; \tag{1.4}$$

$$\frac{\partial U}{\partial \tau} + U\frac{\partial U}{\partial q} - \delta_1\delta_2\frac{\partial W}{\partial q} - \frac{2L_0^2}{L_s^2}\cdot\delta_2\cdot\frac{\partial}{\partial q}\left[\frac{1}{\sqrt{W}}\frac{\partial^2\sqrt{W}}{\partial q^2}\right] = 0; \tag{1.5}$$

Here the new variables and new functions are introduced:

$$W = |\psi|^2; \quad \psi = E(\tau,q)\cdot E_0^{-1}; \quad E_0 = |E(q)|$$

$$\tau = \frac{Z}{L_0}; \quad U = \frac{2L_0}{L_s}\cdot\frac{\partial s}{\partial q}. \tag{1.6}$$

The signs δ_1 and δ_2 correspond to the different types of dispersion and nonlinear permeability. If the waves distribution is sufficiently smooth

$$q^2 >> \frac{2L_0^2}{L_s^2}; \tag{1.7}$$

the role of the last term in eq. (1.5), connected with the small scale diffraction effects contribution to the amplitude evolution, may be negligible. Such disregard permits us to restrict ourselves to eq. (1.5) with the first derivatives only:

$$\frac{\partial U}{\partial \tau} + U\frac{\partial U}{\partial q} - \delta_1\delta_2\frac{\partial W}{\partial q} = 0. \tag{1.8}$$

Equations (1.4) and (1.8) are the fundamental equations of nonlinear geometric optics. Eq. (1.4) represents the transfer equation, while eq. (1.8) represents the nonlinear eikonal equation.

The two functions $W(\tau,q)$ and $U(\tau,q)$ describe the intensity and the phase structure of the field. Such an approach is useful at the "geometric optics" stage of the pulse evolution; its time span depends upon the initial amplitude and phase profiles of the pulse.

The initial envelopes represent the boundary conditions to the set (1.4), (1.8)

$$W\big|_{\tau=0} = W_0(q); \quad U\big|_{\tau=0} = U_0(q). \tag{1.9}$$

The direct construction of the set (1.4), (1.8) exact solutions in the form of explicit functions $W(\tau,q)$ and $U(\tau,q)$ for the arbitrary boundary conditions (1.9) is unknown. The wide classes of exact analytical solutions satisfying conditions (1.9) will be constructed below with the help of special transformations.

It is worthwhile to observe that equations (1.4), (1.8) are analogous to the hydrodynamical Eulerian equations, which describe the arbitrary motion of a one-dimensional compressible gas with a specific heat ratio $\gamma = 2$. In this analogy, the intensity W and the phase derivative U are similar to the hydrodynamical stream density and speed. The significant difference between the set (1.4), (1.8) and the Eulerian equations is connected with the change of the "pressure" sign in the case $\delta_1\delta_2 = +1$. This circumstance leads to the equation type change: the set (1.4), (1.8) in the case $\delta_1\delta_2 = +1$ is an elliptic one. In the opposite case, $\delta_1\delta_2 = -1$, this set, like the usual hydrodynamical equations, is of the hyperbolic type. In spite of this, the likeness of equations permits one to utilize the hydrodynamical methods in nonlinear geometric optics.

For the construction of exact solutions to the set (1.4), (1.8), it is convenient to transform this nonlinear set into a linear one, by regarding W and U as independent variables, and $\tau(W,U)$, $q(W,U)$ as dependent functions:

$$\frac{\partial q}{\partial U} - U\frac{\partial \tau}{\partial U} + W\frac{\partial \tau}{\partial W} = 0, \tag{1.10}$$

$$U\frac{\partial \tau}{\partial W} - \frac{\partial q}{\partial W} + \delta_1\delta_2\frac{\partial \tau}{\partial U} = 0. \tag{1.11}$$

Unlike the initial system, this set of equations is linear. Here the functions $\tau(W,U)$ and $q(W,U)$ are supposed to be single-valued. Equation (1.11) plays the role of the eikonal equation, whereas eq. (1.10) plays the role of the transfer equation in (W,U) space.

It is convenient to reduce the set (1.10) - (1.11) to one second-order differential equation, which expresses the quantities τ and q in terms of some function $f(W,U)$ through relations such as the Legendre transformations. In the geometric optics approximation, this function f contains all the information connected with the nonlinear pulse evolution. This can be accomplished in two different ways, by choosing the function f to satisfy exactly either the eikonal equation or the transfer equation in (W,U) space. Of course, the f functions are in each case different; depending upon the initial shapes of the intensity and frequency profiles one or the other method is the convenient one to utilize.

Transformation I satisfies exactly the transfer equation in (W,U) space. The functions $q(W,U)$ and $\tau(W,U)$ are introduced by the formulae

$$\tau = \frac{\partial f_1}{\partial U}\ ;\ q = -f_1 - W\frac{\partial f_1}{\partial U} + U\frac{\partial f_1}{\partial U}\ . \tag{1.12}$$

The equation for the function f_1 follows from the eikonal equation and has the form:

$$\frac{\partial^2 f_1}{\partial b^2} + \frac{3}{b}\frac{\partial f_1}{\partial b} + \delta_1\delta_2\frac{\partial^2 f_1}{\partial V^2} = 0;\ b = \sqrt{W}\ ;\ V = 2U; \tag{1.13}$$

Transformation II satisfies exactly the eikonal equation in (W,U) space. The functions τ and q are introduced now by the formulae:

$$\tau = -\frac{\partial f_2}{\partial W}\ ;\ q = -U\frac{\partial f_2}{\partial W} - \delta_1\delta_2\frac{\partial f_2}{\partial U}\ . \tag{1.14}$$

The equation for the function f_2 follows from the transfer equation and has the form:

$$\frac{\partial^2 f_2}{\partial b^2} + \frac{1}{b}\frac{\partial f_2}{\partial b} + \delta_1\delta_2\frac{\partial^2 f_2}{\partial V^2} = 0 \tag{1.15}$$

Eq. (1.13) is the five-dimensional axisymmetrical Laplace equation in (b,V) space; eq. (1.15) is the three-dimensional one. These equations are very important in nonlinear geometric optics.

Once the functions f_1 and f_2 are calculated, the description of the field self-action reduces to the formal differentiation procedure in accordance with formulae (1.12) and (1.14). The "inverse" functions $\tau(W,U)$ and $q(W,U)$ are constructed in this approach. The explicit calculation of the functions $W(\tau,q)$ and $U(\tau,q)$ is, as usual, impossible, but the treatment of the inverse dependences $\tau(W,U)$ and $q(W,U)$ permits us to analyse the evolution dynamics. First of all, such treatment uncovers the characteristic evolution moments, connected with the singularities formation during the evolution of initially smooth wave profiles.

I.2. *The Peculiarities and the Eigenfunctions in Nonlinear geometric optics*

1). *The Monotonic Regime of Pulse Evolution.* In this case the monotonic change of pulse slope leads to the appearance of a region with an appreciable field gradient. In this region the derivatives $\frac{\partial W}{\partial q}$ and $\frac{\partial U}{\partial q}$ formally tend to infinity. Such behavior is connected with the presence of a vertical tangent to the pulse profile. If such region of quick intensity growth arose within the energy localization region, the conditions for the origin of such singularity have the form:

$$\left.\frac{\partial q}{\partial W}\right|_{\tau} = 0, \quad \left.\frac{\partial^2 q}{\partial W^2}\right|_{\tau} = 0. \tag{1.16}$$

This singularity is analogous to the "spillover" of a simple wave in hydrodynamics. If the above-mentioned region develops at the waves distribution periphery, the conditions for such singularity to occur are

$$W = 0, \quad \left.\frac{\partial q}{\partial W}\right|_{W} = 0. \tag{1.17}$$

The formation of quick intensity growth region may also occur near the profile maximum. The pulse profile being symmetric, i.e., $W_0(q) = W_0(-q)$, the conditions of such singularity formation are

$$q = 0; \quad \left.\frac{\partial q}{\partial W}\right|_{\tau} = 0; \tag{1.18}$$

unlike the case (1.16), the singularity (1.18) has no hydrodynamical analogy.

Thus, the nonlinear evolution of an initially smooth wave profile in the nonlinear geometric optics approximation is characterized formally by the presence of singularities in the solutions of self-action equations. After such singularities occur, the nonlinear geometric optics equations are no longer valid.

2). *The Nonmonotonic Regime of Pulse Evolution.* Unlike the monotonic regime, the other tendency of wave profile evolution is connected with the nonmonotonic pulse development. In this case the slope of the intensity profile decreases everywhere. Such processes lead to the formation of horizontal bend points on the profile. The conditions for such points to occur are:

$$\left.\frac{\partial W}{\partial q}\right|_{\tau=0} = 0, \quad \left.\frac{\partial^2 W}{\partial q^2}\right|_{\tau=0} = 0. \tag{1.19}$$

After the formation of these bend points, the self-stratification of the profile is developed. The single valuedness of the functions $\tau(W,U)$ and $q(W,U)$ is now broken.

The success in the exact analytical solution of self-action equations (1.13) and (1.15) depends essentially upon the choice of the coordinate system and on the eigenfunctions of the equations associated with this system. The direct utilization of the (b,V) coordinate system produces difficulties in the formulation of boundary conditions. It seems convenient to transform the self-action equations (1.13) and (1.15) by using spheroidal coordinates. In this case, the corresponding eigenfunctions permit us to describe the pulse structure peculiarities during the evolution.

In the (W,U) space, let us introduce the spheroidal coordinates (ε,η) through the substitutions:

$$b^2 = (1+\varepsilon^2)(1-\eta^2), \quad V = \varepsilon\cdot\eta, \quad \varepsilon \geq 0; \ -1 \leq \eta \leq 1. \tag{1.19}$$

The solution to the three-dimensional Laplace equation can be written in terms of oblate spheroidal harmonics of the first (P_n) and second (Q_n) kind:

$$f_2 = \sum_n \{Q_n(\eta)[A_n Q_n(i\ \varepsilon) + B_n P_n(i\ \varepsilon)] + P_n(\eta)[C_n Q_n(i\ \varepsilon) + D_n P_n(i\ \varepsilon)]\}\ . \tag{1.20}$$

The coefficients in this expansion can be determined from the boundary conditions, which are specified on the surface $\varepsilon = 0$. For the initially nonmodulated pulse $(V|_{\varepsilon=0} = 0)$, the boundary conditions have the form:

$$\left.\frac{\partial f_2}{\partial \eta}\right|_{\varepsilon=0} = 0; \quad \left.\frac{\partial f_2}{\partial \varepsilon}\right|_{\varepsilon=0} = -2\eta\cdot q_0(W); \tag{1.21}$$

here $q_0(W)$ is the inverse of $W_0(q)$.

Unlike the expansion (1.20), the solutions of the five-dimensional Laplace equation (1.13) may be represented in a form:

$$f_1 = \sum_n \{T_n(\eta)[A_n T_n(i\,\varepsilon) + B_n R_n(i\,\varepsilon)] + R_n(\eta)[C_n T_n(i\,\varepsilon) + D_n R_n(i\,\varepsilon)]\} \tag{1.22}$$

Here R_n and T_n are Jacobi polinomials and functions:

$$R_0(\eta) = 1; \qquad T_0(\eta) = \frac{1}{2}\left[\frac{\eta}{1-\eta^2} + L\right];$$

$$R_1(\eta) = \eta; \qquad T_1(\eta) = \frac{1}{2}\left[2 - \frac{\eta^2}{1-\eta^2} - 3\cdot R_1(\eta)\cdot L\right];$$

$$R_2(\eta) = 1-5\eta^2; \qquad T_2(\eta) = \frac{1}{2}\left[15\eta - \frac{2\eta}{1-\eta^2} + 3R_2(\eta)\cdot L\right];$$

$$R_3(\eta) = 3\eta-7\eta^3; \qquad T_3(\eta) = \frac{1}{16}\left[16-99\eta^2 + \frac{6\eta^4}{1-\eta^2} - 15R_3(\eta)\cdot L\right]$$

$$L = \frac{1}{2}\,ln\,\frac{1+\eta}{1-\eta} \,. \tag{1.23}$$

The boundary conditions in this case have the form:

$$\left.\frac{\partial f_1}{\partial \varepsilon}\right|_{\varepsilon=0} = 0; \qquad \left.\frac{\partial f_1}{\partial \eta}\right|_{\varepsilon=0} = -f_1 + \frac{1-\eta^2}{2\eta}\,\frac{\partial f_1}{\partial \eta} \,. \tag{1.24}$$

The representation of the solutions of eqs. (1.13) and (1.15) in terms of spheroidal harmonics is convenient when the convex profile is analysed. The concave profile evolution may be described with the help of toroidal harmonics in (W,U) space [2].

I.3. *The Controlled Self-constriction and Self-stratification of Localized Wave Pulses*

The (ε,η) coordinate system permits us to express the complicated initial profiles as sums of eigenfunctions (1.20) and (1.22). The simplest profiles may be described with the help of only one Legendre function in (ε,η) space:

(i) the parabolic intensity profile

$$W_0(q) = 1-q^2, \quad q_0(W) = \eta = P_1(\eta); \tag{1.25}$$

(ii) the bell-like intensity profile

$$W_0(q) = cosh^{-2}q; \quad q_0(W) = \frac{1}{2}\, ln\, \frac{1+\eta}{1-\eta} = Q_0(\eta). \tag{1.26}$$

Thus these profiles, expressed with the help of only one Legendre function, may be represented in the explicit forms (1.25) and (1.26); due to this circumstance, they have been utilized for a long time as standard solutions in nonlinear electrodynamics. However, the more complicated initial distributions $q_0(W)$ associated with linear combinations of eigenfunctions $P_n(\eta)$ and $Q_n(\eta)$ cannot be expressed in the direct form $W_0(q)$; in these cases, the approach connected with the "inverse" dependence $\tau(\varepsilon,\eta)$ and $q(\varepsilon,\eta)$ proves to be extremely useful.

a). *The Pulse Monotonous Self-constriction.* Let us consider a pulse defined on the interval $-1 \leq q \leq 1$ and characterized, contrary to (1.25), by a smooth intensity variation at the edges (Fig. 1):

$$W_0\Big|_{q=1} = W_0\Big|_{q=-1} = 0, \quad \frac{\partial W_0}{\partial q}\Big|_{q=1} = \frac{\partial W_0}{\partial q}\Big|_{q=-1} = 0. \tag{1.27}$$

Let us consider the profile (Fig. 1):

$$q_0 = \eta + \frac{1-\eta^2}{2(\kappa-1)}\, ln\, \frac{1+\eta}{1-\eta} \quad . \tag{1.28}$$

Here κ is a parameter, and the relation between the intensity W_0 and variable η at the surface $\varepsilon = 0$ is:

$$W_0 = W\Big|_{\varepsilon=0} = 1-\eta^2 \ .$$

The function (1.28) describes the one-parameter family of symmetric profiles $W_0(q) = W_0(-q)$. The need for the function (1.28) to be single-valued restricts the range of the parameter κ to values

$$\kappa < 0. \tag{1.29}$$

FIGURE 1. The one-parameter family of initial intensity distributions in the pulse (κ is the parameter).

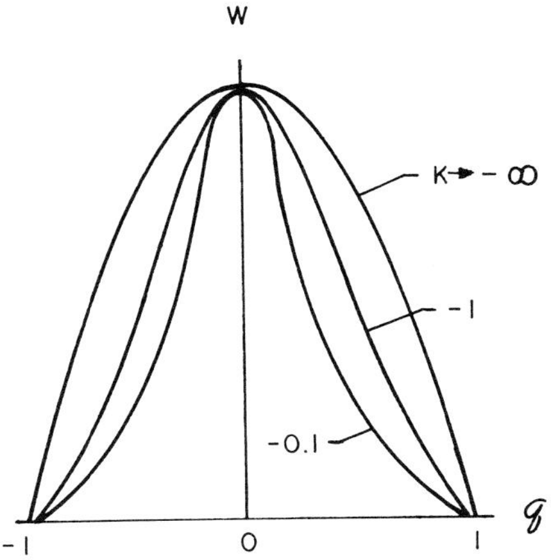

Supposing that (1.29) is henceforth satisfied, let us substitute (1.28) into the boundary condition (1.21). One obtains the following expressions for the coefficients in (1.20):

$$A_1 = \frac{4i}{5(\kappa-1)}, \quad A_3 = \frac{2}{3} A_1, \quad C_0 = \frac{5}{6} A_1,$$

$$C_2 = \frac{-2i(\kappa-2)}{3(\kappa-1)}, \quad D_2 = \frac{\pi i}{2} C_2. \tag{1.30}$$

All other coefficients in (1.20) are equal to zero. Hence, using (1.14) one can obtain the following formulae for the required functions

$$q = (\kappa-1)^{-1} \cdot F_1(\varepsilon;\eta;\kappa); \quad \tau = (\kappa-1)^{-1} \cdot F_2(\varepsilon;\eta;\kappa);$$

$$F_1 = (1+\varepsilon^2+\varepsilon^2\eta^2-\eta^2)L + \frac{\eta\varepsilon^4(1+3\eta^2)}{3(1+\varepsilon^2)(1-\eta^2)} + \frac{\eta(\kappa-1)}{1+\varepsilon^2};$$

$$L = \frac{1}{2}\ ln\ \frac{1+\eta}{1-\eta}; \tag{1.31}$$

$$F_2 = 2\varepsilon\left[\eta L - \frac{1}{1+\varepsilon^2} - \frac{\varepsilon^2(3\eta^2-2)}{(1+\varepsilon^2)(1-\eta^2)}\right]$$

$$+ \frac{\kappa-2}{2}\left[arctan\ \frac{1}{\varepsilon} - \frac{\pi}{2} - \frac{\varepsilon}{1+\varepsilon^2}\right].$$

In these formulae the variables ε and η are related to *W* and *U* by (1.19).

The formulae (1.31) contain all the information about the nonlinear evolution of the pulse (1.28). On the basis of (1.31), we shall now consider the singularities which appear during the evolution of pulse (1.28). With respect to parameter values, two types of singularities are possible.

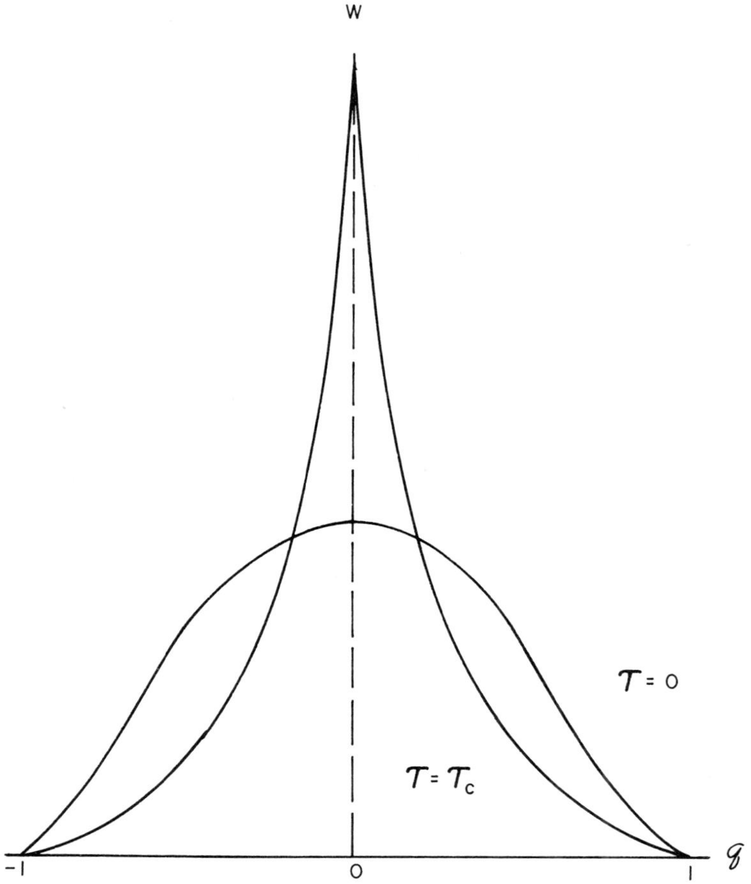

FIGURE 2. The appearance of large intensity gradients near the pulse maximum.

1. *The singularity in the intensity maximum* ($\eta = 0$). The substitution of solution (1.31) into condition (1.18) gives the critical value ε_k of the variable ε [12]:

$$\varepsilon_k^2 = \frac{3}{4}\left[-1 + \sqrt{1 - \frac{4\kappa}{3}}\,\right]. \tag{1.32}$$

The distance of this singularity generation inside the medium is

$$\tau_k = (\kappa - 1)^{-1} \cdot F_2(\varepsilon_k;\ 0;\ \kappa). \tag{1.33}$$

The intensity value at this point is finite, but the intensity derivatives near this point are large (Fig. 2).

2. *The periphery singularity* ($\eta \neq 0$). Such singularity corresponds to quick intensity growth at the pulse periphery. The coordinates of this "spillover" point in (ε,η) space are determined with the help of solution (1.31) and condition (1.16):

$$\varepsilon_k^2 = \left(-\frac{2\kappa}{3}\right)^{\frac{1}{2}}; \quad \eta_k^2 = 10^{-2}. \tag{1.34}$$

In this case, the evolution leads to the generation of a thin intensity spike with a smooth maximum, confined within a region of large intensity gradients (Fig. 3).

The analysis of solution (1.31) shows the essential dependence of the rebuilding tendencies of pulse (1.28) upon the value of the parameter κ. It is convenient to compare the evolution of pulse (1.28) with the evolution of the parabolic pulse (1.25), corresponding to the limit $|\kappa| \to \infty$ in (1.28). The parabolic pulse focusing distance τ_ϕ is equal to

$$\tau_\phi = \frac{\pi}{4}. \tag{1.35}$$

If $|\kappa|$ is large enough ($|\kappa| > |\kappa_c|$), the initial profile (1.28) is not very different from the parabolic one, which has the tendency to self-constriction in the intensity maximum. For smaller $|\kappa|$, the effect of the profile edges leads to a qualitatively

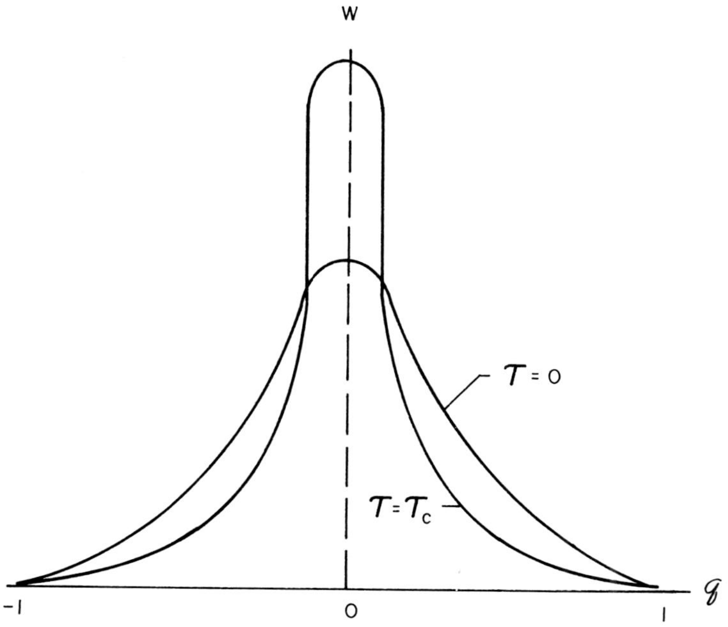

FIGURE 3. The "spillover" appearance at the pulse periphery.

different result of pulse self-action, connected with the increase of the envelope steepness outside the center and with the appearance of "spillover" singularity. The analysis of solution (1.31) shows that $\kappa_c = -235$.

The above-mentioned solutions were analysed with the help of the eigenfunctions of eq. (1.15). The utilization of the eigenfunctions of eq. (1.13) is analogous. One may consider, for example, the initial pulse

$$q\Big|_{\varepsilon=0} = \frac{2\eta}{1-\eta^2} - \frac{3}{2}\kappa\eta - \frac{20 - 30\eta^2 - \kappa}{2} L \quad . \tag{1.36}$$

The function (1.36) is single-valued if the parameter values are limited by the condition $-14 < \kappa < -10$. Substituting the distribution (1.36) into the boundary conditions (1.24), one may calculate the coefficients in the sum (1.22):

$$B_0 = 2 - \kappa, \quad C_1 = \kappa - 10, \quad B_2 = -1.$$

The analysis of singularities for this solution may be performed along the lines of the above-mentioned example.

b). *The Pulse Large Scale Self-stratification.* The monotonous self-constriction of the pulse represents only one possible tendency of pulse self-action. Another nonlinear evolution tendency is connected with the decrease of profile slope at some envelope region and the consequent pulse lamination. Let us consider the two-parameter family of symmetrical initial profiles [2]

$$q\Big|_{\varepsilon=0} = \frac{1-\eta^2}{2} \; ln \; \frac{1+\eta}{1-\eta} + (\kappa-1)\eta + M\eta^3. \tag{1.37}$$

The solution in the form (1.20) shows that the values of the parameters κ and M may stimulate the steepness decrease at any parts of the envelope. In the simple case when the bend point formation occurs quickly ($\varepsilon^2 << 1$) near the profile edges ($\eta^2 \to 1$), one may calculate the bend point coordinates ε_k, η_k with the help of condition (1.19). For instance, if $\kappa = -12$ and $M = 0.7$, one can obtain $\varepsilon_k = 0.45$, $\eta_k = \pm 0.9$. In the (τ, q) space the bend point coordinates are $\tau_k = 0.45$, $q_k = \pm 0.7$. The intensity value at the bend point is $W_k = 0.25$. After the new maximum occurs, pulse lamination develops (Fig. 4).

It is essential to note that the wavelength of such self-stratification does not depend upon high-frequency wavelengths; the pulse lamination is connected with the pulse initial duration and its envelope fine structure.

The above mentioned approach is not restricted to symmetric initial profiles. The asymmetric envelope evolution [12] may lead to a considerable displacement of the pulse maximum, accompanied by pulse self-constriction and consequent "spillover."

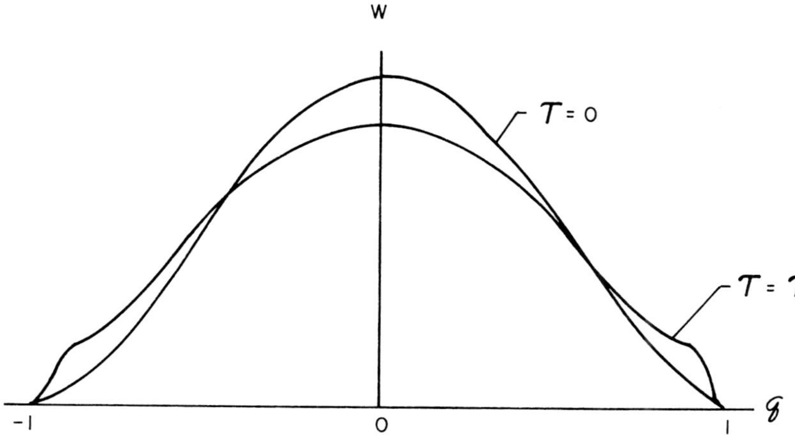

FIGURE 4. The pulse periphery large scale self-stratification (the continuous curve shows the initial intensity profile, the dotted curve illustrates the new maxima origin).

I.4. *The Nonlinear Deformation of the Pulse Phase Structure*

The nonlinear phase velocity perturbations stimulate the development of phase distortion in the pulse. The phase shifts, connected with the intensity envelope, lead to the occurrence of the frequency envelope in the initially monochromatic pulse. The phase distortion takes place even in a nondispersive medium [6], but in such medium the phase shifts are not connected with the growth of intensity maximum. The coupled rebuilding of intensity and phase envelopes may be described with the help of the obtained exact analytical solutions.

The frequency envelope development depends essentially upon the type of pulse front slope growth. Let us consider the one-parameter family of pulses with abrupt fronts [13]:

$$q\Big|_{\varepsilon=0} = \eta - a\eta^2 , \quad V\Big|_{\varepsilon=0} = 0. \qquad (1.38)$$

The solution in the form (1.20), related to condition (1.38), predicts the development of the self-modulation phenomena of

(a)

(b)

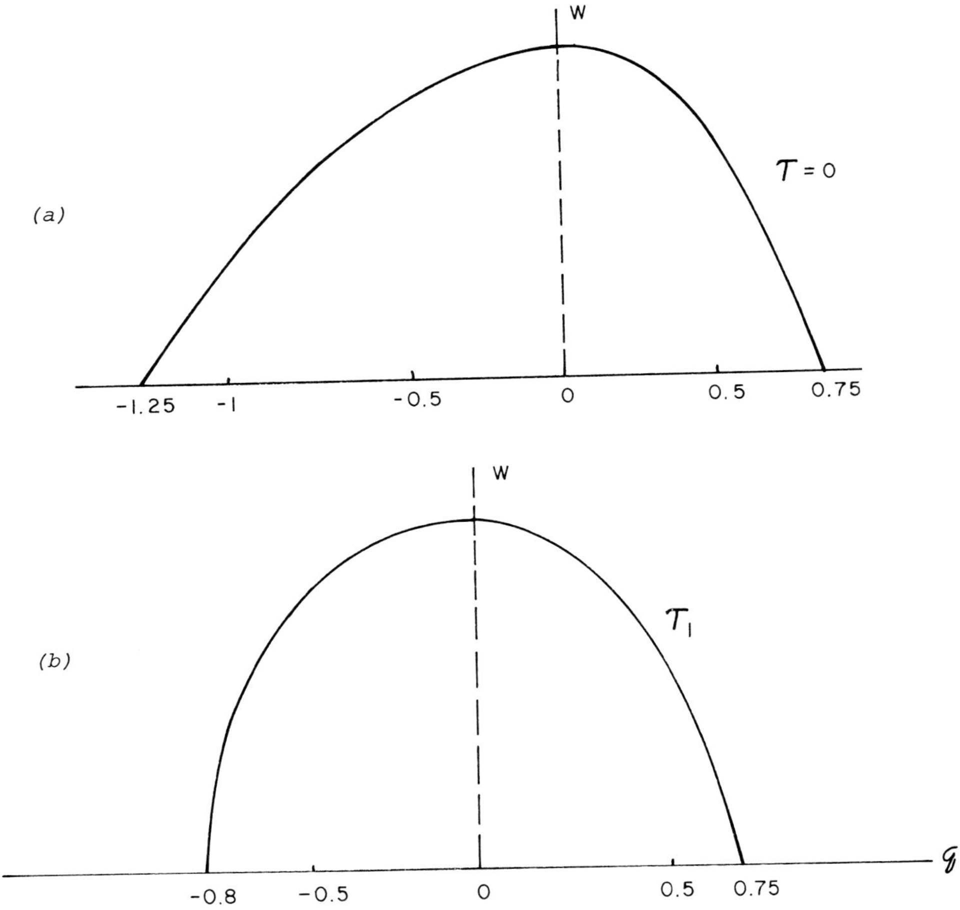

FIGURE 5. The coupled nonlinear deformation of the intensity and the frequency envelopes at the plane $\tau = 0.2$. (a) The initial intensity profile (1.39), a = -0.25. (b) The pulse fronts formation, accompanied by self-constriction. (c) The spectrum broadening (next page).

pulse (1.38) (Fig. 5). Such development contains the simultaneous processes of pulse duration decrease, pulse fronts slope growth, and spectrum broadening, and, in particular, leads to the shock wave formation at the pulse's front. Such shock singularity is determined by condition (1.17).

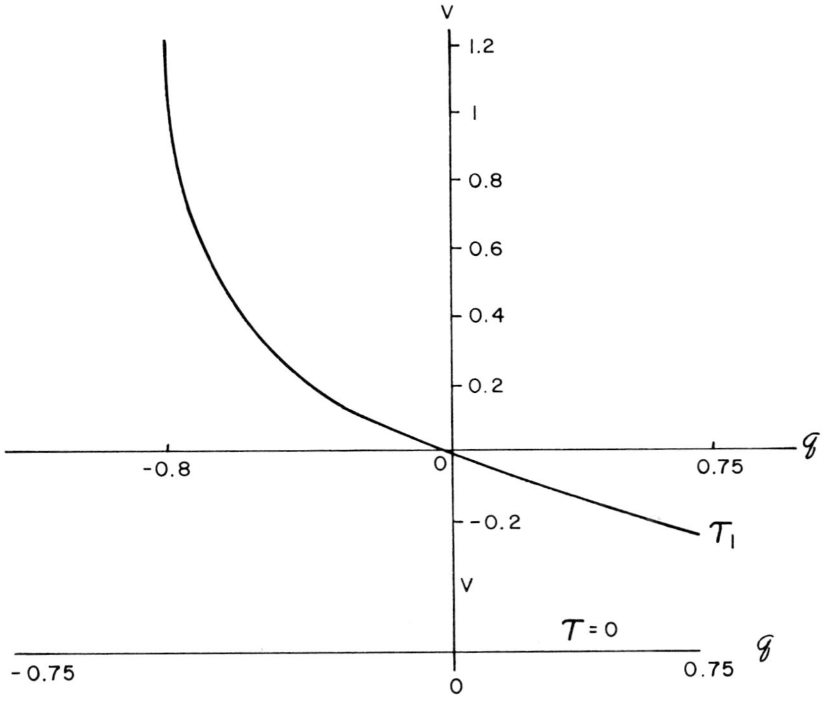

FIGURE 5c.

The spectrum broadening increases together with the intensity envelope gradients. In the abrupt fronts case, the maximum frequency modulation develops at the pulse edges (Fig. 5c). The pulse frequency self-modulation sign is different on the ascending and descending parts of the intensity profile. Near the pulse end, the modulation distribution connected with the intensity gradient tends to zero.

It is worthwhile to notice the possibility of interaction between the initial pulse frequency modulation and the nonlinear development of modulation. The nonlinear effect dependence upon the intensity derivatives may augment or suppress the initial modulation.

The aforesaid analysis is connected with the linear polarization of the intense wave. If the wave has an elliptical polarization, the effect of nonlinear interaction of both polarization components will change the self-action picture. In the isotropic medium, such interaction will lead to the rotation of the polarization ellipse, the ellipse eccentricity and rotation angular frequency Ω being constant. The value of Ω depends upon the polarization components A_1 and A_2 and the nonlinear permittivity of the medium; in particular, in an isotropic plasma one obtains [14]:

$$\Omega = \frac{e^2 A_1 \cdot A_2 \cdot \Omega_e^2}{4m^2 c^2 \omega^3} , \tag{1.39}$$

where Ω_e and ω are the electron Langmuir frequency and the intense wave frequency, respectively. The ellipse rotation frequency Ω is slow ($\Omega << \omega$). Comparison with the electric field vector rotation in the elliptically polarized wave shows that the polarization ellipse revolves in the opposite direction. The characteristic distance L_1, connected with the ellipse nonlinear turn by an angle $\frac{\pi}{2}$ is equal to $L_1 = \pi V_0 \cdot (2\Omega)^{-1}$, where V_0 is the wave group velocity. The ellipse rotation is determined by the local field amplitude values, but these values depend upon the pulse intensity profile deformation (the characteristic scales L_0 of these processes were determined in (1.6)). The comparison of such deformation characteristic scales L_0 and L_1 permits us to evaluate the role of both effects in the pulse nonlinear evolution.

Let us now note that the solutions of eq. (1.15) were represented here in the form of linear combinations of eigenfunctions. On the other hand, the solution of eq. (1.15) in the case $\delta_1\delta_2 = +1$ may be represented [15] as a contour integral in the complex (b, iV) plane:

$$f_2 = -\frac{i}{2}\int_{b-i(V-V_0)}^{b+i(V-V_0)} db_0 \cdot q_0(b)\sqrt{\frac{b_0}{b}} \quad \times$$

$$\times F\left[\frac{1}{2};\frac{1}{2};1;\frac{(V-V_0)^2+(b-b_0)^2}{-4bb_0}\right]. \tag{1.40}$$

Here F is the hypergeometric function; $V_0 = V(q)\big|_{\tau=0}$; $q_0(b)$ is the inverse to $b_0(q)$. Such integral calculation presents serious difficulties, connected with the integration in the complex plane with different initial distributions $q_0(b)$. The integral (1.40) was calculated in [15] for the case $W_0 = (1+q^2)^{-2}$. This result contains no free parameters, similar to the coefficients A_n, B_n in the solution (1.20); therefore, it is difficult to predict, in the framework of eq. (1.40), the effects of varying the initial profile $W_0(q)$. At the same time, the field's dynamic picture is very sensitive to the initial envelope distribution $q_0(b)$. An adequate description of pulse rebuilding demands that the function f_2 construction procedure be very adaptable. This procedure must easily reflect the influence of the different details of the initial distribution. Unlike the first solution of eq. (1.15) in nonlinear waves theory in the form (1.40) [15], the eigenfunctions approach permits us to unify the analytical investigation technique in its application to wide classes of initial distributions. The standard analytical operations with eigenfunctions of multidimensional Laplace equations permit us to separate the influence of each eigenfunction on the general evolution picture. Thanks to sets of harmonics adequate to the type of initial distribution, it becomes possible to separate qualitatively the different phenomena of nonlinear pulse evolution.

II. THE CONTROLLED SPREADING OF LOCALIZED WAVE PULSES

The nonlinear geometric optics approach may be generalized in the case $\delta_1\delta_2 = -1$. In this case, the similarity between nonlinear geometric optics and hydrodynamics is displayed in the effects of pulse fronts formation and initial distribution spreading. In this case, the self-action eq. (1.15) corresponds to the hyperbolic type. However, it is useful to generalize the developed eigenfunctions method to the case $\delta_1\delta_2 = -1$.

II.1. *The Processes of Heterogeneities Smoothening in Nonlinear Pulse Dynamics*

Let us introduce the new variables (ε,η) in the (b,V) space by the formulae $(\varepsilon \le 1)$

$$b^2 = (1-\varepsilon^2)(1-\eta^2); \quad V = -\varepsilon\eta. \tag{2.1}$$

The substitution

$$\varepsilon \to i\varepsilon; \quad \frac{\pi}{2} - \arctan\frac{1}{\varepsilon} \to \frac{1}{2}\ln\frac{1+\varepsilon}{1-\varepsilon}; \tag{2.2}$$

transfers the solutions (1.20), (1.22) to the corresponding solutions associated with the hyperbolic case. The further analysis of these solutions is quite similar to the case $\delta_1\delta_2 = +1$.

For example, let us consider the evolution of initial profile (1.37). Unlike the case $\delta_1\delta_2 = +1$, the central maximum of this pulse will decrease during the evolution. This decrease will be accompanied by spreading of the pulse. If the maximum decrease is quick enough, the peculiar self-stratification becomes possible [2] (Fig. 6). In particular, if the parameters κ and M in the initial profile (1.37) are equal to $\kappa = 1$, $M = -0.25$, the critical value $\varepsilon = \varepsilon_k$, connected with condition (1.19), is reached in the pulse center $\eta = 0$, $\varepsilon_k = 3^{-1}$. The intensity value at the bend point is $W_k = 0.91$. The development of a minimum is accompanied by two new maxima. After these maxima originate, their nonlinear evolution may lead to their partial self-constriction and their relative height growth. Similarly to the case $\delta_1\delta_2 = +1$, the pulse large scale lamination in the case $\delta_1\delta_2 = -1$ depends upon the pulse duration and the initial envelop derivatives.

The pulse front formation, connected with its profile smoothening, may be similarly analysed with the help of the solutions

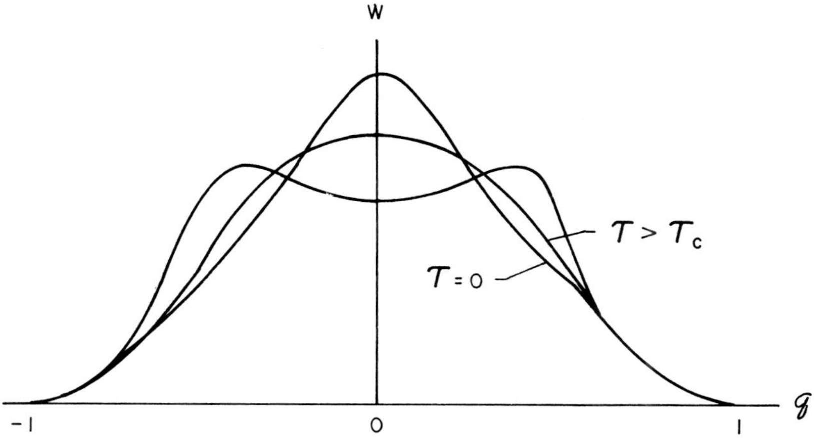

FIGURE 6. The pulse self-stratification, accompanied by spreading.

of eq. (1.15) after the substitution (2.1). But it is useful to discuss here another approach to this problem.

II.2. *The Riemann Waves Analogy in Dispersive Media*

It is typical for the aforesaid solutions of eq. (1.15), that the profiles of intensity W and modulation U cannot be represented during the evolution in the form of a simple running wave

$$U = U(W). \tag{2.3}$$

For example, the varying intensity values in the pulse maximum are accompanied by the constant zero value of the pulse self-modulation near this maximum ($V = 0$). However, in the case $\delta_1\delta_2 = -1$, one may obtain a new class of exact solutions of eqs. (1.4) - (1.5), analogous to Riemann waves in hydrodynamics. Namely, in this case condition (2.3) is fulfilled and the set (1.4) - (1.5) may be written in the form:

$$\begin{aligned} &\frac{\partial W}{\partial \tau} + U \cdot \frac{\partial W}{\partial q} + W \cdot \frac{\partial U}{\partial W} \cdot \frac{\partial W}{\partial q} = 0; \\ &\frac{\partial U}{\partial W}\left(\frac{\partial W}{\partial \tau} + U \cdot \frac{\partial W}{\partial q}\right) + \frac{\partial W}{\partial q} = 0. \end{aligned} \tag{2.4}$$

These equations are compatible with condition (2.3) if the following condition is fulfilled:

$$\left.\frac{\partial W}{\partial \tau}\right|_W + \theta(W) \left.\frac{\partial W}{\partial q}\right|_W = 0; \tag{2.5}$$

here $\theta(W)$ is an arbitrary function. Eq. (2.5) shows that the W values are constant along the characteristics in the (τ, q) plane, determined with the help of equations [14]:

$$\frac{\partial q}{\partial \tau} = \theta, \quad q = \theta \cdot \tau + P(\theta), \tag{2.6}$$

where $P(\theta)$ is an arbitrary function. After substitution of (2.6) into the set (2.4) one obtains

$$(U + \theta) \frac{\partial U}{\partial W} + 1 = 0, \qquad U + \theta + W \frac{\partial U}{\partial W} = 0. \tag{2.7}$$

This set has an integral

$$(U + \theta)^2 = W. \tag{2.8}$$

The solutions of set (2.4) may be written in the form:

$$U = C - \frac{2\theta}{3}, \quad W = \left(C + \frac{\theta}{3}\right)^2, \tag{2.9}$$

where C is an arbitrary constant. The obtained solution shows that if the intensity and modulation initial distributions are connected via the relation

$$W\big|_{\tau=0} = \frac{1}{4}(3C - U)^2, \tag{2.10}$$

then the evolution of these distributions is described by formulae (2.9).

The function $P(\theta)$ of (2.6) is determined with the help of an initial profile $U\big|_{\tau=0} = U_0(q)$,

$$q\big|_{\tau=0} = P(\theta). \tag{2.11}$$

Formulae (2.9) describe the family of "obvious" solutions,

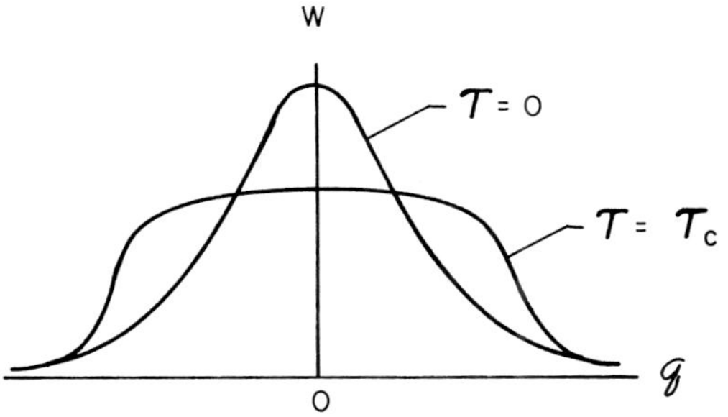

FIGURE 7. The "spillover" origin during the modulated pulse spreading.

depending upon the parameter C. The function $P(\theta)$, connected with the initial pulse envelops, describes the evolution of the "running wave" (2.3).

This solution is analogous to Riemann's solution in one-dimensional nonstationary wave theory in hydrodynamics. It describes the generation of "spillover" inside the nonlinear medium. For example, in the case of symmetrical initial pulse modulation (Fig. 7):

$$U_0(q) = \frac{aq^2}{1+q^2}\ ; \quad W_0(q) = \frac{1}{4}\left[3C - aq^2(1+q^2)^{-1}\right]^2\ ; \qquad (2.12)$$

and the function $P(\theta)$ of (2.6) is equal to:

$$P(\theta) = \left[\frac{3C - 2\theta}{3(1-C) + 2\theta}\right]^{\frac{1}{2}} . \qquad (2.13)$$

Now the equality (2.6) may be considered as a cubic equation determining the function $\theta(\tau,q)$. This analysis shows that the evolution of pulse (2.12) is connected with the pulse spreading and the slope growth at the profile periphery. These effects lead to "spillover." The coordinates of the "spillover" region in (τ,q) plane are determined, as usual, from the analysis of cubic eq. (2.6). In particular, if the constants values in (2.12) are $C = 2\times3^{-1}$, $a = 2$, then the "spillover" point coordinates are $q_c = 1.5$, $\tau_c = 1.6$.

The peculiarity of these Riemann-like solutions is connected with the dispersive properties of the medium and the symmetric character of pulse evolution. Unlike the dispersionless medium model, the "spillover" regions form simultaneously on both fronts of the pulse.

Let us note that the aforesaid nonlinear solutions are not traditional in mathematical physics. The evolution equation is of elliptical type in the case $\delta_1\delta_2 = +1$, while in the case $\delta_1\delta_2 = -1$ is of hyperbolic type. As usual, the elliptic equations solutions are represented with the help of eigenfunctions, and the hyperbolic equations solutions are constructed in the framework of the method of characteristics. However, the nonlinear geometric optics approach developed here leads to the elliptic self-action equations with boundary conditions of Cauchy type; such boundary conditions are associated with hyperbolic problems.

Such a mixed situation stimulates the unusual mathematical technique: the elliptical equation (1.15) ($\delta_1\delta_2 = +1$) is solved with the help of characteristics (1.40) in the complex plane. At the same time, the hyperbolic equation (1.15) ($\delta_1\delta_2 = -1$) is solved with the help of eigenfunctions. The "uncorrect" character of the problem stimulates the generation of singularities in the solutions of elliptic equations. It is essential that such singularities, connected with the origin of large field gradients, arise during a finite evolution time. After such singularities appear, the nonlinear geometric optics equations lose their usefulness. However, the theory which predicts the time and position of the appearance of such singularities seems to be of considerable theoretical and experimental interest. Such theory outlines the role of initial amplitude-phase envelopes during the pulse evolution and shows the possibility of controlled rebuilding of these pulses.

III. THE CRITICAL REGIMES OF THE NONSTATIONARY EVOLUTION OF INTENSE PULSES

The nonlinear geometric optics solutions describe the large scale evolution of localized pulses. If the pulse is short enough, the utilization of more rigorous solutions of the nonlinear Schroedinger equation is necessary. The wide interest in the properties of intense short pulses stimulates the attempts of theoretical predictions and controlled development of such pulses. Moreover, the limits to the use of nonlinear geometric optics must be analysed in the framework of such rigorous solutions.

The nonstationary solutions of the nonlinear Schroedinger equation are analysed here with the help of inverse scattering method [11]. The critical values of the wave amplitudes, characterizing the threshold of production of self-constricted high intensity field region, are calculated for different initial amplitude profiles. The threshold value dependence upon the initial slope of pulse fronts proves to be essential, even if the pulse maximum amplitude remains constant. Unlike the quasi-stationary soliton-like profiles, the above-mentioned regions are formed during the nonstationary supercritical (or critical) evolution regime inside the nonlinear medium. The sensitivity of such regime dynamics to the initial amplitude profiles is shown. The considerable enlargement of limits of nonlinear geometric optics utilization in the supercritical regime is illustrated.

III.1. *The Threshold Character of Pulse Evolution and the Critical Regime Origin*

Let us utilize here some results of the inverse scattering problem for a certain linear differential operator connected with eq. (1.1). Namely, the pulse asymptotic behavior may be described with the help of a special function $a(\xi)$, connected with the scattering matrix for the set of equations

$$\frac{\partial U_1}{\partial q} + i\xi U_1 = i\sqrt{\frac{\chi}{2}}\ E(q,0)U_2 ;$$
$$\frac{\partial U_2}{\partial q} - i\xi U_2 = i\sqrt{\frac{\chi}{2}}\ E^*(q,0)U_1 ; \tag{3.1}$$

here ξ is some real parameter; the wave amplitude $E(q,0)$ describes the initial pulse profile

$$E(q,0) = E_0 \cdot f(q) ,$$

E_0 is the maximum value of the function at the surface $\tau = 0$. The solutions of set (3.1), connected with conditions

$$U_1\Big|_{q\to-\infty} = exp(-i\xi q), \qquad U_2\Big|_{q\to-\infty} = 0,$$

determine the scattering matrix in the form $(q \to \infty)$

$$U_1 = a(\xi) \cdot exp(i\xi q) ,$$
$$U_2 = b(\xi) \cdot exp(i\xi q). \tag{3.2}$$

In order to analyse the asymptotic behavior of some initial amplitude profile $f(q)$ it is necessary to calculate the scattering matrix (3.2) connected with this profile.

It is convenient to introduce the new function:

$$V(q,\xi) = V(q)exp(iq\xi), \quad V(\infty;\xi) = a(\xi),$$

which permits one to reduce the system (3.1) to an integral equation

$$V(q,\xi) = 1 - \frac{\chi}{2}\int_{-\infty}^{q} dy\ \kappa(q,y)V(y,\xi), \tag{3.3}$$

where

$$\kappa(q,y) = f(y)e^{-i\xi q}\int_{y}^{q} d\eta\ f(\eta)e^{i\xi\eta} .$$

Let us rewrite eq. (3.3) in the form

$$V(q,\xi) = 1 - \left(\frac{\chi}{2}\right)^n \cdot V_{2n}(q,\xi), \tag{3.4}$$

where

$$V_m(q,\xi) = \int_{-\infty}^{q} dq_1 f(q_1) e^{i\xi q_1} \int_{-\infty}^{q_1} dq_2 f(q_2) e^{-i\xi q_2}$$

$$\ldots \int_{-\infty}^{q_{m-1}} dq_m f(q_m) e^{i\xi q_m} . \tag{3.5}$$

The signs in the exponential factors in (3.5) are alternating. The expansion (3.5) permits one to utilize the perturbation theory in the case $|\chi| << 1$, associated with the important problem of evolution of intense short pulses. Namely, the pulse duration T_0 is limited by the condition (1.3):

$$T_0^2 << \left| \frac{V_\omega}{2KV_0^2} \cdot \frac{n_0}{\Delta n} \right| . \tag{3.6}$$

The function $a(\xi)$ has a peculiarity at the point $\xi = 0$, connected with the amplitude maximum of the symmetric profile $f(q)$ [9]. It is essential that in the case $\xi = 0$ one may calculate the sum of all the terms in (3.4) and obtain the formula

$$|a(0)|^2 = \cos^2\left[\sqrt{\frac{\chi}{2}} \int_{-\infty}^{\infty} f(q)\,dq\right], \tag{3.7}$$

where χ is arbitrary. The critical value of the parameter χ in (3.7), connected with the above-mentioned singularity origin, may be calculated from (3.7) directly:

$$\chi_c = \frac{\pi^2}{2} \left| \int_{-\infty}^{\infty} f(q)\,dq \right|^{-2} . \tag{3.8}$$

It is convenient to analyse the field structure sensitivity to the initial profile with the help of two types of symmetrical amplitude profiles:

$$f(q) = \begin{cases} \left(1-q^2\right)^{\frac{\nu}{2}} , & |q| \le 1 , \\ 0 , & |q| \ge 1 . \end{cases} \tag{3.9}$$

The difference between these profiles near the distribution maximum ($q \to 0$) is negligible. However, these profiles are distinguished essentially at the periphery: the case $\nu = 1$ corresponds to the parabolic intensity profile with abrupt fronts, whereas the case $\nu = 2$ corresponds to smooth fronts. The profile dynamics is very sensitive to this difference in the derivatives.

Let us calculate the critical value χ_c connected with the central peculiarity origin. After the substitution of profiles (3.9) into (3.8) one obtains

$$\chi_c = 2\sqrt{2\pi} \, \frac{\Gamma\left(\frac{\nu+3}{2}\right)}{\Gamma\left(\frac{\nu+2}{2}\right)} , \tag{3.10}$$

where Γ is the gamma function. Parameter χ_c determines the threshold amplitude E_c values in the cases $\nu = 1$ and $\nu = 2$. These values for the parabolic and the bell-like profiles are $\chi_c = 1$ and $\chi_c = 1.4$, respectively. The bend point effect on the initial intensity profile in the case $\nu = 2$ stimulates the self-constriction of the pulse central part only; therefore, in this case the critical amplitude value is higher than in the case $\nu = 1$.

The aforesaid analysis is connected with the pulse asymptotic state. However, these formulae determine neither the distance of such state formation, nor the characteristic scale of localized region near the maximum. The time-space picture of the formation of these states during the evolution may be described in the framework of the direct solution of eq. (1.1). In such solution, the parameter χ_c values separate the subcritical ($\chi < \chi_c$, $E_0 < E_c$) and the supercritical ($\chi > \chi_c$, $E_0 > E_c$) evolution regimes.

It is worthwhile to notice that the classical problem connected with the rectangular pulse evolution is described by the particular case of formulae (3.9) and (3.10) related to the value $\nu = I$ $(f(q) = 1)$. The critical amplitude E_c may be calculated from (3.10): $E_c^2 = \pi^2(2\chi_c)^{-1}$; this value was found independently in [9].

III.2. *The Controlled Formation of Spike-like Pulses in the Supercritical Evolution Regime*

The waves interference, connected with nonlinear phase shifts, changes the picture of field evolution. The results of interference depend upon the evolution regime: Figs. 8, 9, and 10 are related to the subcritical, critical and supercritical regimes, respectively. The interference effect display is shown in Fig. 8 in the case of parabolic profile ($\nu = 1$); one can see the partial nonlinear compensation of the dispersive spreading of the pulse. Unlike these phenomena, the critical and supercritical evolution regimes lead to the gradual development of a narrow region of a field localization at the pulse maximum. Unlike the quasi-stationary soliton-like waves distribution, the above-mentioned localized region is formed as the result of the initial profile nonstationary evolution inside the medium. The wave energy cumulation at the central maximum creates the conditions for a peculiarity origin in the region $\xi \to 0$. The dynamics of central peculiarity origin is shown in Figs. 9 and 10. After the peculiarity origin, the wave amplitude changes in the peculiarity region may become negligible. This effect permits one to connect such a peculiarity with the nonlinear spike pulse formation.

The important role of the wave profile periphery regions during this profile nonlinear rebuilding is shown in Fig. 10. The bend points availability on the initial profile stimulates this profile self-stratification and new maxima development, even though the bend points are located in the regions of small amplitude values at the profile's edges.

The high intensity narrow region establishment at the pulse center shows the possibility of a peculiar electromagnetic shock wave formation. Unlike the electromagnetic shock wave on the pulse front in a dispersionless medium [6], these shock waves are developed near the pulse center; moreover, the formation of these waves is connected with the growth of the pulse maximum.

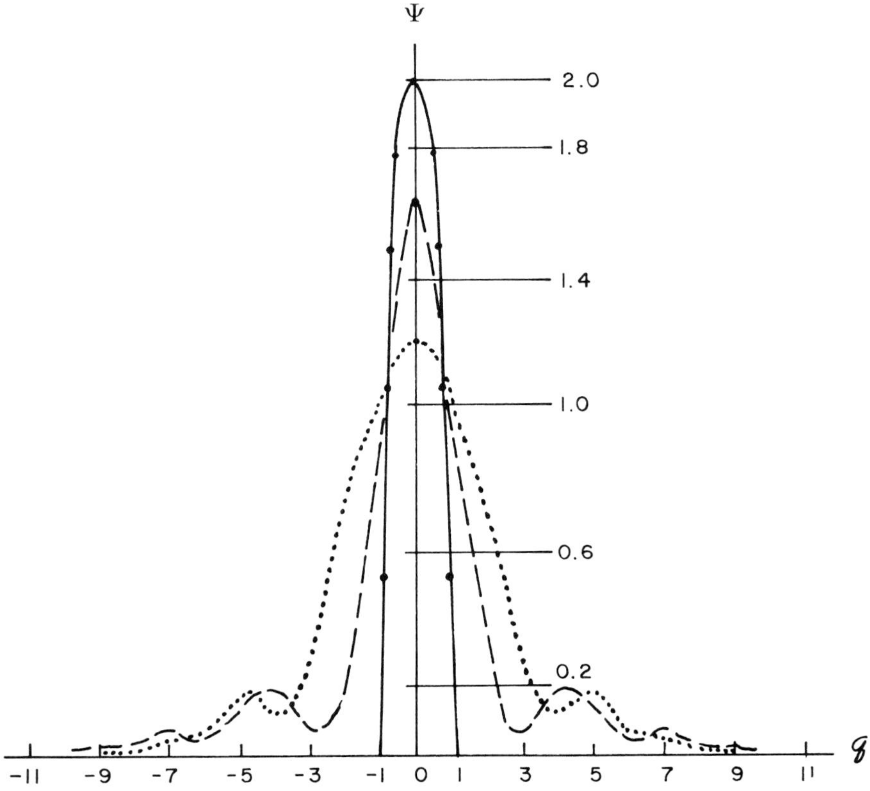

Figure 8a.

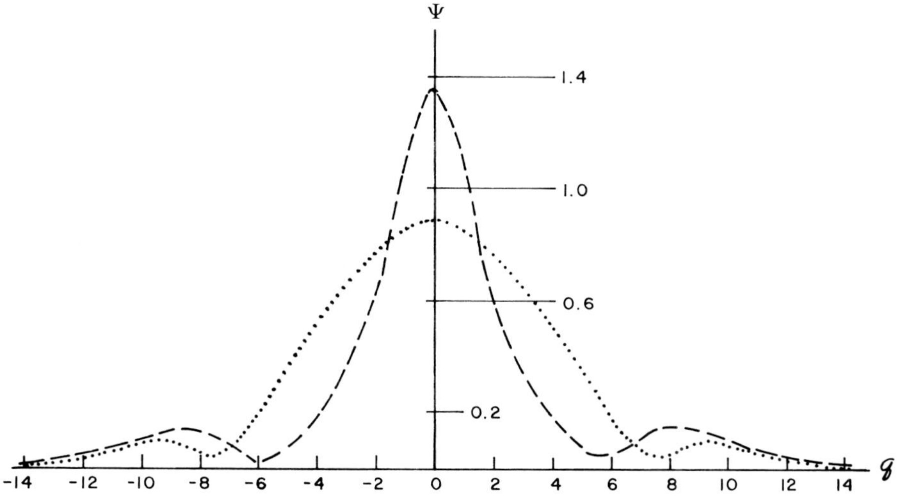

Figure 8b.

FIGURE 8. The dimensionless wave amplitude distribution during the initial parabolic profile subcritical evolution. (a) and (b) relate to the values $\tau_1 = 0.5$; $\tau_2 = 1$. The continuous lines relate to a linear theory picture, the dotted lines show the nonlinear distortions of this picture.

One can recognize the qualitative similarity between the predictions of nonlinear geometric optics and the exact solutions of the nonlinear Schroedinger equation, related to the origin of the spike-like region at the pulse center and the periphery regions self-stratification. However, the critical and supercritical evolution description is for arbitrary values of the parameter χ. On the other hand, the nonlinear geometric optics approximation may be utilized if the following condition is fulfilled:

$$|\chi| >> 1. \tag{3.11}$$

This circumstance permits one to determine the limits of utilization of such an approximation by comparing the geometric optics

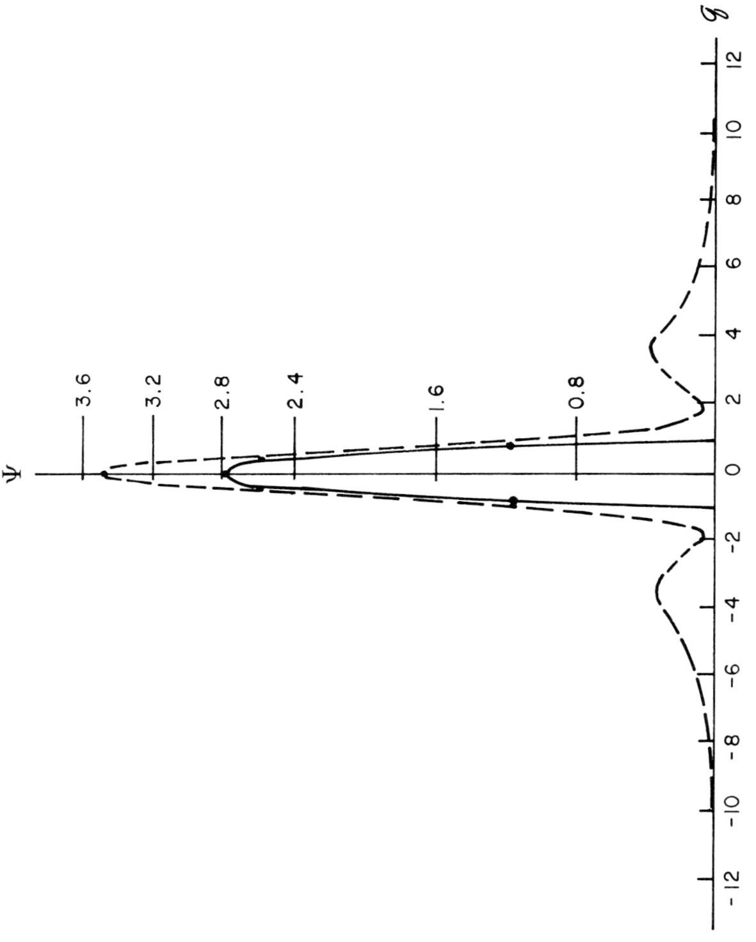

Figure 9a.

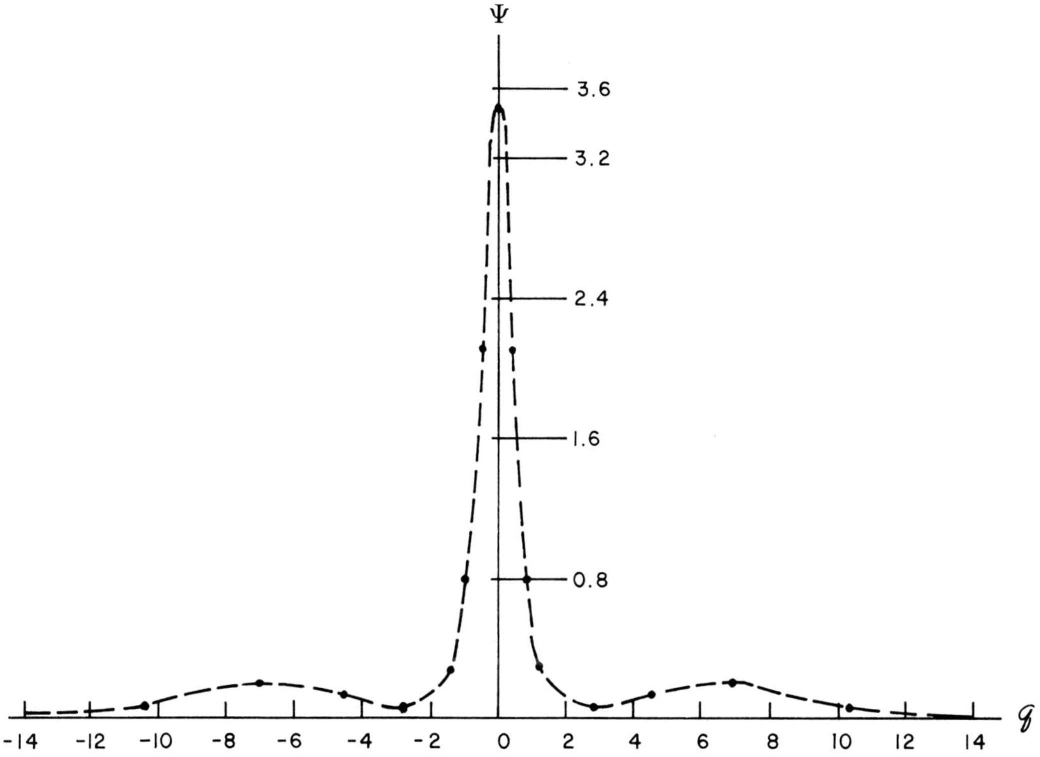

Figure 9b.

Figure 9. The critical regime of self-constriction of parabolic initial amplitude profile ($E_o E_c^{-1} = 1$; $\chi = 1$). *The curves (a) and (b) relate to the values* $\tau_1 = 0.5$; $\tau_2 = 1$.

(a)

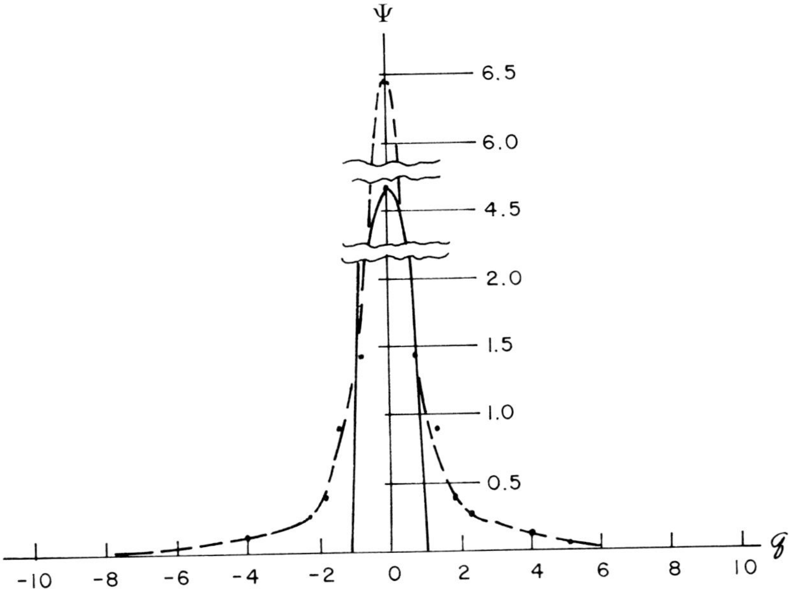

FIGURE 10. The stimulation of new amplitude maxima development due to turning point existence at the initial amplitude profile, during the nonlinear evolution in the supercritical regime ($E_0 \cdot E_c^{-1} = \sqrt{2}$; $\chi = 1$). The curves (a) and (b) relate to the values $\tau_1 = 0.2$; $\tau_2 = 0.4$ (see next page).

solutions with the solutions of eq. (1.1). The transition to geometric optics in the linear theory of wave propagation is connected with the condition that the medium parameters be slowly varying. Unlike this condition, the sensitivity of nonlinear pulse dynamics to the initial profile shows that it is necessary to carry out such comparison separately for each profile. This comparison verifies the condition of applicability of nonlinear geometric optics in the form (3.11).

Let us perform such comparison in the case of a parabolic initial profile (3.9) with $\nu = 1$. In this case, the nonlinear geometric optics solution is very simple. This solution predicts the monotonic formation of the high intensity region, characterized by large values of intensity gradients, near the pulse

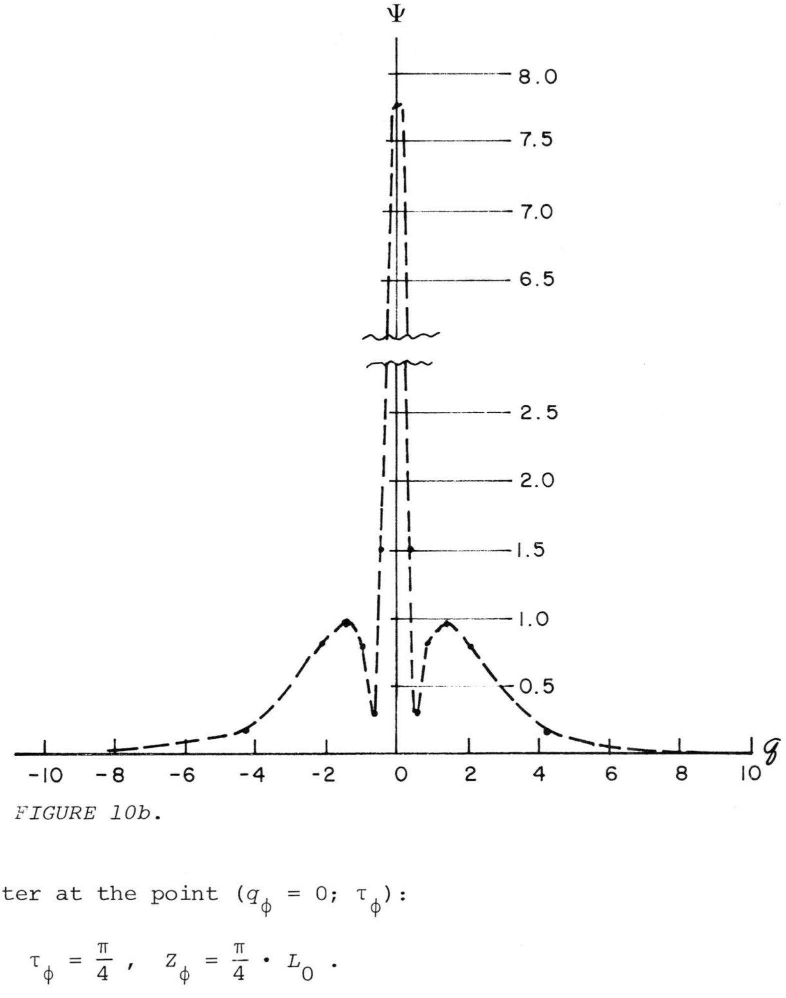

FIGURE 10b.

center at the point ($q_\phi = 0$; τ_ϕ):

$$\tau_\phi = \frac{\pi}{4}, \quad Z_\phi = \frac{\pi}{4} \cdot L_0 .$$

On the other hand, in the framework of the exact solution of eq. (1.1), the formation of the quasi-stationary high intensity region during the supercritical evolution regime ($E_0 \cdot E_c^{-1} = \sqrt{2}$) in the case $\chi = 1$ may be pinpointed from the point $Z_s = L_s \cdot \tau_s$. Therefore, the characteristic distance of narrow high intensity region formation inside the medium may be satisfactory determined in the framework of nonlinear geometric optics, even under the condition $\chi = 1$.

It is worthwhile to notice that one and the same solution, characterized by a given value of the parameter χ and by a given envelop shape, may describe the evolution of the different pulses whose amplitude E_0 and duration T_0 are connected by the relation $E_0^2 \cdot T_0^2 = \text{constant}$. This dependence shows the possibility of nonlinear pulse evolution modelling and controlled nonstationary rebuilding inside the nonlinear medium.

IV. TWO-DIMENSIONAL NONSTATIONARY EVOLUTION OF INTENSE ANISOTROPIC WAVES

The simplest examples of self-action of two-dimensional waves are connected with the evolution of waves localized in the axisymmetric region inside the nonlinear medium. Such problems, associated with nonlinear waveguides and plasma waves collapse [16], [17], stimulated the interest in more complicated field structures, in particular, in field distributions without axial symmetry. The self-action of such anisotropic distributions is distinguished by special peculiarities. The numerical analysis of two-dimensional anisotropic profiles of Langmuir waves in a plasma and high-frequency electromagnetic waves in a nonlinear dielectric [18] show the tendency of such profiles to increase their asymmetry during the evolution.

The nonlinear dynamics of such fields is analysed here with the help of elliptic amplitude-phase distributions, the ellipse eccentricity being arbitrary. The dynamics of an axisymmetrical profile may be considered as a particular case of these general results.

The importance of controlled waves evolution in two-dimensional problem is obvious, but the mathematical technique of exact solutions construction, developed in one-dimensional problem, is not suitable here. (Nonlinear geometric optics in (W,U) space may be utilized in two-dimensional problem in the form of perturbations theory only [19].) Unlike the one-dimensional problem, the self-similar solutions of the nonlinear Schroedinger equation will be utilized below. These solutions show the tendency, dependent upon the initial amplitude-phase profiles, of such profiles self-constriction onto a point or a narrow high-intensity region. The oscillating evolution regime is also possible. These tendencies, connected with the central profile region rebuilding, are analysed more rigorously in the framework of a variational approach to the two-dimensional nonlinear Schroedinger equation.

Such equation determines the nonstationary evolution of a dimensionless amplitude A:

$$i \frac{\partial A}{\partial \tau} + \frac{\partial^2 A}{\partial p^2} + \frac{\partial^2 A}{\partial q^2} + \chi |A|^2 A = 0. \tag{4.1}$$

Let us consider, with the help of eq. (4.1), the evolution of Langmuir and ion-acoustic waves in a plasma. The meaning of the dimensionless function A and of the variables p, q, τ and the parameter χ in each of these problems is as follows [23].

a). *Langmuir waves:*

$$\tau = \frac{3}{2} \Omega_e t , \quad p = \frac{x}{z_\infty} , \quad q = \frac{y}{z_\infty} , \quad \chi = \frac{|E_0|^2}{12\pi NT} , \quad A = \frac{E}{E_0} ; \tag{4.2}$$

here E_0 is the initial value of the amplitude maximum [25].

b). *Ion-acoustic waves:*

$$\tau = \frac{2}{3} \frac{c_s^2}{\omega z_\infty^2} \left(\frac{kc_s}{\omega} \right)^4 , \quad \chi = \frac{2}{9} \frac{c_s^2}{\omega^2 z_\infty^2} \left(\frac{kc_s}{\omega} \right)^4 |A_0|^2 ; \tag{4.3}$$

here A_0 is connected with the maximum value of the electron density N perturbation ΔN_0 ($A_0 = \frac{\Delta N_0}{N}$) in the initial distribution; Ω_e is electron Langmuir frequency, z_∞ the Debye radius, T the plasma temperature, c_s the sound velocity in the plasma, ω and k the frequency and the wave vector of the ion-acoustic wave.

IV.1. *The Self-similar Types of Nonlinear Evolution Regime*

Let us consider the Gauss-like form of initial amplitude A distribution:

$$A(p,q)\Big|_{\tau=0} = A_0 \cdot exp\left[- \frac{p^2}{2a_0^2} - \frac{q^2}{2b_0^2} + i\,F_{10}p^2 + i\,F_{20}q^2 \right] \tag{4.4}$$

where we assume that the intensity $|A|^2$ is constant along the contour of the ellipse

$$\frac{p^2}{a_0^2} + \frac{q^2}{b_0^2} = 1.$$

The constants F_{10} and F_{20} describe the initial phase structure of the profile, a_0 and b_0 are the dimensionless values of the ellipse half-axes x_0 and y_0 ($a_0 = x_0 \cdot z_\infty^{-1}$; $b_0 = y_0 \cdot z_\infty^{-1}$); the ellipse eccentricity is arbitrary.

Let us construct the self-similar solutions of eq. (4.1), which satisfy condition (4.4), in the Gauss-like form:

$$A(p,q,\tau) = \frac{A_0}{\sqrt{f_1(\tau) \cdot f_2(\tau)}} \cdot exp\left\{ - \frac{p^2}{2a_0^2 f_1^2(\tau)} - \frac{q^2}{2b_0^2 f_2^2(\tau)} + iF_1(\tau)p^2 + iF_2(\tau)q^2 + i\Phi(\tau) \right\}. \tag{4.5}$$

This self-similar solution conserves the elliptical symmetry during the evolution; the time-dependent parameters of the distribution are described by the dimesionless functions $f_1(\tau)$, $f_2(\tau)$, $F_1(\tau)$, $F_2(\tau)$, $\Phi(\tau)$. The initial values of f_1 and f_2 are:

$$f_1\Big|_{\tau=0} = f_2\Big|_{\tau=0} = 1. \tag{4.6}$$

The functions F_1, F_2 and Φ are connected with the nonlinear deformation of phase structure. Let us consider here the simple case of initially unmodulated profiles:

$$F_1\Big|_{\tau=0} = F_2\Big|_{\tau=0} = \Phi\Big|_{\tau=0} = 0. \tag{4.7}$$

It is convenient to introduce the new variable S and new functions U and V:

$$S = 2\tau,\ U = a_0 f_1,\ V = b_0 f_2\ . \tag{4.8}$$

After substitution of (4.5) into equation (4.1) one obtains:

$$F_1 = -\frac{1}{2U}\frac{dU}{dS}\ ,\quad F_2 = -\frac{1}{2V}\frac{dV}{dS}\ , \tag{4.9}$$

$$\frac{d\Phi}{dS} = -\frac{M}{2UV} - \frac{1}{2U^2} - \frac{1}{2V^2}\ ,\ M = \chi\, a_0\, b_0\ , \tag{4.10}$$

$$\frac{d^2U}{dS^2} + \frac{M}{U^2V} = \frac{1}{U^3}\ ,\quad \frac{d^2V}{dS^2} + \frac{M}{UV^2} = \frac{1}{V^3}\ . \tag{4.11}$$

These equations must be solved with the initial conditions

$$U\Big|_{S=0} = a_0,\quad V\Big|_{S=0} = b_0,\quad \frac{dU}{dS}\Big|_{S=0} = \frac{dV}{dS}\Big|_{S=0} = 0. \tag{4.12}$$

The set (4.9) is the exact result of this substitution; during the construction of eqs. (4.10), (4.11), the first terms of the exponent expansion in the nonlinear term $|A|^2$ in eq. (4.1) are utilized. This expansion permits one to simplify the analysis of the competition of dispersive and nonlinear effects near the profile center; the geometric optics limit in this problem, connected with the disregard of the terms on the right-hand side of eq. (4.11), was discussed in [2].

Let us concentrate our attention on the set (4.11). This set has two conservation laws:

$$U^2 + V^2 = C + \varepsilon_0 S^2\ , \tag{4.13}$$

$$\left(\frac{dU}{dS}\right)^2 + \left(\frac{dV}{dS}\right)^2 + \frac{1}{U^2} + \frac{1}{V^2} - \frac{2M}{UV} = \varepsilon_0\ , \tag{4.14}$$

where the constants C and ε_0 are determined from the initial conditions (4.6), (4.7):

$$C = a_0^2 + b_0^2 \ , \ \varepsilon_0 = \frac{1}{a_0^2} + \frac{1}{b_0^2} - \frac{2M}{a_0 b_0} \ . \tag{4.15}$$

It is worthwhile to introduce the new variables (R,ϕ) in (4.13), (4.14):

$$U = R \cos \phi \ , \ V = R \sin \phi \ . \tag{4.16}$$

The waves evolution may be considered, due to this substitution, as a mechanical problem of some point moving in the (R,ϕ) plane; the variable S is analogous to the time variable, and eq. (4.14) is analogous to the energy conservation law. Now one can rewrite the set (4.13), (4.14) in the form:

$$\begin{aligned} &R^2(S) = R_0^2 + \varepsilon_0 S^2 \ ; \ R_0^2 = C; \\ &Q(S) = U(\phi) \ ; \ Q(S) = \int_{S=0}^{S} R^{-2}(S')dS' \ ; \end{aligned} \tag{4.17}$$

$$U(\phi) = \int_{\phi_0}^{\phi} d\phi_1 \left[\varepsilon_0 R_0^2 - \frac{1}{\cos^2 \phi_1} - \frac{1}{\sin^2 \phi_1} + \frac{2M}{\cos \phi_1 \cdot \sin \phi_1} \right]^{-\frac{1}{2}} \ . \tag{4.18}$$

The initial value of the function $\phi(S)$ is

$$\phi_0 = \phi(S) \Big|_{S=0} = \arctan \frac{b_0}{a_0} \ ; \tag{4.19}$$

Thus, the problem of construction of self-similar solutions of the two-dimensional nonlinear Schroedinger equation is reduced to the study of the dimensionless equation (4.18), which describes the qualitatively different types of initial profiles nonlinear rebuilding.

IV.2. *The Monotonic and Non-monotonic Regimes of Localized Waves Rebuilding*

The different results of waves nonstationary evolution, described in the framework of eq. (4.18), are connected with the distribution's geometry and wave spectrum properties. The occurrence of these possibilities is determined by the roots of denominators in (4.18).

The function R(S) in (4.17) decreases monotonically during the evolution, if $\varepsilon_0 < 0$. The characteristic time of distribution collapse is

$$S_c = R_0(-\varepsilon_0)^{-\frac{1}{2}} . \tag{4.20}$$

In the opposite case $\varepsilon_0 > 0$ the function $R(S)$ increases.

Other evolution possibilities are connected with the form of initial distribution, expressed by the value ϕ_0. The function ϕ varying from the initial value ϕ_0 depends upon the sign of the second derivative

$$\left.\frac{d^2\phi}{dS^2}\right|_{S=0} = \phi''_0 = \frac{1}{a_0^2 b_0^2} \cdot \frac{1-\mu^2}{1+\mu^2} \cdot \left(\frac{1+\mu^2}{\mu} - M\right), \quad \left.\frac{d\phi}{dS}\right|_{S=0} = 0. \tag{4.21}$$

In the case $\phi''_0 > 0$ ($\phi''_0 < 0$) the function $\phi(S)$ begins to increase (decrease) during the evolution; the limits of its variation are determined by the roots of the equation

$$\varepsilon_0 R_0^2 - \frac{1}{\cos^2\phi} - \frac{1}{\sin^2\phi} + \frac{2M}{\sin\phi \cdot \cos\phi} = 0; \tag{4.22}$$

the value $\phi = \phi_0$ is a root of this equation.

If eq. (4.22) has no other positive roots, the function $\phi(S)$ changes from $\phi = \phi_0$ to $\phi = \phi_c = \frac{\pi}{2}$ if $\phi''_0 > 0$, or to $\phi = \phi_c = 0$ if $\phi''_0 < 0$. In these cases the distribution self-constriction leads to formation of the narrow high intensity region, stretched out along the x-axis in the case $\phi_c = 0$. Such rebuilding time may be determined from eq. (4.18) (Fig. 11):

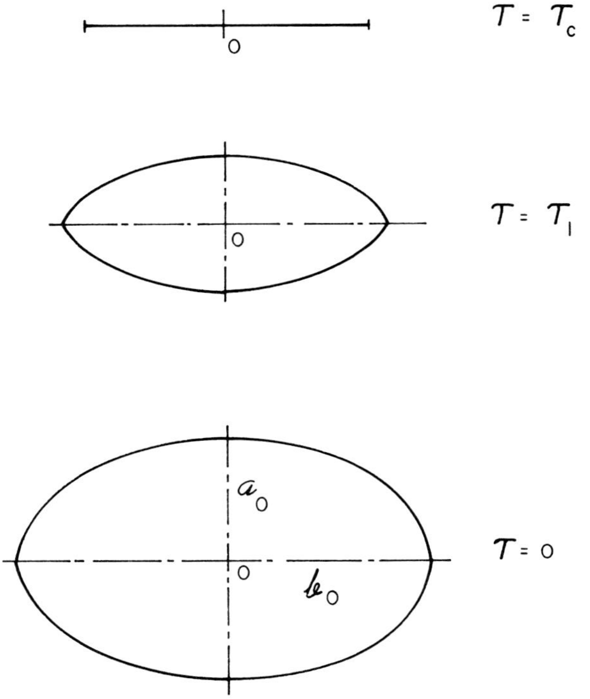

FIGURE 11. The monotonic evolution of two-dimensional wave anisotropic distribution.

$$Q(S_c) = U(0). \tag{4.23}$$

The characteristic length of the narrow region is

$$\rho_c = 2 \cdot R(S_c). \tag{4.24}$$

A similar analysis applies to the case $\phi_c = \frac{\pi}{2}$.

If eq. (4.22) has the positive root $\phi = \phi_1$, distinguished from ϕ_0, the function $\phi(S)$ oscillates between the values corresponding to values $\phi = \phi_0$ and $\phi = \phi_1$, and the profile self-action oscillates with time. The period of these oscillations is not constant (Fig. 12).

In the case $\varepsilon_0 = 0$ the distribution oscillations have a periodic character. The function $U(V)$ varies between the values

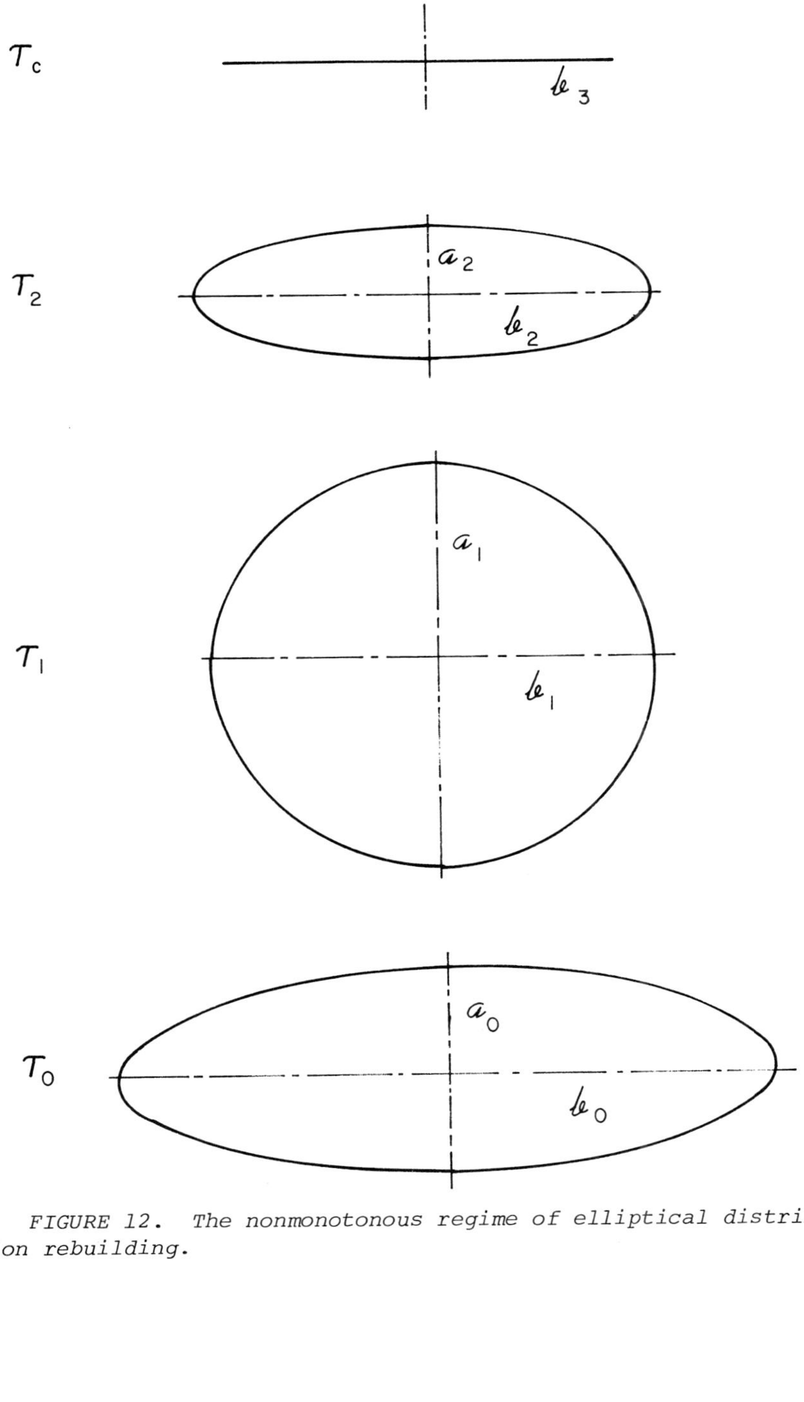

FIGURE 12. The nonmonotonous regime of elliptical distribution rebuilding.

$R_0 cos\ \phi_0$ and $R_0 cos\ \phi_1$ ($R_0 sin\ \phi_0$ and $R_0 sin\ \phi_1$). The oscillations period S_p is

$$S_p = 2R_0^2 \cdot U(\phi_1). \tag{4.25}$$

The abovementioned regimes are determined by the initial profiles. Similar solutions may be utilized in the problem of laser beam stationary self-focusing, governed by equation (4.1) [16]; the τ coordinate in this case is measured along the beam axis.

The universal character of the dimensionless self-similar solutions that we have obtained stimulates efforts to generalize these solutions by relaxing the paraxial approximation. This generalization may be carried out in the framework of the variational approach, which is typical for wave problems in one-dimensional situations [22].

IV.3. *The Variational Approach to the Two-dimensional Nonlinear Schroedinger Equation*

Unlike the paraxial approximation, the influence of both central and periphery parts of profile (4.5) on its self-similar evolution regime may be described with the help of a variational method. In this case, the functions $f_1(\tau)$, $f_2(\tau)$, $F_1(\tau)$, $F_2(\tau)$ and $\Phi(\tau)$ in (4.5) must satisfy a special condition, which may be obtained from the functional [24]:

$$L(A,A^*) = \int_{-\infty}^{+\infty}\int_{-\infty}^{+\infty} dp\ dq \int_{-\infty}^{+\infty} d\tau \left[\nabla_\perp A \cdot \nabla_\perp A^* + i\left(A\,\frac{\partial A^*}{\partial \tau} - A^*\,\frac{\partial A}{\partial \tau}\right) - \frac{\chi |A|^4}{2} \right] . \tag{4.26}$$

This functional has an extremum if the function is the solution of eq. (4.5). After the substitution of (4.5) into (4.26), integration over dp and dq, and utilization of standard Lagrange conditions, one obtains the equations governing the abovementioned functions. The calculations show that the relations between the pairs of functions U, F_1 and V, F_2 coincide with (4.9). However, the relations connected with the amplitude distribution parameters $U(S)$ and $V(S)$ and the phase shift $\Phi(S)$ are now changed:

$$\frac{d\Phi}{dS} + \frac{3M}{8UV} - \frac{1}{2U^2} - \frac{1}{2V^2} = 0, \tag{4.27}$$

$$\frac{d^2U}{dS^2} + \frac{M}{4U^2V} = \frac{1}{U^3}, \quad \frac{d^2V}{dS^2} + \frac{M}{4UV^2} = \frac{1}{V^3} \quad . \tag{4.28}$$

The comparison of eqs. (4.27) and (4.28) with the paraxial approximation equations (4.10) and (4.11) shows that the form of equations is the same, but the coefficient M is replaced by $\frac{3M}{4}$ in (4.10) and by $\frac{M}{4}$ in (4.11). The obtained solutions show the essential dependence of nonlinear dynamics upon the fine details of initial structure of the waves profile. The substitution $M \to \frac{M}{4}$ into formulae (4.27), (4.28) permits one to calculate the characteristic time values of the monotonic and nonmonotonic collapse in two-dimensional problems and the period of the peculiar nonlinear oscillating regime.

V. CONCLUSIONS

It is worthwhile to point out the main achievements and difficulties of the analytical theory of nonstationary nonlinear rebuilding of localized wave pulses in dispersive media.

1). The hydrodynamical analogy in nonlinear electromagnetic wave theory seems to be very useful. This analogy permits us to develop nonlinear geometric optics and to construct wide classes of exact analytical solutions, describing the large-scale evolution of localized pulses inside the medium. Due to special transformations, the nonlinear self-action problem is reduced, in the one-dimensional case, to a linear one in the intensity-modulation space. Such an approach permits one to construct the exact analytical solutions in nonlinear geometric optics in a very simple form, represented by means of the eigenfunctions of multi-dimensional Laplace equations, in the aforesaid space. Due to such method, the analysis of nonlinear rebuilding of pulses is reduced to simple algebraic operations on the eigenfunctions, that are well known in mathematical physics. Such simplification permits one to analyse the qualitative effects during the nonlinear evolution, to unify the analytic techniques, and to utilize the same approach in problems of pulse spreading.

2). The obtained solutions characterize the essential dependence of the dynamics of pulse rebuilding upon the initial amplitude-phase profiles. This dependence shows the possibility of controlled pulse rebuilding, connected with pulse self-constriction and self-stratification, its front formation and spectrum broadening inside the medium.

3). The family of exact nonstationary solutions is extended in the fremework of the nonlinear Schroedinger equation. The utilization of the inverse scattering method permits one to analyse the critical and supercritical regimes in the evolution of different localized profiles. Such regimes are connected with threshold values of initial pulse amplitude that depend upon the initial profile shape. Unlike the quasi-stationary soliton-like distributions, the considered nonstationary rebuilding in the supercritical regime leads to a spike-like pulse formation inside the medium. The possibility of controlled evolution of short pulses, connected with the pulse self-constriction and self-stratification, is illustrated. The important role of periphery parts of the initial profile during its nonlinear evolution is pointed out. The limits of utilization of nonlinear geometric optics are substantially expanded, within the framework of the aforesaid solutions.

4). The monotonous and oscillating regimes of self-constriction of anisotropic distributions of Langmuir waves and ion-acoustic waves in plasma are analysed in the framework of exact self-similar solutions of the two-dimensional nonlinear Schroedinger equation. The variational approach to the two-dimensional nonlinear Schroedinger equation is outlined.

The obtained results permit us to outline the following areas of further theoretical development.

I). Due to considerable similarity of the tendencies of nonlinear pulse rebuilding among different wave fields, it is worthwhile to analyse these tendencies in the framework of one universal dimensionless eq. (1.1) or (4.1). Such similarity shows the possibility of prediction and modelling of one instance of wave evolution with the help of another instance, if both these instances are described in the framework of these equations. This possibility may be especially fruitful, if one of these instances is well studied in optical and radiophysical experiments.

II). The previous analysis is connected with a conservative medium. Previous attempts to calculate the common action of nonlinear dissipation or amplification of waves, accompanying the pulse amplitude-phase deformation, were connected with perturbation theory only [20]. The elaboration of a more rigorous theory, containing the nonlinear effects in the real and the imaginary parts of the dielectric permittivity simultaneously, seems to be very actual.

III). The aforesaid examples of exact solutions of the nonlinear Schroedinger equation in one-dimensional and two-dimensional problems may be considered as a first stage of a more complicated problem: the analysis of the three-dimensional nonstationary nonlinear Schroedinger equation. The original approach [7], based on the qualitative theory of differential equations, is well known in the theory of nonlinear oscillations and seems to be very useful in the stationary limit of this problem. This approach permits one to stay away from the "shortened" equations (1.1), (4.1) and to analyse the rigorous nonlinear electrodynamics equations. The attempts based on a variational approach to this problem also seem to be quite useful.

The analytical theory of nonlinear localized waves began to form in connection with the analysis of nonlinear wave processes during recent years. The first successes of this theory in radio physics and optics stimulated the study of problems in the adjacent areas of acoustics, plasma physics, and wave electronics. The first obtained solutions became useful for peculiar "nonlinear intuition" promotion in different fields of physics. We hope that the general interest in nonlinear waves in distributed systems will stimulate further developments in the theory of intense localized waves.

ACKNOWLEDGMENTS

The author is very grateful to Professor E. Wolf for his attention to the main results of this analysis. The author thanks Professor E. Velikhov and Professor V. Letokhov for their kind criticism.

REFERENCES

1. Ginzburg, F.L., and Rukhadze, A.A., *The Waves in the Magnetoplasma*, Nauka, Moscow, (1975).
2. Shvartsburg, A.B., *Geometrical Optics in Non-linear Waves Theory*, Nauka, Moscow, (1976).
3. Shvartsburg, A.B., Soviet Physics - Uspekhi, *113*, 735 (1974).
4. Silin, V.P., *The Parametrical Action of the High Power Radiation on a Plasma*, Nauka, Moscow, (1973).
5. Shvartsburg, A.B., *J.E.T.P. 70*, 947 (1976).
6. De-Martini, F., Townes, C.H., Gustafson, T.K. and Kelley, P.L., *Phys. Rev. 164*, 312 (1967).
7. Eleonsky, V.M., and Silin, V.P., *J.E.T.P. 63*, 532 (1972).
8. Lugovoi, V.N., and Prokhorov, A.M., Soviet Physics - Uspekhi *111*, 194 (1973)
9. Manakov, S.V., *J.E.T.P. 65*, 1392 (1973).
10. Zakharov, V.E., and Shabad, A.B., *J.E.T.P. 61*, 118 (1971).

11. Gardner, C.S., Green, J., Kruskal, M. and Miura, R., *Phys. Rev. Lett. 19*, 1095 (1967).
12. Shvartsburg, A.B., Opto-Electronics *8*, 393 (1976).
13. Shvartsburg, A.B., *Phys. Lett. 50A*, 208 (1974).
14. Gurevich, A.V., and Shvartsburg, A.B., *Nonlinear Theory of Radiowaves Propagation in the Ionosphere*, Nauka, Moscow, (1973).
15. Lighthill, M.J., *Proc. Roy. Soc. 299 A*, 28 (1967).
16. Vorob'ev, V.V., Radiophysics, *XIII*, 1905 (1970).
17. Gibbons, J., Thornhill, S., Wardrop, M. and ter Haar, D., *J. Plasma Phys. 17*, 153 (1977).
18. Sodha, M.S., Tripathi, V.K. and Nayyar, V.P., Optics Communications *9*, 381 (1973).
19. Shvartsburg, A.B., Optics Communications, *X*, 127 (1974).
20. Shvartsburg, A.B., Radiophysics *XIX*, 1775 (1976).
21. Schubert, M., and Wilhelmi, B., *Einführüng in Die Nichtlineare Optik*, Leipzig (1971).
22. Whitham, G.B., *Linear and Non-linear Waves*, Wiley, New York, (1974).
23. Kako, M., Supplement of the Progress of Theoretical Physics, *55*, 120 (1975).
24. Letokhov, V.S., *J.E.T.P. 50*, 1148 (1966).
25. Shvartsburg, A.B., Plasma Physics *21*, (1979).
26. Emery, V.J., *J. Math. Phys. 11*, 1893 (1970).
27. Rudenko, O.V. and Soly'an, S.I., *The Theoretical Foundations of Nonlinear Acoustics*, Nauka, Moscow, (1975).

THE NONLINEAR RESONANT PLASMA-WAVE INTERACTION IN A COLLISIONAL MAGNETOPLASMA

A.B. Shvartsburg

IZMIRAN
Academgorodok
Moscow Region, USSR

New possibilities of controlled evolution of power high frequency electromagnetic radiation in wave electronics and plasma radiophysics are considered. These possibilities are based on the specific sensitivity of the absorption, refraction, polarization and energy flux direction of the wave in the vicinity of Langmuir and electron gyro-resonance upon the frequency of electron collisions in a magnetoplasma. The thermal variations of this frequency due to plasma Joule heating may essentially change the resonant regime of wave propagation, even though this frequency is less than other characteristic magnetoplasma frequencies. Such sensitivity determines the wide classes of thermal interactions between the heating waves and the resonant wave. The complex hysteresis dependence of plasma dispersive properties upon the wave intensity due to Langmuir resonant heating induces the coupled rebuilding of radiation parameters and thermal lamination of the magnetoplasma layer.

I. INTRODUCTION

The peculiarity of resonant interaction between electromagnetic waves and collisional plasma is connected with the considerable influence of collisions on plasma dispersive properties at the resonant frequencies. The essential sensitivity of these interactions to electron temperature shows the possibility of controlled development of nonlinear wave phenomena due to plasma

ISBN 0-12-709660-4

heating. Recently, the interest in collisional effects was restricted by diffusion and heat conductivity processes in a plasma and by Joule absorption of electromagnetic waves. Other wave phenomena in plasma electrodynamics, connected with refraction and polarization of waves in a plasma, were usually considered in a collisionless approximation. In the framework of these ideas the collisions, usually assumed to be sufficiently infrequent, were responsible only for the small corrections to the principal collisionless effects, such as the phase shifts in the wave propagating through the dissipative medium (Weinstein, 1957), the slow collisional relaxation of solitons in a plasma [35], and the perturbation of the wave's group velocity due to its weak attenuation in a plasma [18]. During the last decade, the achievements in the theory of plasma oscillations attracted attention to important types of plasma instabilities that depend essentially upon collisions: the parametric instability [32] and the drift-dissipative instability in a heterogeneous plasma [11].

The motives for these theoretical and experimental investigations were two-fold: the interest in the effects themselves, and the possible use of these effects for controlled plasma-wave interactions. The possibilities of such interactions in a collisional plasma, based on the thermal variations of electron collision frequency and Joule absorption of the waves, were illustrated in the radiowave crossmodulation effect long ago. However, the crossmodulation was only the first example of this class of phenomena. Afterwards, the appearance of intense radiation sources in radio and optical frequency ranges stimulated interest in the whole complex of plasma heating phenomena under laboratory and cosmic conditions [8]; similar effects in a solid body were described by Vladimirov et al. [40]. The search for new possibilities in controlled plasma-waves interactions is presently of considerable theoretical and practical interest in connection with problems of wave electronics, of energy transfer through a plasma, and of thermal processes in plasma dynamics.

The range of such possibilities is substantially extended in resonant situations, when the electromagnetic wave frequency is sufficiently close to plasma characteristic frequencies. Unlike the nonresonant case in which the collisions are mainly responsible for the waves' Joule absorption, here the dependence of wave refraction, polarization and energy flux direction upon collisions near resonances is also important. Such resonant effects may be very sensitive to the frequency of electron collisions, even though these frequencies are much lower than other characteristic plasma frequencies. This sensitivity leads to possibilities of jump-like transitions between the temperature states of plasma due to resonant Joule heating. These transitions are accompanied by discontinuous exchanges of refraction, polarization and absorption of the heating wave itself. Moreover, the relatively short time needed to establish the electron temperature attracts attention to the possibilities of its utilization in nonlinear phenomena such as the controlled rebuilding of intense radiation fluxes in a plasma (it is essential that the aforesaid effects not be connected with slow processes of ionization redistribution, such as diffusion). The frequency of electron elastic collisions may be modified by the propagating wave itself or by other waves. Thus, the wide class of resonant wave processes connected with wave interaction and self-action via thermal effects can by analyzed; such radiophysical phenomena in a collisional magnetoplasma are, in fact, discussed below. The first section of this review is devoted to the different actions of resonant Joule heating in a plasma, the second and third sections are respectively connected with the thermal self-action of high-frequency electromagnetic waves in the vicinity of plasma resonances and with the thermal coupling of such waves.

II. RESONANT THERMAL EFFECTS IN PLASMA RADIOPHYSICS

The thermal phenomena connected with plasma-wave interaction in a collisional magnetoplasma are amplified considerably, when the frequency of the electromagnetic wave ω is close to the electron Langmuir frequency Ω_e. These effects are caused by a temperature dependence of the effective collisions frequency $\nu_e(T_e)$ and may occur even at constant ionization N_e. When calculating the range of frequencies of electron elastic collisions, responsible for such effects, it is convenient to write the expression for the complex refractive index $n-i\chi$ for high-frequency waves in the form

$$(n-i\chi)^2 = 1 - \frac{2v(1-v-is)}{2(1-is)(1-v-is)-u\,sin^2\alpha \pm \sqrt{u^2 sin^4\alpha + 4u\,cos^2\alpha(1-v-is)^2}} \tag{1.1}$$

Here n is the refractive index, χ is the absorption coefficient,

$$s = \frac{\nu_e(T_e)}{\omega}, \quad v = \frac{\Omega_e^2}{\omega^2}, \quad u = \frac{\omega_H^2}{\omega^2},$$

ω_H is the electron gyrofrequency, α is the angle between the direction of wave propagation (z) and the magnetic field $\underline{H}$. The vectors $\hat{z}$ and $\underline{H}$ are assumed to lie in the same plane and the signs "+" and "-" in (1.1) correspond to the ordinary and extraordinary polarization. The neglect of ions motion restricts the wave frequency ω with the unequality $\omega \gtrsim \omega_H(m_e/M_i)^{\frac{1}{2}}$ where m_e and M_i are the electron and ion masses. The effective frequency of collisions ν_e is usually determined by means of the formula

$$\nu_e(T_e) = \frac{4\pi m_e}{3N_e T_e}\int_0^\infty v^4\nu(v)f_0(v)dv\,. \tag{1.2}$$

Here $\nu(v)$ is the frequency of collisions between the electrons and the heavy particles, and $f_0(v)$ is the electron distribution function.

Let us analyze the travelling wave in a transparent region, determined by the condition $n^2 > \chi^2$. Herein, the Langmuir resonance ($v = 1$) for the extraordinary wave field in the magnetoplasma permits one to analyze the series of thermal phenomena far from the point of wave reflection $v = 1 - \sqrt{u}$. The temperature dependence of the frequency of electron elastic collisions ν_e may be usually written in the form

$$\nu_e = \nu_0 f^m , \quad f = T_e/T_{e0} . \tag{1.3}$$

Here T_{e0} and ν_0 are the unperturbed values of electron temperature and electron collision frequency, while the value of $T_e (T_e \geq T_{e0})$ is connected with the electrons heating. The power m value in (1.3) depends upon the degree of plasma ionization: if the electron-ion collisions (ν_{ei}) are dominating, m is equal to -1.5; in the opposite case, when the electron-neutral collisions (ν_{em}) are dominating, m varies from $m = 1$ for oxygen and nitrogen [22] to $m = 1.5$ for the noble gases [2]. These dependences permit one to analyze the temperature effects in a collisional plasma due to thermal variations of the frequency of collisions. In the following, it is convenient to utilize the dimensionless value of this frequency:

$$q = \frac{\nu_e}{\nu_c} , \quad \nu_c = \frac{\omega_H \sin^2\alpha}{2\cos\alpha} . \tag{1.4}$$

In the resonant case ($v = 1$) the value $q = 1$ relates to phase synchronism between the ordinary and extraordinary waves in a magnetoplasma [23].

I.1. The Temperature Anomaly of Langmuir Resonant Wave Absorption

The analysis of Joule heating shows the anomalous character of resonant absorption χ, connected with the occurrence of a nonmonotonic dependence of absorption upon the electron temperature and the effective collision frequency for some range of plasma parameters. A nonmonotonic character of absorption as a function of temperature indicates the possibility of a threshold for enhancement in heating and anomalous dissipation in the plasma, i.e., the critical temperature value T_c corresponding to the threshold of the effect is associated with an extremum of the function $\chi(T_e)$.

Let us restrict ourselves to the case $q \leq 1$. In this region, the changes of absorption coefficient χ are shown in Fig. 1. The maximum value χ_m of this coefficient in a limited range of angles, when $\frac{2\nu_e}{\omega_H} < \tan^2\alpha << 2$, is obtained at the point [31]:

$$q_0 = \frac{2\sqrt{3u}\ |\cos\alpha|}{1 + 3u\cos^2\alpha} . \tag{1.5}$$

In the region $0 < q \leq q_0$ the value of the absorption coefficient increases from $\chi = 0$ to $\chi = \chi_m$, whereas in the region $q_0 \leq q \leq 1$ it decreases from $\chi = \chi_m$ to $\chi = \chi_0$, where:

$$\chi_0 = \left(\frac{P^{-\frac{1}{2}} - P^{-1}}{2} \right)^{\frac{1}{2}} , \quad P = 1 + (u\cos^2\alpha)^{-1} .$$

The critical temperature value is determined from the equation

$$q(T_c) = q_0 .$$

In this case, the change of collision frequency of the particles on heating may lead, depending on the degree of plasma ionization, to threshold effects of two kinds: (a) In a slightly ionized plasma, when the effective frequency of electron collisions increases on heating $\left(\frac{\partial \nu_e}{\partial T_e} > 0 \right)$, a decrease may occur in

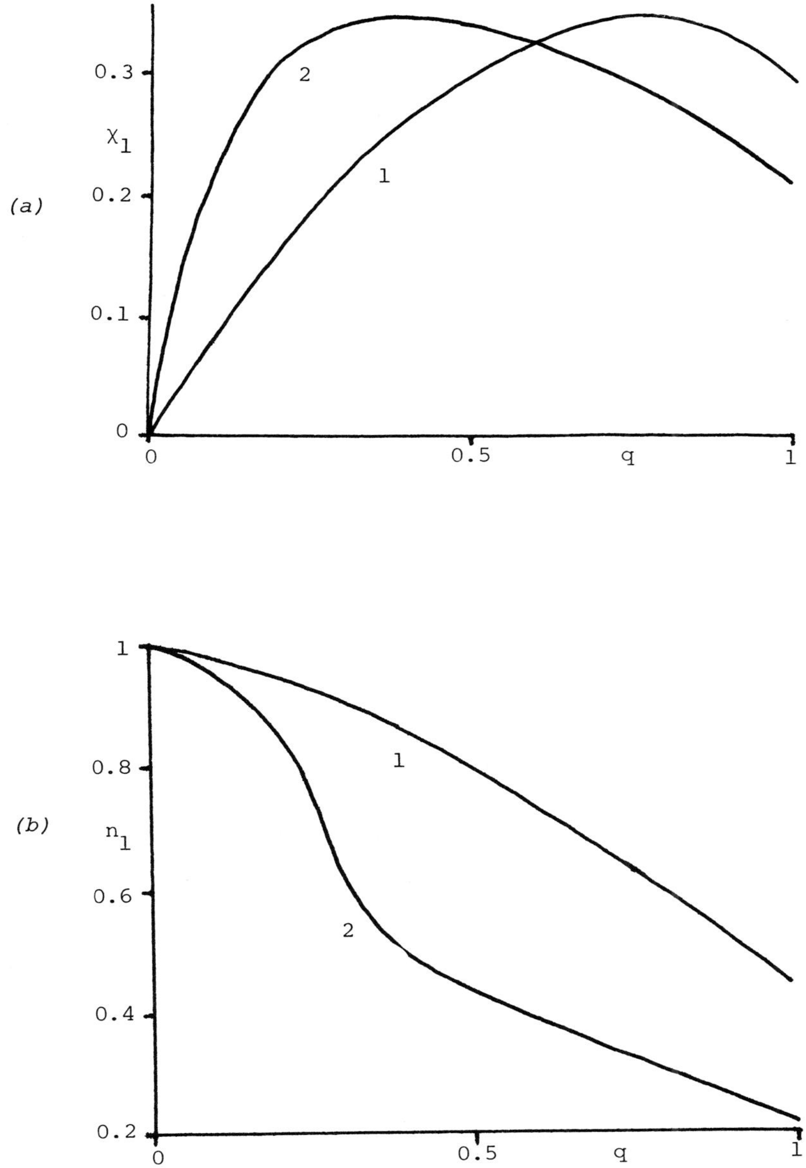

FIGURE 1. The effect of heating on resonant anomalous Joule absorption (1a) and refractive index (1b) of the extraordinary wave in a magnetoplasma (curves 1 and 2 correspond to the values $u\cos^2\alpha = 0.1$ and $u\cos^2\alpha = 10^{-2}$, respectively).

the value of $\chi(T_e)$, when $T > T_c$. (b) When the ionization is high, a region of positive slope $\frac{\partial \chi}{\partial T_e} > 0$ may occur on the $\chi(T_e)$ curve in the region $T > T_c$, in the case of a decreasing frequency of Coulomb collisions, connected with the heating $\frac{\partial \nu_e}{\partial T_e} < 0$.

Using the dependence $\nu_e(T_e)$, Fig. 1a may be applied to both cases (a) and (b). Notice that the change of absorption for the given effect may be considerable; thus, in the region $q < 1$, for the parameter value $u \cos^2 \alpha = 0.01$, a relative increase in the absorption during the Coulomb collisions $\left(\frac{\partial \nu_e}{\partial T_e} < 0 \right)$ will make $\Delta\chi/\chi_m = 0.4$ for a heating from $q = 1$ to $q = q_0 = 0.35$. In this case the temperature will increase from $T = T_0$ to $T_c = 2T_0$. The same Fig. 1a for a slightly ionized plasma shows a considerable relative enhancement $\Delta\chi/\chi_m = -0.4$ for heating from $q = q_0 = 0.35$ to $q = 1$; herein the temperature will increase from $T_e = T_c$ to $T_e = 3T_c$.

The heating in the vicinity of Langmuir resonance will lead to a considerable change of the refractive index of the resonant wave (Fig. 1b). Thus, it is seen that the magnetoplasma layer, transparent on condition that $q = 0.2$ (curve 2) and $n^2 >> \chi^2$, becomes non-transparent ($n^2 \approx \chi^2$) for the same wave due to a four fold growth of collision frequency.

The above analysis relates to the limited values of the angle α in the range $\frac{2\nu_0}{\omega_H} << \tan^2 \alpha << 2$. In the opposite case, when α tends to $\frac{\pi}{2}$, the resonant dependence $\chi = \chi(\nu_e)$ may be more complicated. Thus, the resonant extraordinary wave propagation across the magnetic field may be accompanied by two absorption maxima χ_1 and χ_2. In this case the "critical collision frequency" ν_c tends to infinity, and it is convenient to utilize the dimensionless parameters $s = \nu_e/\omega$. Herein, the absorption maxima are located at the points

$$s_{1,2} = \frac{1}{2\sqrt{3}} \left(1 \pm \sqrt{1 - 12u} \right) . \qquad (1.6)$$

There is an absorption minimum χ_3 between those two maxima, at

$$s_3 = \sqrt{u} \quad . \tag{1.7}$$

These effects may occur if the plasma is sufficiently ionized ($u < 12^{-1}$); in the case $u = 12^{-1}$, these three points are merged into one point $s_1 = s_2 = s_3$; in the case $u > 12^{-1}$ the graph $\chi = \chi(s)$ has one maximum at the point $s = s_3$.

The difference between the maximum values of absorption coefficient $\chi_1 = \chi_2 = 0.35$ and its minimum value χ_3 increases due to a decrease in the magnetic field on condition that $u < 12^{-1}$. Simultaneously the first maximum in the vicinity of the point s_1 is narrowing. The depth of the minimum χ_3 may be very considerable: thus, $\chi_1/\chi_3 = 3.5$ in the case $u = 0.001$.

The temperature dependence $\chi = \chi(q)$ in the abovementioned situation is single-valued. However, the inverse dependence $q = q(\chi)$ in some ranges of values q is not single-valued (Figs. 1,2). This circumstance may become a source of difficulties in the determination of collision frequency by means of measurement of resonant wave attenuation.

This analysis relates to a simplified case, when the wave vector of the plane wave propagating through the magnetoplasma and the magnetic field are disposed in one plane. In a more general case the possibilities of resonant absorption are extended due to angular anisotropy of the effect.

I.2. The Anisotropy of the Resonant Absorption

The comparison of Fig. 1a and Fig. 2 illustrates the difference in absorption of the waves, propagating inside some cone around the magnetic field ($\tan^2\alpha \ll 2$) and across the magnetic field $\left(\alpha = \frac{\pi}{2}\right)$. The angular dependence of such resonant heating effects is obvious. However, the essential heating effects are usually realized in the intense narrow wave beams. In this case the picture of plasma-wave interaction is complicated, because the rays of the beam form different angles with the magnetic field. Herein the resonant condition along each ray will depend upon the ray direction.

The general analysis of such oblique waves propagation is rather complicated [3]. Let us analyze the simplified example of Joule interaction between the axisymmetrical beam of extraordinary waves and the magnetoplasma. Let us suppose that the beam's axis is orthogonal to the boundary of the plasma layer, and the magnetic field is parallel to this boundary. Herein, the peripheric rays of the beam may be characterized by the projections of their wave vectors on the beam's axis, (K_z) and on a direction orthogonal to this axis (P_0). Let us consider the particular case when the projection P_0 is orthogonal to the magnetic field $\underline{H}$. In this geometric scheme the resonant condition may be written as

$$v = 1 - P_0^2 . \tag{1.8}$$

Even in this simplified case [20] the formulae for the refractive index n and the absorption coefficient χ of the extraordinary wave are complicated:

$$n_1 \chi = \sqrt{\frac{\pm M + \sqrt{M^2 + N^2}}{2}} , \tag{1.9}$$

where

$$M = -(\beta_1 F + \beta_2 \gamma_2)\Delta^{-1}, \quad N = (\beta_1 \gamma_2 - \beta_2 F)\Delta^{-1}, \quad \Delta = 2(\beta_1^2 + \beta_2^2),$$

$$\beta_1 = P_0^2(1 - s^2) - u - 2s^2, \quad \beta_2 = s(u + s^2 - 1 - 2P_0^2),$$

$$\gamma_1 = (1 - P_0^4)(u + 2s^2), \quad \gamma_2 = 2s(1 - P_0^2)(P_0^2 - u - s^2),$$

$$F = (1 - P_0^2)[2(u + s^2) + P_0^2(u + 2s^2)].$$

In the special case $P_0 = 0$ (plane wave), formula (1.9) leads to the expression (1.7). The angular anisotropy of absorption of the wave beam is shown in Fig. 2 for the case $u = 0.001$. The

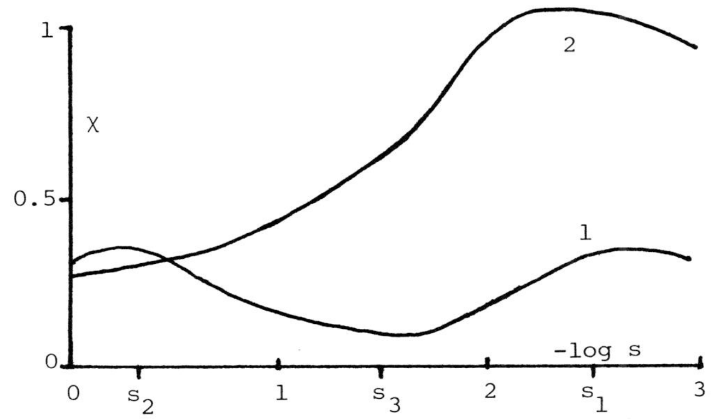

FIGURE 2. The angular anisotropy of resonant absorption of the narrow wave beam, propagating across the magnetic field ($u = 10^{-3}$). Curves 1 and 2 correspond to the axial and the periphery rays respectively; the angular width of the beam is 10°.

comparison of curves 1 and 2 shows that in a heterogeneous magnetoplasma the resonant absorption of the peripheric rays ($V = 1 - P_0^2$) is much more pronounced than the absorption of the paraxial rays ($V = 1$), even when the angle θ between these rays is small ($\theta = 10°$). According to Fig. 2, the absorption of such oblique rays may increase threefold in the range ($0.15 < s < 0.45$). This difference is connected with the essential anisotropic character of the resonant absorption effect.

This phenomenon may be important in the problem of interaction of a narrow conical wave beam with a heterogeneous collisional magnetoplasma. If the beam propagates along the electron density gradient, the resonant condition for each ray is realized at a different depth inside the plasma and the attenuation rate for these rays is also different. Thus, in a plasma layer the peripheric rays of the beam may be attenuated much more than the axial rays. This effect may be utilized for the formation of the angular divergence of the wave beam. It is worthwhile to note that the narrowing of the beam due to intensified dissipation of the periphery rays, connected with the specific profile of the radial heterogeneity of absorption in an isotropic medium was discussed by Kogelnik [17]. Unlike this, the above resonant effect in a magnetoplasma does not require any special distribution of dissipation. The sensitivity of attenuation to the collision frequency and the magnetic field permits us to compare the angular dependence of dissipation with the action of a peculiar aperture, governed by the thermal and magnetic effects over a wide range of plasma parameters.

In this way, a nonmonotonic temperature dependence of absorption may lead to a regime of anomalous absorption or enhancement of plasma heating in a time of the order of the time of the heating:

$$\tau_T = (\delta\nu_0)^{-1}, \quad \nu_0 = \nu^0_{em} + \nu^0_{ei} \tag{1.10}$$

where δ is the mean energy fraction, transferred by elastic collisions of electrons with heavy particles. Unlike the case of an isotropic plasma, in which the maximum absorption is also possible in the vicinity of plasma resonance, the anomaly which we consider here is connected with the high frequency ($\omega >> \nu_0$) and with the action of the magnetic field. In an inhomogeneous plasma such an effect is localized in the region $|1 - v| \lesssim s$. The reflection of the ordinary wave from such a plasma layer depends upon the collision frequency ν_0 in a very thin region, close to the resonant region. These properties of resonant plasma-wave interaction may be utilized for the determination of the fine structure of heterogeneous plasma distribution [13]. The threshold character, the magnitude of the effect and the short time required for it to occur may be of interest for plasma heating experiments, when the high-frequency wave dissipation is governed by the plasma temperature. The useful application of such effects may be connected with the possibility of selective heating of the localized resonant regions inside the heterogeneous plasma.

I.3. The Thermal Deviation of the Wave Energy Flux in a Plasma

The phase and group velocities of the wave may depend upon the frequency of electron collisions in the series of resonant situations. Thus, the collisions determine the condition of the phase synchronism between the ordinary and extraordinary modes in a magnetoplasma in the vicinity of Langmuir resonance [23]. Another example of such dependence is connected with the penetration of the gyroresonant extraordinary wave into the collisional magnetoplasma in the vicinity of the point of reflection $u = 1$, $v = 2$. If the collisions are rare enough ($s << 1$), one can derive from eq. (1.1) the refractive index and the absorption coefficient of this wave in the form

$$n = \chi = \sqrt{\frac{s}{2}} \quad . \tag{1.11}$$

However, these effects do not change the direction of the wave vector $\underline{k}$. Unlike these phenomena, the direction of wave energy flux may depend upon the collision frequency. This flux in a magnetoplasma is not directed along the wave vector $\underline{k}$. The angle between these vectors is determined by the plasma density and by the magnetic field, and leads to the deviation of rays of the high-frequency radiowaves [24] and whistlers (Mullaly, 1962) in the ionospheric plasma and to specific cross-modulation effects [37]. However, in the vicinity of Langmuir resonance the dependence of this deviation upon the collision frequency may become essential. In this case the thermal variations of the collision frequency lead to complicated changes of energy flux direction, because the changes of each component are different.

Let us consider the energy flux of the extraordinary wave in

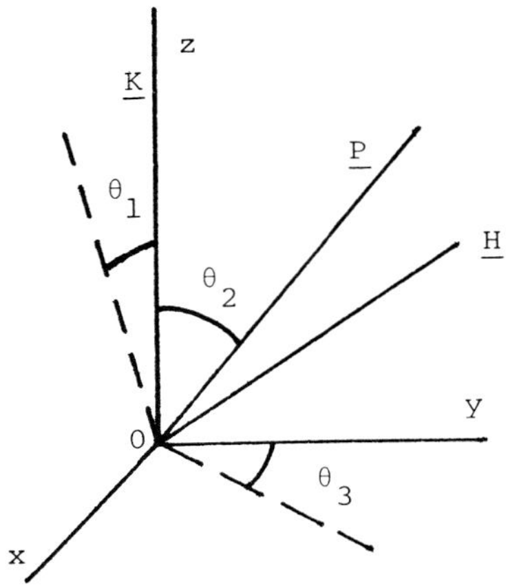

FIGURE 3. Orientation of the Poynting vector P in a magnetoplasma.

a transparent plasma region ($v = 1$). It is convenient to introduce a coordinate system with the z-axis directed along the wave vector $\underline{k}$ (Fig. 3); the y-axis is parallel to the $(\underline{k}, \underline{H})$ plane, the x-axis is orthogonal to this plane. The Poynting vector direction is determined in the case $q \leq 1$ by the angles $\theta_1, \theta_2, \theta_3$ [24]:

$$\tan\theta_1 = -\frac{1 - T}{\tan\alpha}, \quad \tan\theta_2 = \frac{1 - T}{\kappa \tan\alpha}, \quad \tan\theta_3 = -\kappa,$$

$$\kappa = -\frac{q}{1 + \sqrt{1 - q^2}}, \quad T = \frac{\ell}{\sqrt{1 + \ell^2}},$$

$$\ell = \sqrt{u}\cos\alpha\left(q + \frac{\tan^2\alpha}{2\kappa}\right). \tag{1.12}$$

During the thermal change of the parameter q from $q = 0$ to $q = 1$ the angle θ_3 varies monotonically. Unlike this tendency, the variation of angles θ_1 and θ_2 may be more complicated; in some ranges of plasma and radiation parameters, nonmonotonic variations of these angles are possible (Fig. 4). Similarly to

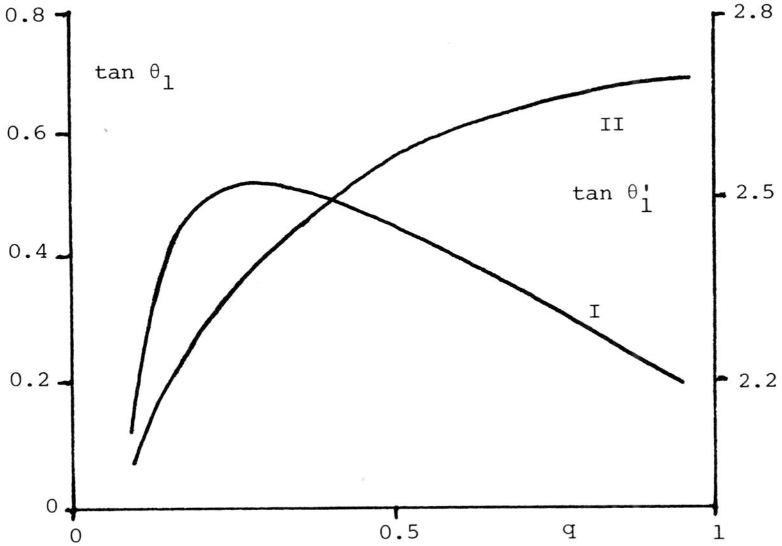

FIGURE 4. The thermal focusing of the energy flux of the resonant extraordinary wave in a magnetoplasma (u = 0.1). Curves I and II and the axes θ_1 *and* θ_1' *relate to the values* $\tan^2\alpha = 0.1$ *and* $\tan^2\alpha = 1$*, respectively.*

Fig. 1, this dependence may be utilized in both cases of electron-ion and electron-neutral collisions.

Let us consider in more detail the energy flux connected with the propagation of the extraordinary wave across the magnetic field in a plasma. If the plasma is sufficiently rarefied ($u >> s_0^2$), the angle θ_2 between the Poynting vector $\underline{P}$ and the wave vector $\underline{k}$ (Fig. 3) ($\theta_1 = \theta_3 = 0$) may be obtained by means of Maxwell equations in the form:

$$\tan^2\theta_2 = \frac{u}{u^2 + s^2} \quad . \qquad (1.13)$$

This formula, similar to eq. (1.12), describes the monotonic deviation of Poynting's vector via the thermal variations of the collision frequency s.

The temperature dependence of those angles shows the possibility of controlled deviation of the extraordinary wave energy flux due to plasma heating by the field of the wave itself or by another source of heat. This effect may occur in the phenomena connected with thermal self-focusing or cross-focusing of waves in a plasma. In particular, the modulation of the heating wave amplitude at the boundary of the plasma layer leads to an angular scanning of the energy flux direction inside the plasma. This effect attracts attention to the possibility of angular selection of resonant signals having the same frequency but different intensity.

I.4. Gyroresonant Negative Absorption and Negative Radiation Temperature

The plasma-wave interaction in a collisional magnetoplasma may be substantially amplified in the vicinity of gyroresonance ($\omega \approx \omega_H$). In this range of frequencies the maximum and the shape of the resonant dependence $\chi = \chi(\omega)$ are very sensitive to the kinetic effects connected with the electron distribution function $f_0(v)$. One may illustrate such dependence by means of components of the conductivity tensor σ_{ij} in a plasma:

$$\sigma_{ij} = C \int_0^{\infty} f_0(v) \frac{\partial}{\partial v} \left[\tilde{\sigma}_{ij} v^3 \right] dv; \tag{1.14}$$

here C is a normalization constant; the Hermitean tensor $\tilde{\sigma}_{ij}$ describes the electron motion in the wave field in a collisional magnetoplasma; the components of this tensor depend upon the magnetic field and may be written, on condition that $\nu^2 << \omega_H^2 \approx \omega^2$, in the form:

$$\tilde{\sigma}_{\perp} = \frac{\sigma_0}{1 + X^2}, \quad \tilde{\sigma}_{\wedge} = -i\tilde{\sigma}_{\perp}, \quad \tilde{\sigma}_{||} = \sigma_0, \quad \sigma_0 = \frac{V\omega_H}{8\pi s}.$$

The parameter X is essential in various gyroresonant effects in collisional magnetoplasma:

$$X^2 = \frac{(1 - \sqrt{u})^2}{s_0^2} . \tag{1.15}$$

The combination of different velocity dependences $\nu(v)$ and distribution functions $f_0(v)$ characterizes the variety of Joule effects.

In a simple case, when the electron distribution is Maxwellian and the electron-ion collisions are dominating, formula (1.14) permits one to calculate the kinetic corrections in the vicinity of gyroresonance (Shkarofsky, 1961).

However, formula (1.14) predicts not only the quantitative corrections to conductivity, but also the new qualitative effect based on the appearance of negative conductivity for some ranges of wave frequencies. In such a situation, the Joule current in the wave field and the electric vector of this field are directed opposite to each other. This possibility is connected with the velocity dependence of frequency of collisions $\nu(v) = \nu_0 (v/v_{Te})^h$. The condition for negative conductivity to originate is obvious from (1.14):

$$h \geq 3 . \tag{1.16}$$

These values of the quantity h may be realized in slightly ionized gases Ar and Kr due to the Ramsauer effect [2]. Another possibility of negative conductivity occurrence is connected with the well-known condition of an "inverse" distribution function in some range of electron velocities.

A negative conductivity signals the appearance of waves amplification (negative absorption). In a tenuous plasma the absorption coefficient of the gyroresonant wave may be written as [26]:

$$\chi = \chi_0 \frac{(h+3)X^2 + 3 - h}{3(1+X^2)^2} , \quad \chi_0 = \frac{\nu}{2s} (1 + \cos^2\alpha), \tag{1.17}$$

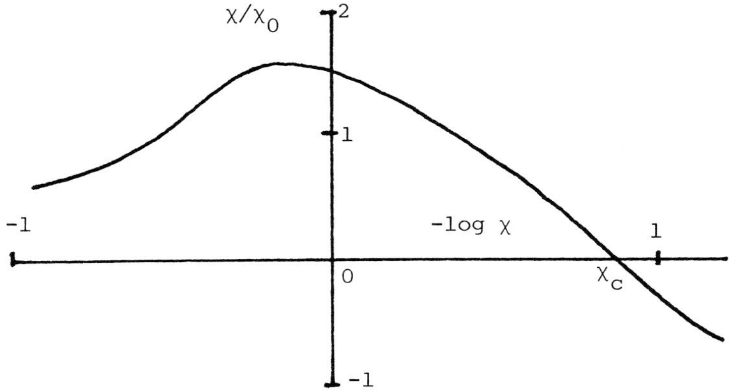

FIGURE 5. The transition from positive to negative gyroresonant absorption due to plasma heating and magnetic field variation (the parameter X is given by eq. (1.15)).

where the parameter X is determined from eq. (1.15). The graph of this dependence illustrates the region of negative conductivity (Fig. 5).

This kinetic phenomenon is interesting in view of the possibility of generation and amplification of radiation in such plasmas. It is important to note that the amplification is wide-banded, and that the structure of formula (1.17) shows the possibilities of thermal and magnetic regulation of the waves' resonant evolution.

The negative absorption leads to a considerable difference between the radiation temperature T_r and the electron temperature T_e, defined by the average electron energy. Herein, the radiation temperature may become negative. In the vicinity of the gyroresonance ($X \to 0$), the radiation temperature is governed by the equation:

$$\frac{T_r}{T_e} = \frac{3}{3 - h} \quad ; \tag{1.18}$$

accordingly, the value of T_r is negative under condition (1.16).

The radiation intensity also displays an important peculiarity in the region $h \geq 3$. It is convenient to write the expression for the gyroresonant ($X \to 0$) radiation intensity that escapes per unit solid angle from unit a area of surface of the homogeneous plasma in the absence of reflection at the plasma boundary in the form:

$$\frac{I\omega_H}{B(\omega_H, T_e)} = \frac{3}{3-h}\left\{1 - exp\left[\frac{h-3}{3}\;\frac{\omega_H \chi_0 L}{c}\right]\right\}, \qquad (1.19)$$

where the coefficient χ_0 is as given in eq. (1.17), and $B(\omega_H, T_e)$ is the blackbody radiation emitted by a Maxwellian plasma with average electron energy $\frac{3}{2}KT_e$ (here K is Boltzmann's constant). Because of the effect of negative absorption the radiation intensity may exceed that of blackbody, that is, $I\omega_H > B(\omega_H, T_e)$.

The analysis of the amplification action of such slightly ionized plasmas advances the problems of these systems stability, connected with the quick growth of electron density fluctuations due to negative conductivity. During the low frequency amplification in an isotropic plasma ($\omega << \nu$), the competition of the negative absorption and the electrons diffusion will lead to quasi-stationary periodical distributions of electron density [4]. However, in a magnetoplasma the diffusion limitation of the fluctuations growth may lead to a more complicated picture of density perturbations due to the anisotropic character of diffusion and to the arising of curl diffusion currents. Unlike this, the series of self-stratification processes in a collisional magnetoplasma, related to the case of "normal" absorption ($\chi > 0$), will be considered below.

II. THERMAL DISCONTINUITIES AND PLASMA LAMINATION DUE TO RESONANT HEATING

The electron temperature increase due to plasma Joule heating is usually proportional to the dissipated part of the energy of the electromagnetic field. The dependence of this part upon the frequency of electron collisions may lead to a complicated rebuilding of plasma parameters and wave characteristics. In particular, some heating regimes may lead to the origin of a double-valued electron temperature dependence from the wave intensity. This hysteresis effect may stimulate the nonlinear jump of wave field structure. If the thickness of the plasma layer along the direction of wave propagation is large enough, the plasma heat conductivity may lead to a nonlinear periodic distribution of temperature and dispersive properties of the wave in the initially homogeneous plasma.

One of the first examples of hysteresis of the temperature dependence upon the wave amplitude was connected with the non-resonant low-frequency ($\omega \ll \nu$) heating of fully ionized plasma [1]. However, the possibilities of such phenomenon may be substantially enlarged due to resonant situations. These self-action effects have a threshold character; the threshold amplitude values depend upon the electron collision frequency. The coupling hysteresis jumps of plasma temperature and the waves refraction and absorption may occur in a collisional magneto-plasma in the transparent region. These problems show an interesting analogy with the Van der Waals equation in thermodynamics and with the nonlinear pendulum model in mechanics. The peculiar resonant self-lamination of plasma due to transfer phenomena, connected with the aforesaid discontinuities, may occur if the plasma layer is sufficiently thick.

For simplicity, let us suppose that the ion-molecular collisions ν_{im} are frequent enough:

$$\nu_{im} >> \delta\nu_{ei} .$$

In this case the ion temperature T_i is close to the molecular one; thus, the variations of ion temperature will be neglected below. Moreover, the thickness of the plasma layer along the direction of wave propagation is supposed to be limited: $L << L_\chi$; here L_χ is the characteristic dissipation length

$$L_\chi = \frac{c}{\omega\chi} ; \tag{2.1}$$

therefore, the spatial attenuation of the wave amplitude due to wave absorption will be neglected here.

Under the above conditions, the thermal self-action phenomena may be analyzed by means of the energy balance equation:

$$\frac{3}{2} K N_e (T_e - T_{e0}) \delta\nu_e = \frac{1}{2} \sigma_{ij} E_i E_j^* . \tag{2.2}$$

Here $\{\sigma_{ij}\}$ is the plasma conductivity tensor, and the electric components E_i, E_j of the wave field depend upon the wave polarization. In the resonant case it is convenient to rewrite eq. (2.2) in the dimensionless form:

$$f-1 = a^2 \Phi(f) ; \tag{2.3}$$

here $a^2 = E_x^2 E_p^{-2}$; the axes orientation corresponds to that of Fig. 3; the characteristic plasma field E_p is determined as

$$E_p^2 = \frac{3m_e \delta K T_{e0} (\omega^2 + \nu_e^2)}{e^2} \phi ,$$

$$\phi = \left[cos^2\beta + \frac{sin^2\beta}{2} \left(\frac{\omega^2 + \nu_e^2}{(\omega-\omega_H)^2 + \nu_e^2} + \frac{\omega^2 + \nu_e^2}{(\omega+\omega_H)^2 + \nu^2} \right) \right]^{-1} , \tag{2.4}$$

δ is the mean energy fraction transmitted during the electron-ion collisions, and β is the angle between the vectors $\underline{E}$ and $\underline{H}_0$.

The function $\Phi(f)$ depends upon the plasma parameters and the wave mode. Let us examine some resonant situations ($v = 1$ and $u = 1$) by means of this function.

II.1. The Nonlinear Jumps of Wave Refraction in a Collisional Magnetoplasma

The phase structure of the electromagnetic wave field in a collisional magnetoplasma in the vicinity of Langmuir's electron resonance may become very sensitive to electron collisions. The resonant polarization ($v = 1$) of the extraordinary wave is described by means of the formula:

$$E_y/E_x = \kappa = -q\left[1 + \sqrt{1-q^2}\,\right]^{-1}, \tag{2.5}$$

$$E_z = E_x\ [\kappa\sqrt{u}\ \cos\alpha - s - i]\,[u(1-is) + is(1-is)^2 - u\cos^2\alpha\,]^{-1}. \tag{2.6}$$

These general relations permit us to analyze the collisional corrections to the Faraday effect in a magnetoplasma [10]; this effect is also possible in a "linear" theory. Unlike this, let us concentrate attention on some peculiar effects, which are possible due to intense fields only.

Let us consider the propagation of the extraordinary wave across the magnetic field. The polarization structure of this wave may be written in the form

$$E_y = 0;\ \frac{E_z}{E_x} = \rho\ exp(-i\phi),\ \rho = \left[\frac{u}{s^2 + (u+s^2)^2}\right]^{1/2};$$

$$\phi = arctan\left(s + \frac{u}{s}\right). \tag{2.7}$$

The variations of the collision frequency may lead to variations of the amplitude ρ and the phase ϕ of polarization. Herein the dependence of the phase shift ϕ upon the collisions is non-monotonic.

Now let us calculate the function of self-action $\Phi(f)$ connected with our problem:

$$\Phi(f) = 1 + \frac{u}{s^2 + (u+s^2)^2} \quad . \tag{2.8}$$

In some domain of intensity values the occurrence of S-like dependence of temperature upon the intensity a^2 is possible. This region is described by means of the set of equations (2.3) and

$$\frac{\partial \Phi}{\partial f} = 0 \ . \tag{2.9}$$

The positive roots of eq. (2.9) determine the coordinates of the turning points of the S-like curve $f = f(a^2)$. The corresponding critical values of amplitude $a = a_c$ may be calculated from eq. (2.7). Herein, in the range $a_{c1} \le a \le a_{c2}$, eq. (2.7) has three positive roots; the middle root is unstable.

Let us consider here the rarefied ($u >> s_0^2$), fully ionized plasma ($s = s_0 f^{-3/2}$). In this case eq. (2.9) has the form

$$Af^4 = f - 1. \tag{2.10}$$

The real roots of this equation are different on condition that $A < 0.1$. Herein one of these roots is located in the interval $1 < f \le \frac{4}{3}$. However, only two states, corresponding to the upper and lower branches of the S-like curve are stable. Transitions from a low temperature state to a high temperature one and vice versa are shown by arrows, a_{c1} and a_{c2} being the critical amplitudes corresponding to these transitions. The temperature discontinuity Δf is equal to $\Delta f = f_1 - f_c$ where the values f_1 and f_2, connected with the threshold amplitude a_c, are the positive roots of equation

$$f = f(a_{c1}) . \tag{2.11}$$

This temperature discontinuity is responsible for the appearance of nonlinear jumps of polarization amplitude and phase:

$$\Delta\rho = \rho(f_c) - \rho(f_1), \quad \Delta\phi = \phi(f_c) - \phi(f_1). \tag{2.12}$$

The common action of both these effects leads to deformation of the shape of the polarization ellipse and to rotation of its axes.

The complex of such hysteresis phenomena is shown in Fig. 6. It is interesting to note the considerable magnitude of the temperature discontinuities: so, in the case $u = 0.01$, $s_0 = 0.001$ the transition from the lower branch to the upper one leads to a temperature increase of five times ($f_{22}/f_{12} = 5$) (Fig. 6a, curve 1). The direction of the Poynting vector of the extraordinary wave depends upon the parameter θ_2 according to formula (1.13). Therefore, the temperature jump is accompanied by a jump in the energy flux direction (Fig. 6a, curve θ_2). This phenomenon shows the possibility of decreasing the energy flux from some region inside the plasma, if the wave intensity reaches the threshold value. Moreover, heating stimulates the simultaneous jump-like variations of refractive index n and absorption coefficient χ:

$$n,\chi = \left\{ \frac{1}{2} \left[\pm \frac{u^2}{u^2 + s^2} + \sqrt{\frac{u^2}{u^2 + s^2}} \right] \right\}^{\frac{1}{2}} . \tag{2.13}$$

Fig. 6b shows the hysteresis character of energy dependence $n(a^2)$ and $\chi(a^2)$ in the field of the extraordinary wave. One notes the considerable plasma enhancement due to the transition $f_{12} \rightarrow f_{22}$. These effects will also lead to the exchange of the refractive properties of the plasma layer.

The considerable magnitude and the relatively short characteristic times τ_T (see eq. 1.10) of such threshold effects attract attention to the possibility of utilizing these phenomena for the controlled evolution of intense wave fluxes in plasma. It is worthwhile to point out that the described hysteresis phenomena develop in a transparent magnetoplasma in the field of the travelling electromagnetic wave in the vicinity of Langmuir

(a)

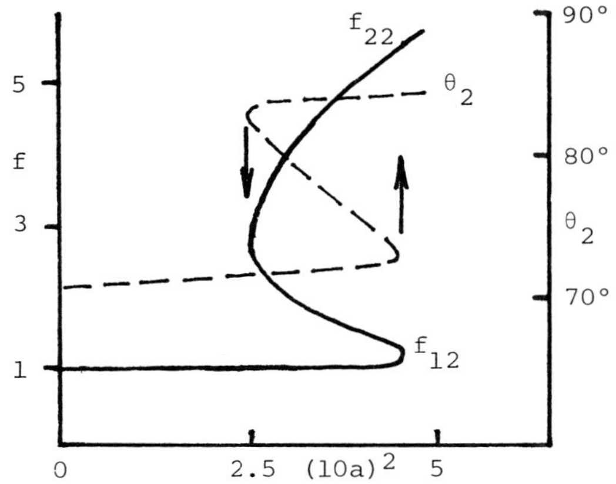

(b)

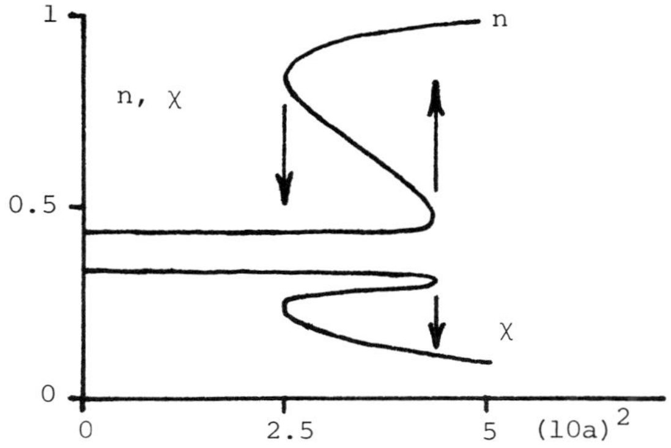

FIGURE 6. Nonlinear threshold phenomena connected with the hysteresis of dependence of plasma properties upon the wave intensity. (a) The jumps of the electron temperature f and of the angle θ_2 between the wave vector and the Poynting vector due to plasma Joule heating. (b) The thermal discontinuities of the refractive index and the absorption coefficient of the resonant wave in a magnetoplasma. The transitions are indicated by arrows.

resonance. Unlike this, the hysteresis phenomena in the vicinity of electron gyroresonance will be considered below.

II.2. The Hysteresis Character of Wave Absorption due to Plasma Gyroresonant Heating

Let us consider the S-like dependence of wave absorption upon the wave amplitude in the vicinity of electron gyroresonance. Such phenomenon is possible due to plasma heating by means of the extraordinary wave propagating along the magnetic field. The considerable absorption of this mode was formerly analyzed in connection with gyroresonant cross-modulation effects [14]. Unlike this, the peculiar self-action of this wave will be considered below.

During the longitudinal propagation of the aforesaid wave the electron temperature is governed by the dimensionless equation (2.3):

$$\Phi(f) = s_0^{-2}[X^2 + f^{-3}(1 + pf^2)^2]^{-1} \tag{2.14}$$

Here the parameter p characterizes the plasma ionization degree:

$$p = \frac{\nu_{em}^0}{\nu_{ei}^0}; \tag{2.15}$$

the dimensionless amplitude a is connected with the resonant wave amplitude E_0; the parameter X was determined in (1.15); the m order value in eq. (1.3) for electron-neutral collisions is chosen for simplicity as $m = 0.5$.

One finds from (2.14) that the occurrence of an S-like graph $\chi = \chi(a^2)$ corresponds to the region of three positive roots in eq. (2.14). The boundaries of this interval are determined by means of eq. (2.9); in our problem this condition has the form

$$X^2 f^4 + (1 + pf^2)[2pf - pf^2 - 2f + 3] = 0. \tag{2.16}$$

It is convenient to separate in this equation two limit cases, connected with either the exact gyroresonance ($X = 0$) in the partially ionized plasma ($p \neq 0$), or the fully ionized plasma ($p = 0$) in the vicinity of gyroresonance ($X \neq 0$). In the case of exact resonance ($X = 0$), eq. (2.16) degenerates into a cubic one; the threshold effects may arise on condition that $p < 12^{-1}$. In the opposite case ($X \neq 0$, $p = 0$) the threshold values are determined by two positive roots of the equation:

$$X^2 f^4 = 2f = 3. \tag{2.17}$$

If the parameter X in (2.17) satisfies the condition

$$\omega^2 < \frac{\nu_0^2}{16(1 - \sqrt{u}\,)^2} , \tag{2.18}$$

then the roots $f = f_1$ and $f = f_2$ are different; herein the lesser of these roots is located inside the interval $1 < f \leq 2$. The intermediate case ($X \neq 0, p \neq 0$) in eq. (2.16) may be solved numerically.

The temperature discontinuity from $f = f_1$ to $f = f_2$ leads to a hysteresis jump in the resonant absorption coefficient χ of the extraordinary wave. In the simple case of a tenuous transparent plasma, the simultaneous calculation of the refractive index and of the absorption coefficient leads to the expression

$$\chi = \left\{ \frac{1}{2} \left[- A + \sqrt{A^2 + \frac{v^2 f^3}{s_0^2 (1 + X^2 f^3)^2}} \right] \right\}^{\frac{1}{2}} ,$$

$$A = 1 - \frac{X v f^{3/2}}{1 + X^2 f^3} . \tag{2.19}$$

The temperature-dependent absorption coefficient χ is shown in Fig. 7 for the case $u = 0.95$, $s_0 = 0.2$.

The origin of the discontinuities (Fig. 8) in the case ($X^2 << 1$; $f_0 = 1$) may be understood qualitatively by means of an energy balance equation. The portion of dissipated energy

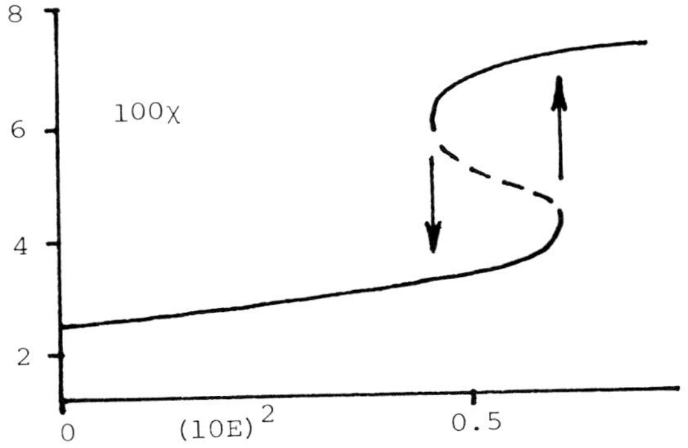

FIGURE 7. The discontinuity in the hysteresis dependence of the absorption coefficient χ *upon the electric field* $E(\omega = \omega_H)$.

Q_1 of the wave E increases with the growth of electron temperature ($Q_1 \sim f^{-3/2}$). On the other hand, the fraction of the energy Q_2, which is transferred from electrons to ions, decreases ($Q_2 \sim \delta\nu_e(T_e - T_{e0}) \sim f^{-1/2}$). If the electric field is sufficiently intense, the electrons cannot transfer to the ions a considerable fraction of the wave field energy, dissipated in the plasma [1]. The electron temperature T_e will increase, and the frequency of collisions will decrease and eventually the condition $x^2 \ll f^{-3}$ ($f > 1$) will be violated. This is the reason for the occurrence of the second state, i.e., the high temperature one.

In the particular case connected with the isotropic plasma ($u = 0$), condition (2.18) leads to the inequality $\omega < 0.25\nu_e$. Unlike this nonresonant low-frequency case, analyzed by Altshuler [1], the aforesaid resonant phenomena in magnetoplasma are developed in the high frequency ranges ($\omega \sim \Omega_e, \omega_H$). The analogous effects in the semi-conductors were described by Vladimirov et al. [40]. Moreover, the abovementioned plasma phenomena are associated with the travelling waves in a homogeneous plasma. Unlike this, the hysteresis effects in a quasi-static electric field will be considered below.

II.3. Discontinuities of Nonlinear Dielectric Permittivity in a Heterogeneous Plasma

The quasi-static electric field in the vicinity of electron Langmuir resonance may become responsible for considerable exchanges of plasma dielectric properties. This effect is connected with electron density perturbations due to electric field. The growth of the electric field amplitude in the vicinity of the reflection point inside the heterogeneous plasma ($\varepsilon(z) \to 0$) leads to the onset of the sharp maximum of the averaged high-frequency potential U [9]:

$$U = \frac{e^2 E^2(z)}{4\pi m\omega^2} \quad .$$

The redistribution of electron density $N(z)$ due to the averaged force

$$\underline{F} = -\nabla U$$

may be described by means of the Boltzmann-like formula:

$$N(z) = N_0(z)\, exp\left(- \frac{U}{2KT_{e0}}\right) \quad . \qquad (2.20)$$

Such density perturbation stimulates the change of resonant absorption. Herein this absorption depends upon the characteristic thickness ℓ of the resonant layer. The space dispersion may be neglected on condition that [11]:

$$\frac{eE_m}{m_e\omega} \ll \nu_e \ell, \quad s \gg \left(\frac{r_D}{\ell}\right)^{2/3} \quad .$$

Here r_D is the Debye electron radius, and E_m is the maximum amplitude value. Let us consider the restricted range of the wave amplitude:

$$E^2 << E_c^2 = \frac{8\pi m_e \omega^2 KT_{e0}}{e^2} . \tag{2.21}$$

In the vicinity of Langmuir resonance ($v = 1$) the amplitude values are determined by the collision frequency. The real part of the dielectric permittivity in a layer with linear profile of electron density may be written as

$$\varepsilon_r = -\frac{z}{\ell} + \frac{E^2}{E_c^2} . \tag{2.22}$$

In the following , let us assume that the imaginary part of the dielectric permittivity $Im\varepsilon = s$ does not depend upon the coordinate z.

On the other hand, the equation

$$(\varepsilon_r + is)E(z) = E_0 \tag{2.23}$$

permit us to eliminate the quantity E^2 from eq. (2.22). Herein the nonlinear equation, governed by the real part of the dielectric permittivity, may be written in the form

$$\frac{z}{\ell s} + \frac{\varepsilon_r}{s} - \frac{\eta}{1 + \left(\frac{\varepsilon_r}{s}\right)^2} = 0, \tag{2.24}$$

where the dimensionless parameter

$$\eta = \frac{E_0^2}{E_c^2 \, s^3} \tag{2.25}$$

describes the nonlinear effect. The analysis of eq. (2.24), similarly to eqs. (2.3) and (2.17), shows that the dependence $\varepsilon_r = \varepsilon_r(\eta)$ is single-valued, if the wave amplitude is small enough:

$$\eta < \eta_c = \frac{8}{3\sqrt{3}} .$$

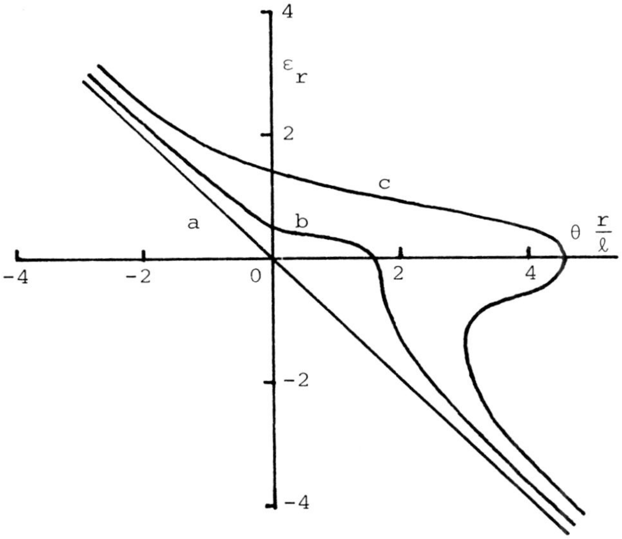

FIGURE 8. The hysteresis of the nonlinear resonant dependence of the dielectric permittivity in a heterogeneous plasma; curves (a), (b), (c) correspond to the dimensionless field intensities $\eta \to 0$, $\eta = \eta_c$, $\eta = 3\eta_c$, respectively.

Unlike this, in the region $\eta \geq \eta_c$ the single-valuedness is violated and the dependence $\varepsilon_r = \varepsilon_r(\eta)$ has a hysteresis behavior (Fig. 8).

Analogous discontinuous transitions between negative and positive values of the dielectric permittivity due to the electric field may also occur in a homogeneous plasma [19].

As a result of such nonlinear effect, the electron density distribution in the vicinity of the Langmuir resonance acquires the shape of a step. This density perturbation, connected essentially upon the frequency of electron collisions, leads to the origin of a particular plasma layer with a dielectric permittivity value close to zero. The occurrence of such a layer characterizes the peculiar "self-screen" resonant effect in a heterogeneous plasma. This effect may be interesting in problems connected with heterogeneous plasma diagnostics, with the resonant effects in limited plasmas with diffusive boundaries, with the radiation acceleration of plasma clots in waveguides, and with the treatment of controlled band-filters.

II.4. Kinetic Effects and Thermal Solitons in a Plasma

The spatial distributions of plasma temperature, connected with the low-temperature and the high-temperature branches of the aforesaid hysteresis curves, may be analyzed by means of the heat-conductivity equation. The exact analytical treatment of the energy balance equation reveals domains of monotonic and periodic solutions. Herein the periodic solutions predict the occurrence of peculiar quasi-stationary temperature waves in region of S-like electron temperature dependance from the intensity of the distributed heating wave (Figs. 6,7). It is important that such oscillations arise from stationary and spatially homogeneous heat source distributions due to transfer phenomena. In this case, the temperature oscillations have a nonlinear character and their shape, amplitude and space scale are given by plasma characteristics and by the heating source.

The periodic structure of the temperature distribution may be analyzed in connection with the different examples of hysteresis discontinuities described above. For simplicity, let us consider the model of resonant heating in the fully ionized magnetoplasma. The longitudinal heat conductivity of the electrons is much higher than the transversal conductivity. Thus, the energy balance consideration leads to the dimensionless heat-conductivity equation in the form

$$\frac{\partial}{\partial \varepsilon}\left(f^{5/2}\ \frac{\partial f}{\partial \varepsilon}\right) = f^{-3/2}\ F(f), \quad F(f) = 1 - f + \Phi(f). \tag{2.26}$$

Here ε is a dimensionless coordinate: $\varepsilon = z/L_T$; the constant L_T is the characteristic "heat conductivity length"

$$L_T = \left(\frac{1.6 T_{e0}}{m_e \delta \nu_0^2} \right)^{\frac{1}{2}} . \tag{2.27}$$

The numerical coefficient in (2.27) is the result of an exact kinetic calculation. The multiplication of both sides of eq. (2.26) by the quantity $f^{5/2} \frac{\partial f}{\partial \varepsilon}$ permits one to obtain the solution of eq. (2.26) in the form

$$\varepsilon(f, f_0) = \frac{1}{\sqrt{2}} \int_{f_0}^{f} dx\, x^{5/2} [\Theta(x) - \Theta_0]^{-\frac{1}{2}} . \tag{2.28}$$

The function $\Theta(x)$ may be written as

$$\Theta(x) = \Theta_0 + \int_{f_0}^{x} yF(y)\, dy. \tag{2.29}$$

Here f_0 is the minimum value of the temperature in the periodic profile; this value is located on the low-temperature branch of the corresponding S-like curve. The dimensionless length of the spatial period is $L = 2\varepsilon(f_0; f_m)$. The integration of eq. (2.26) is necessary only for the determination of the temperature profile shape of its spatial period. One may see that the amplitude values of such temperature profile may be determined directly from the S-like temperature curve $f = f(a^2)$. Herein these extrema height may be changed substantially, due to kinetic effects in the gyroresonant absorption (Fig. 9). If plasma lamination develops across the magnetic field (2.8), the characteristic length L_T of eq. (2.27) must be multiplied by the small factor $\nu_e \omega_H^{-1}$, on condition that $\nu_e^2 \ll \omega_H^2$. Such case of lamination due to Langmuir resonant radiation is shown on Fig. 10.

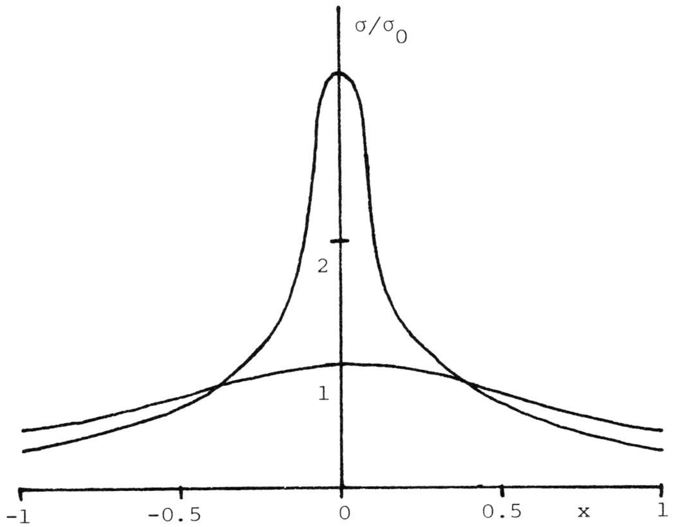

FIGURE 9. The narrowing and amplification of the gyroresonant wave absorption due to kinetic effects. Curves I and II correspond to velocity-independent and velocity-dependent frequencies of elastic electron collisions, respectively.

Outside the double-valued region the spatial temperature distribution is monotonic, inside this region the distribution is periodic. This region is limited by the critical value a_c of the heating wave amplitude. Both temperature distributions are stable against temperature perturbations, with the exception of the region close to the critical points. Small perturbations in the vicinity of these points may lead to the change of the character of distribution from monotonic to periodic. These self-stratification phenomena are possible in the vicinity of both gyroresonance (Fig. 7) and Langmuir resonance (Fig. 6). Herein the above results, based on assumption about the constancy of the heating wave amplitude are valid if the period of the temperature profile $L \sim L_T \varepsilon$ is much smaller than the

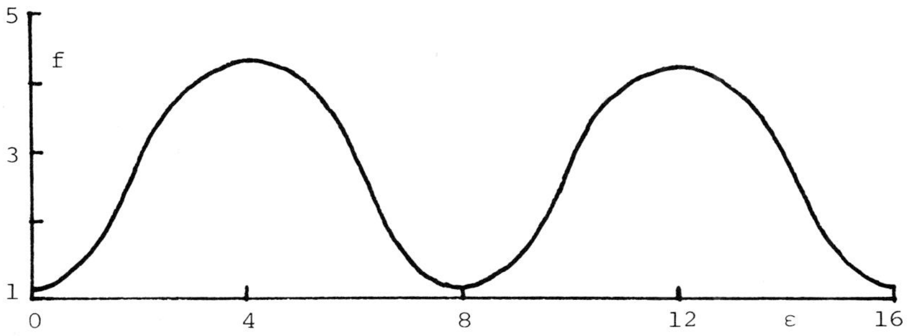

FIGURE 10. The thermal self-lamination in a plasma due to hysteresis dependence of electron temperature f upon the resonant heating wave intensity ($v = 1$; $u = 0.01$; $s_0^2 = 10^{-3}$; $a^2 = 3 \times 10^{-2}$).

characteristic absorption length L_χ. This condition restricts the plasma electron temperature according to the inequality ($\varepsilon \gtrsim 1$):

$$\frac{\varepsilon^2 T_{e0}}{m_e c^2} \ll \frac{\delta s_0^2}{\chi^2} . \tag{2.30}$$

The establishment of the periodic temperature profile stimulates the slow process of plasma diffusion due to temperature gradient.

III. RESONANT WAVE COUPLING IN A COLLISIONAL PLASMA

The possibilities of resonant wave coupling in a plasma may be extended substantially due to collisional effects. One of the first examples of such wave interaction was illustrated in the resonant triple splitting effect [7]. This effect, connected with the partial conversion of the ordinary mode into an extraordinary one in a magnetoplasma, is very sensitive to the electron collision frequency in the Langmuir resonance region ($v = 1$).

The amplitude coefficient R of this mode conversion may be written in the form [10]:

$$R \sim \begin{cases} exp(1 - q)^{3/4}, & q \leq 1; \\ 1 & q \geq 1. \end{cases} \quad (3.1)$$

The heating of the slightly ionized plasma, connected with the growth of the parameter q, may substantially increase the effectiveness of such conversion. Fig. 1b illustrates the considerable temperature dependence of the refraction index n and of the absorption coefficient χ in the field of the extraordinary wave in a resonant homogeneous plasma ($v = 1$). A similar effect may also be obtained in the field of the ordinary wave; herein the temperature value

$$f_0 = q_0^{-1} \quad (3.2)$$

corresponds to the polarization degeneration phenomenon, connected with the rapproachment of the complex refractive indexes of these modes. In a heterogeneous plasma the values n depend upon the profile of the collision frequency. The display of this effect in the lower ionosphere is illustrated in Fig. 11; the numbers along the curve $n_1 = n_2$ correspond to the fulfillment of condition (3.2). This situation shows the possibility of the controlled transformation of the modes in the middle wave range due to heating effects [26]. Fig. 11 relates to the value of the angle α between the magnetic field and the wave vector $\underline{k}$. such possibility of forced wave synchronism is not restricted to small values of the angle α only. Thus, the temperature dependence of the coefficient of wave conversion may be utilized for the local analysis of the temperature perturbations in the heterogeneous plasma.

The triple splitting effect attracted attention to the phenomena of waves' resonant coupling in dissipative media. Some examples of such phenomena will be described below. Unlike the

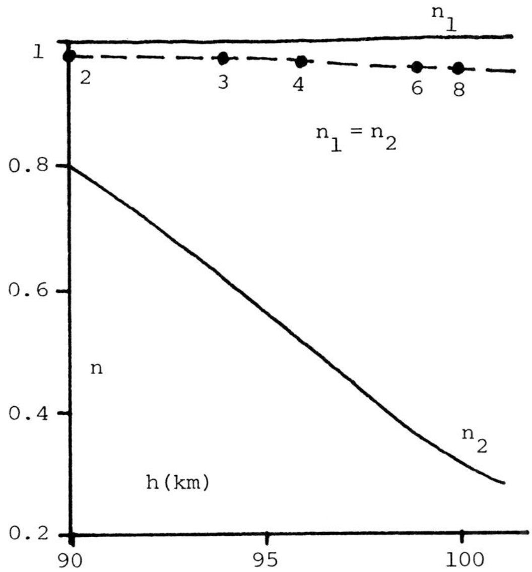

FIGURE 11. The resonant ($v = 1$) polarization degeneration of the extraordinary (n_1) and ordinary (n_2) modes in the ionospheric plasma due to electron heating. The solid curves show the resonant values of refraction indexes n_1 and n_2 in an unperturbed plasma ($f = 1$); the dotted curve illustrates the thermal merging of these modes; the numbers along the dotted curve characterize the local temperature increase ($f > 1$) needed for these modes' merging.

well-known nonlinear interactions of waves in a collisionless plasma [38], the role of collisions in the following effects is fundamental.

III.1. The Thermal Excitation of Parametric Instabilities in a Plasma

The peculiar temperature dependence of the parametric instability threshold in a partially ionized plasma is connected with the different tendencies of the thermal perturbations of electron-ion and electron-neutral collisions. The electron heating will lead to an increase of the electron-neutral collision frequency and to a decrease of the electron-ion collision frequency. The dependence of parametric instability threshold amplitude upon the frequency of electron collisions shows the possibility of thermally influencing such instability development. The competition of the aforesaid tendencies in a partially ionized plasma reveals the peculiar ranges of heating wave intensity responsible for the parametric instability excitation. Unlike the resonant character of plasma-wave parametric interaction, the frequency of the heating wave may be chosen over a wide range of values.

Let us examine an isotropic collisional plasma. It is convenient to utilize the formula for parametric instability threshold amplitude E_c in the form [21]:

$$E_c^2 = \frac{16\pi N_e \nu_e (T_e + T_i)}{\omega} , \quad \omega \approx \Omega_e . \tag{3.3}$$

Here ν_e is the sum of electron-ion and electron-neutral collision frequencies. It is convenient to utilize below for the parameter m, connected with the temperature dependence of electron-neutral collisions, the value $m = 0.5$; other possible m values lead to qualitatively similar results, but the calculations are rather complicated. Let us assume that the frequency ν_{im} of ion-neutral collisions satisfies the condition $\nu_{im} >> \delta\nu_{ei}$. In this case, the ion temperature may be considered as a constant, and formula (3.3) may be rewritten in the form:

$$E_c^2 = E_{c0}^2 \, Z(f) , \quad E_{c0}^2 = \frac{16\pi N_e T_{e0} \nu_{em}^0 (1+r)(1+q)}{\Omega_e} ,$$

$$Z(f) = \frac{(f+r)(f^2+q)}{f^{3/2}(1+r)(1+q)} , \quad Z(f)\Big|_{f=1} = 1 . \tag{3.4}$$

Here the parameters r and q are connected with the plasma ionization degree and the initial inequality of electron and ion temperatures:

$$r = \frac{T_i}{T_{e0}} \quad , \quad q = \frac{\nu^0_{ei}}{\nu^0_{em}} \quad ;$$

E_{c0} is the instability threshold amplitude in an unperturbed plasma, and the role of thermal perturbations is described by the dimensionless function $Z(f)$.

In a high temperature region ($f^2 >> q$), the function $Z(f)$ increases monotonically. If the f values are not too high, the function $Z(f)$ will have an extremum at the point $f = f_0$, which is determined by means of the equation:

$$f_0^3 + \frac{r}{3} f_0^2 - \frac{q}{3} f_0 - rq = 0. \tag{3.5}$$

The possible types of curves $Z(f)$ are shown in Fig. 12. These types are connected with the values of the plasma parameters q and r. If the function $Z(f)$ increases near the value $f = 1$ of the variable f, the condition $\left.\frac{\partial Z}{\partial f}\right|_{f=1} > 0$ leads to the condition:

$$1 + \frac{r - q}{3} - rq > 0. \tag{3.6}$$

This case is illustrated by the curves a and c of Fig. 12. The opposite case is illustrated by the curves b and d. The thermal change of instability threshold may be considerable: thus, in the case $r = 0.5$, $q = 100$ the curve b reaches its minimum at the value $f_0 = 6.3$, where $Z(f_0) = 0.4$.

Fig. 12 shows the appearance of the limited ranges of parametric instability threshold values due to growth of heating wave intensity. The descending parts of the curves in Fig. 12 illustrate this threshold decrease due to plasma heating; the opposite effect is described by the ascending parts of these curves. Therefore, the heating wave intensity scanning may lead to the controlled development of parametric instability, the

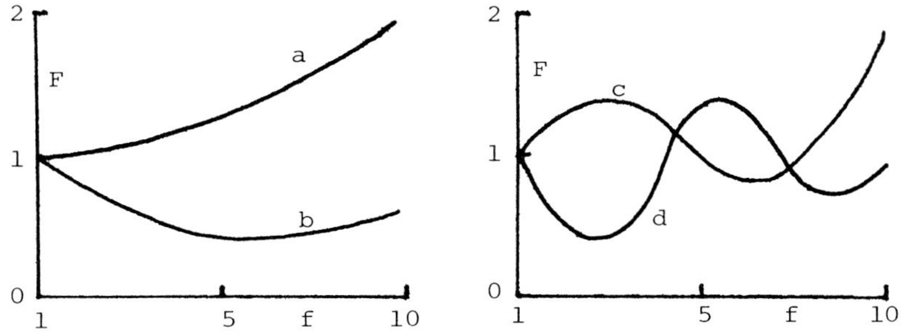

FIGURE 12. Parametric instability threshold function Z(f).

resonant wave intensity being constant.

Moreover, the same effect shows the possibility of separate determination of the frequencies of both types of electron elastic collisions, ν_{em} and ν_{ei}. Such possibility is connected with the indication of parametric instability threshold amplitude in the field of the radiowave, falling on resonant plasma region. The different observations of the beginning of such instability development permit one to write a set of equations similar to (3.3). The unknown quantities in these equations are ν^0_{em}, ν^0_{ei}, T_{e0} and T_i. The observation of an anomalous attenuation of reflected resonant signal, accompanied by the determination of plasma temperatures, attracts attention to the possibility of the separate determination of the frequencies ν^0_{em} and ν^0_{ei}. The appearance of resonant wave instability in a heterogeneous plasma due to another wave heating action may be valid, in particular, in ionospheric modification experiments [39]. The different altitude dependences $\nu^0_{em}(z)$ and $\nu^0_{ei}(z)$ and the utilization of the incoherent scattering technique for the temperature profile determination show the wide range of radiowave frequencies responsible for the aforesaid phenomena.

II.2. Wave Mixing due to Plasma Joule Heating

The nonlinear effects due to velocity-dependent collision frequencies may lead to a significant coupling of the electromagnetic waves in a plasma. The comparison with the waves interaction in a collisional plasma, connected with the nonlinear character of the Lorentz force in the wave field shows the possibility of classifying the above processes in conformity with the number of interacting waves. Thus, one may analyze the third-order, the fourth-order and so on processes, connected with the velocity-dependent frequency of collisions. Let us, for simplicity, consider a spatially uniform electron plasma, where the wavelengths of all interacting waves are assumed to be much larger than the mean free path of electrons. Perturbations of the density N_0 can then be neglected, and the equation of momentum may be utilized

$$\frac{\partial \underline{j}}{\partial t} - \frac{N_0 e^2}{m_e} \underline{E} = -\nu(f)\underline{j} \ . \tag{3.7}$$

Here the current $\underline{j}$ and the wave $\underline{E}$ characterize the result of interaction of the intense waves $\underline{E}_1$ and $\underline{E}_2$; the field $\underline{E}$ is supposed to be weak and its self-action is neglected here.

The effect of intense waves $\underline{E}_1$ and $\underline{E}_2$ (with frequencies ω_1 and ω_2) is described by means of the nonlinear collisional term in eq. (3.7). Let us consider the resonant interactions of these waves in isotropic plasma, satisfying the following types of resonant conditions:

$$\omega_1 - \omega_2 = \Omega, \tag{3.8}$$

or

$$2\omega_1 - \omega_2 = \Omega_e. \tag{3.9}$$

The situation (3.8) is connected with the excitation of one longitudinal plasma wave by means of two transverse waves. Such excitation may become considerable in the presence of an external constant electric field $\underline{E}_0$ [35], if the waves $\underline{E}_1$ and $\underline{E}_2$ are propagating along the direction $\underline{E}$. A Taylor expansion of the right-hand side of eq. (3.7) yields

$$\frac{\partial \underline{j}_L}{\partial t} - \frac{N_0 e^2}{m_e} \underline{E}_L = -\nu_0 \underline{j}_L - \frac{2\nu'}{N^2 e^2} \sigma E_0 (\underline{j}_1 \, \underline{j}_2), \quad \nu' = \frac{\partial \nu}{\partial f}, \qquad (3.10)$$

where $\underline{j}_L = \underline{j}(\Omega_e)$, $\underline{E}_L = \underline{E}(\Omega_e)$, $\underline{j}_{1,2} = \sigma(\omega_{1,2})\underline{E}_{1,2}$, and the non-resonant terms in the Taylor expansion are omitted. Eq. (3.10) shall be combined with the Maxwell equation:

$$\underline{j}_L + \varepsilon_0 \frac{\partial \underline{E}_L}{\partial t} = 0. \qquad (3.11)$$

If the current $\underline{j}_L$ in (3.11) is eliminated by means of (3.10), one finally obtains the equation for slowly varying amplitude E_L (i.e., $\frac{\partial E_L}{\partial t} << \Omega_e E_L$) in a standard form:

$$\frac{\partial E_L}{\partial t} + \frac{E_L}{\tau} = A_\chi E_1 E_2 . \qquad (3.12)$$

Here the coupling coefficient A_χ may be written in the form

$$A_\chi = \frac{i\Omega_e \nu'}{N_0 \omega_1 \omega_2} \frac{\sigma E_0}{T}, \quad \sigma = \frac{\Omega_e^2}{\nu_e} . \qquad (3.13)$$

This effect is connected with third-order processes. A simple example of fourth-order resonant effects, connected with the condition (3.9), was analyzed by Kaw [16] in relation to optical frequency mixing in semiconductors. By means of two laser fields E_1 and E_2 one may thus generate the large current at frequency $2\omega_1 - \omega_2$. The theory, based on the same eq. (3.7), predicts the possibility of generating such harmonics in a plasma, and, in particular, in the ionospheric plasma, if the lasers are replaced by two high-power sources of high frequency radiowaves. The analysis of the ionospheric plasma effects (Wilenskij, 1954) and the semiconductors phenomena [34] permit one to obtain the following expression for the electric field $E_3 exp[i(2\omega_1 - \omega_2)t]$ at the mixed frequency $2\omega_1 - \omega_2$:

$$E_3 = 3iP\, E_1^2 E_2^* \left\{ \omega_1^2 \omega_2 \left(1 - \frac{\Omega_e^2}{\omega_1^2}\right)^2 \left(1 - \frac{\Omega_e^2}{\omega_2^2}\right) \right.$$

$$\left. \times \left[\left(2\omega_1 - \omega_2\right)^2 - \Omega_e^2 - \frac{i\nu_0 \Omega_e^2}{2\omega_1 - \omega_2} \right] \right\}^{-1} . \tag{3.14}$$

Here the value of P in a fully ionized plasma is

$$P = -\frac{3e^4}{8m^3} \int \frac{\nu(v)}{v^2} f_0(v)dv . \tag{3.15}$$

If the resonant condition (3.9) is fulfilled, the ratio E_3/E_1 may be increased to the value of about $10^{-4}\Omega_e\omega_2^{-1} \approx 10^{-5}$. In the ionospheric plasma this figure might be within the limits of what one can observe by means of a sensitive satellite antenna [41]. This effect shows an interesting possibility of analysis of collisional processes in a plasma. It is worthwhile to note that such effect is possible only due to the energy dependence of the collision frequency $\nu(f)$.

It is interesting to compare the coupling coefficient A_χ of eq. (3.13) with the coefficient A_L, which is accounting for the nonlinearity of the conventional Lorentz force. Thus, in the third-order interaction (3.12) this coefficient may be written in the form

$$\underline{A}_L = -\frac{e}{2m_e} \frac{\Omega_e}{\omega_1\omega_2} \underline{K}_L . \tag{3.16}$$

In deriving formula (3.12), the wavelengths of all interacting waves were assumed to be much larger than all the other characteristic scales of this plasma-wave interaction; unlike this, the quantity $\underline{A}_L$ of eq. (3.16) depends upon the Langmuir wavelength.

The comparison of the quantities (3.13) and (3.16) shows that the collisionless generation may be rather weak, if the wavelength of the Langmuir plasma wave is large enough:

$$K_L r_D << \frac{2V_\sim}{v_T}, \quad V_\sim = \frac{eE_0}{m_e \Omega_e}, \tag{3.17}$$

where r_D is the Debye radius and v_T is the electron thermal speed. Therefore, the resonant collisional excitation may be very effective in the long wave range of the plasma oscillations spectrum.

III.3. The Generation of Stationary Electromagnetic Fields due to Self-Action of High-Frequency Waves

The nonlinear interaction of two electromagnetic waves in a plasma may stimulate the generation of combined frequencies. The analogous effects, accompanying the self-action of one wave, may lead to the doubling of frequency and to the origin of static electric and magnetic fields. In a high frequency electromagnetic field such phenomena are connected with the variable and constant components of the nonlinear electron current. The nonlinear electron motion due to Lorentz force action depends upon the polarization of the wave $\underline{E}$ and the variable magnetic field $\tilde{H}$ of the wave. The components of the nonlinear current may be calculated by means of Maxwell's equations and of the equations of electron motion in a collisional magnetoplasma:

$$\frac{d\vec{v}}{dt} + \nu\vec{v} = \frac{e}{m} (\underline{E} + \frac{1}{c} [\vec{v}\times\tilde{H}]). \tag{3.18}$$

Let us consider, for simplicity, the transversal, circularly polarized plane wave propagation along the z-axis; herein, the electrons are moving in the (x,y) plane.

The longitudinal stationary electric field E_z arises from the displacement of the electrons from the immobile ions. The balance of the forward Lorentz force and the return electrostatic force indicates that the field E_z has the form

$$E_z = -\frac{1}{c}[\vec{v}\times\tilde{H}]_z .$$

Such field E_z due to the high frequency pumping wave E_0 in the isotropic collisional plasma may be calculated by means of the formula [15]:

$$E_z = \frac{e}{m}\frac{E_0^2}{\omega^2}\left[\frac{\nu}{\omega}\left(\varepsilon_1\frac{d\varepsilon_2}{dz} - \varepsilon_2\frac{d\varepsilon_1}{dz}\right) - \frac{1}{2}\frac{d}{dz}\left(\varepsilon_1^2+\varepsilon_2^2\right)\right] . \quad (3.19)$$

Here the quantities ε_1 and ε_2 denote the real and the imaginary parts of the complex dielectric permittivity of the plasma. The effective electron collision frequency ν is obtained from the frequency of electron-ion collisions ν_{ei} and a characteristic frequency ν_c, connected with the radiation reaction of the pumping wave:

$$\nu_c = \frac{2e^2\omega^2}{3mc^3} .$$

To get a feeling for the magnitude of the longitudinal field we let $\omega = 10^{15}\,rad/s$, $\nu_c = 10^6$ Hz; in the vicinity of the Langmuir resonance $|\omega - \Omega_e| \approx 10^8\,rad/s$, one obtains $E_z = 10^{-7}E_0$. This self-consistent, time-independent electric field is shown to exist in a collisional cold electronic plasma.

A qualitatively similar effect predicts the generation of a static magnetic field in a plasma, due to the aforesaid wave. This effect is connected with the magnetic dipole produced by the motion of charged particles due to wave's field. The magnetic momentum of the particle, created by this motion, may be written as

$$\underline{\mu} = \frac{e}{2c}[\underline{r}\times\underline{v}]; \quad \underline{r} = \frac{i\underline{v}}{\omega} ; \quad (3.20)$$

The electron speed $\underline{v}$ is governed by eq. (3.18); the vector $\underline{r}$

characterizes the electron trajectory. The sum of the momenta (3.20) in the unit plasma volume permits one to calculate the plasma magnetization. This effect may be amplified in the vicinity of gyroresonance, due to a constant external magnetic field H_0 in the plasma. However, the gyroresonant increase of the effect is restricted due to collisions. Thus, the extraordinary wave, which is propagating along the magnetic field H_0 in a resonant regime $(\omega = \omega_H)$, will lead to an additional static magnetic field ΔH:

$$\frac{|\Delta H|}{H_0} = \frac{E_0^2}{H_0^2} \frac{\Omega_e^2}{16\pi\nu^2} . \tag{3.21}$$

For simplicity, the corrections connected with the final plasma pressure are here neglected.

Unlike the spontaneous generation of magnetic field due to turbulent motion of conducting fluid, the aforesaid perturbation of the magnetic field is connected with the electrons oscillations in the high-frequency pumping wave field, the plasma ions being immobile. Herein, the dependence of the perturbation ΔH upon the electron collision frequency shows a possibility for the thermal regulation of such phenomena. In particular, in the region of ambiguous dependence of electron temperature upon the wave intensity E_0^2 in the vicinity of gyroresonance, the perturbation ΔH may have a hysteresis character.

IV. CONCLUSION

It is worthwhile to note here the main achievements and problems in the theory of resonant interaction of intense electromagnetic waves with a collisional magnetoplasma. The peculiarity of such interaction is connected with the essential dependence of resonant waves refraction, absorption, and polarization upon the frequency of electron elastic collisions, even when this frequency is less than other characteristic magnetoplasma frequencies. This dependence may cause two groups of phenomena:

1) The controlled propagation of a resonant wave due to thermal variations of the electron collision frequency in the field of another heating wave; such phenomenon may lead to plasma enhancement, to the deviation of the wave energy flux, and to variations of the threshold of parametric instability.

2) The self-action of the resonant wave, including the aforesaid effects, connected with plasma heating in the field of the intense resonant wave itself. Moreover, some specific phenomena, associated with the hysteresis dependence of plasma dispersive properties upon the intense wave amplitude, are typical for such thermal self-action.

The threshold character, the strength of the effects and the short time required for it to occur, may be of interest for several applications of these resonant phenomena to plasma radiophysics and wave electronics. If the plasma layer is sufficiently homogeneous, the possibilities of controlled band-filters treatment or angular selection of coherent pulses with different intensity may be very promising. On the other hand, experiments with heterogeneous plasma layers attract attention to the possibilities of selective heating of the resonant region inside the plasma layer, the self-screening of an intense wave in a plasma, and the radiation acceleration of plasma clots in waveguides.

The achievements in the theory of resonant electromagnetic waves interaction with a collisional plasma permit us to outline the following fields of further development of such theory:

a) The low-frequency plasma heating in the vicinity of ion gyroresonance ($\omega \to \Omega_H$). These phenomena are similar qualitatively to the high-frequency ones in the vicinity of electron gyroresonance ($\omega \to \omega_H$); however, the absorption coefficient, connected with an ion gyroresonant wave propagating along the magnetic field in a rarefied plasma ($\Omega_e^4 << \nu_e^2\Omega_H^2$),

$$\chi_i = \frac{\Omega_e}{\sqrt{2\nu_e \Omega_H}} ,$$

is much greater than the coefficient of electron gyroresonant absorption χ_e: $\chi_i/\chi_e = \sqrt{\frac{M_i}{m_e}} >> 1$. Such high effectiveness of ion gyroresonant heating attracts attention to the corresponding hysteresis phenomena and the problem of "run-away" ions.

b) The thermal excitation of different parametric instabilities in a magnetoplasma. The magnetic field extends substantially the variety of interacting wave modes in a plasma. Correspondingly, the possibilities of temperature influence on the threshold amplitudes of these phenomena are extended too. Moreover, the electron temperature determines the velocities of some interacting waves (for instance, the Langmuir waves) and even the possibilities of existence of some waves (the ion-acoustic waves). Moreover, the temperature variations of the electron collision frequency may modify the spectra of drift-dissipative oscillations in a heterogeneous plasma.

c) The turbulence phenomena in a collisional magnetoplasma. The wave-wave interactions in a weakly turbulent plasma may be described by means of a characteristic "turbulent frequency of collisions" ν_t. Such an approach may be useful in the description of wave dissipation near the reflection point [10] or of turbulent heat-conductivity [6]. However, the refusal of the assumption about weak turbulence impedes essentially the corresponding analysis. Therefore, the search of turbulent oscillations spectra of the collisional magnetoplasma is presently very important.

ACKNOWLEDGMENT

The author is very grateful to Professors V.V. Migulin and A.A. Rukhadze for their kind attention to this chapter.

REFERENCES

1. Altshuler, S., *J. Geophys. Res. 68*, 4707 (1963).
2. Bekefi, G. and Hirshfield, J.L., *Phys. Fluids 4*, 173 (1961).
3. Beynon, V.J.G., *Proc. Roy. Soc. 59*, 97 (1947).
4. Brodskij, V.B., *The Ukrainian Phys. Journal 23*, 597 (1978).
5. Crawford, F.W., Sears, D.M. and Bruce, R.L., *J. Geophys. Res. 75*, 7326 (1970).
6. Dum, S.T., *Phys. of Fluids 21*, 945 (1978).
7. Ellis, G.R., *J. Atm. Terr. Phys. 3*, 263 (1953).
8. Fejer, G.J., *Rev. of Geophys. and Space Phys. 17*, 135 (1979).
9. Gaponov, A.V. and Miller, M.A., *Soviet Phys. - J.E.T.P. 34*, 242 (1958).
10. Ginzburg, V.L., "The Propagation of Electromagnetic Waves in a Plasma" Nauka, Moscow, (1967).
11. Ginzburg, V.L. and Rukhadze, A.A., "The Waves in a Magneto-plasma" Nauka, Moscow, (1975).
12. Gildenburg, V.B., *Soviet Phys. - J.E.T.P. 46*, 2156 (1964).
13. Helliwell, R.A., *Phil. Trans. Roy. Soc. 280A*, 137 (1975).
14. Hibberd, F.H., Radio Science *69D*, 25 (1965).
15. Hon-Ming Lai and Yau-Wa Chan, *Phys. Rev. Lett. 35*, 1226 (1975).
16. Kaw, P.K., *Phys. Rev. Lett. 21*, 539 (1968).
17. Kogelnik, H., Applied Optics, *4*, 1562 (1965).
18. Kuzelev, M.V. and Rukhadze, A.A., Radiophysics, *22* (1979).
19. Mironov, V.A., Radiophysics *12*, 1765 (1969).
20. Mitjakov, N.A., Radiophysics *2*, 159 (1959).
21. Nishikawa, K., *J. Phys. Soc. Japan 24*, 1152 (1968).
22. Phelps, A.V. and Pack, J.L., *Phys. Rev. Lett. 3*, 340 (1959).
23. Ratcliffe, D.A., "Magneto-Ionic Theory and Its Applications to the Ionosphere" Cambridge, England, (1953).
24. Scott, J., *Proc. IEEE 38*, 1057 (1950).
25. Shvartsburg, A.B., *Soviet Phys. - Uspekhi 113*, 735 (1974).
26. Shvartsburg, A.B., *Phys. Lett. 57A*, 435 (1976).
27. Shvartsburg, A.B., Plasma Physics *21*, 663 (1979).
28. Shvartsburg, A.B., Physica Scripta (in press)(1979).
29. Shvartsburg, A.B., Korobeinikov, V.G., Deminov, M.G. and Razmadze, A.V., *Planet. Space Sci. 27*, 129 (1979).
30. Shvartsburg, A.B. and Zulina, E.M., Physik Solariterrestris *4*, 93 (1977).
31. Shvartsburg, A.B., Space Science Reviews (in press)(1979).
32. Silin, V.P., "The Parametrical Action of High Power Waves on Plasma" Nauka, Moscow, (1973).
33. Stenflo, L., Physica Scripta *2*, 50 (1970).
34. Stenflo, L., *Journal of Geophys. Res. 76*, 5349 (1971).
35. Stenflo, L., Plasma Physics *14*, 713 (1972).
36. Tanaka, S. and Mitani, K., *J. Phys. Soc. Japan 19*, 1376 (1964).

37. Tewari, D.P., Radio Science *13*, 1041 (1978).
38. Tsytovich, V.N., "Non-linear Effects in Plasma" Nauka, Moscow, (1967).
39. Utlaut, W. and Cohen, R., Science *174*, 245 (1971).
40. Vladimirov, V.V., Volkov, A.F. and Meilikhov, E.Z., "The Plasma of Semiconductors" Nauka, Moscow, (1979).
41. Weyl, G., *Phys. of Fluids 13,* 1802 (1970).

LAGRANGIAN METHODS
IN NONLINEAR PLASMA WAVE INTERACTION

F.W. Crawford

Institute for Plasma Research
Stanford University
Stanford, California

Analysis of nonlinear plasma wave interactions is usually very complicated, and simplifying mathematical approaches are highly desirable. The application of averaged-Lagrangian methods offers a considerable reduction in effort, with improved insight into synchronism and conservation (Manley-Rowe) relations. This chapter indicates how suitable Lagrangian densities have been defined, expanded, and manipulated to describe nonlinear wave-wave and wave-particle interactions in the microscopic, macroscopic and cold plasma models. Recently, further simplifications have been introduced by the use of techniques derived from Lie algebra. These and likely future developments are reviewed briefly.

I. INTRODUCTION

Progress in science is often materially aided by the development and application of a specialized mathematical formulation which provides both a compact description and improved insight into the phenomena of interest. The merits of such formulations, and their integrative quality over a wide variety of special cases, must be balanced against the risks of unfamiliarity to the tyro and evolution towards impenetrable generality.

The averaged-Lagrangian method to be described here [1,2] commends itself for application to nonlinear plasma wave problems not only because the formulation is compact and contributes

ISBN 0-12-709660-4

insight, but also because Lagrangian techniques are familiar to most plasma physicists from their education in classical or quantum mechanics.

The distinguishing integral formulation may also have been encountered in the context of variational methods, in which functional differentiation of an integral over a Lagrangian density, L, relates that density to a differential equation of interest. For example, we have

$$L = \int_a^b L\left(y\ ,\ \frac{dy}{dx}\ ,\ x\right) dx\ , \tag{1}$$

$$L + \delta L = \int_a^b L\left(y + \varepsilon\eta\ ,\ \frac{d}{dx}\ (y + \varepsilon\eta)\ ,\ x\right) dx \qquad (\varepsilon << 1)\ , \tag{2}$$

where $\eta(a) = 0 = \eta(b)$, which leads to $\delta L = O(\varepsilon^2)$ if

$$\frac{\partial L}{\partial y} - \frac{d}{dx}\left(\frac{\partial L}{\partial(dy/dx)}\right) = 0\ . \tag{3}$$

To be specific, if the differential equation were the Sturm-Liouville equation,

$$\frac{d}{dx}\left(p\ \frac{dy}{dx}\right) + (q + \lambda r)y = 0, \tag{4}$$

a suitable Lagrangian density would be

$$L = p\left(\frac{dy}{dx}\right)^2 - (q + \lambda r)y^2 \tag{5}$$

where p, q, r are functions of x, and λ is a constant.

It may be remarked in passing that variational integrals have proved useful for predicting plasma resonant frequencies accurately in complicated geometries. We shall not deal with this topic here, but will concentrate on the multiple time- and space-scale features of the averaged-Lagrangian method. The foregoing comments serve, however, to emphasize that the starting point of any Lagrangian analysis must be a Lagrangian density leading to the differential equations of the relevant plasma model. The

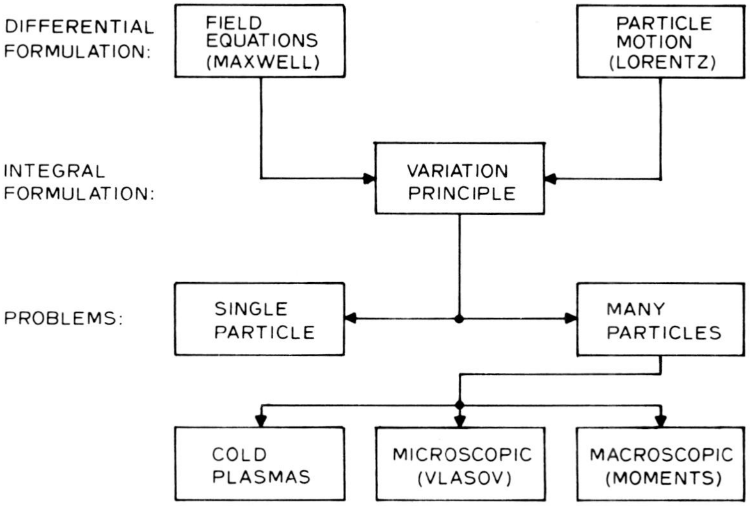

FIGURE 1. Lagrangian formulation.

range covered is illustrated by Figure 1.

This figure expresses the fact that our basic equations are the Maxwell equations for the electromagnetic field,

$$\nabla \times \underline{E} = -\frac{\partial \underline{B}}{\partial t}, \quad \nabla \cdot \underline{D} = \rho, \quad \underline{D} = \varepsilon_0 \underline{E},$$

$$\nabla \times \underline{H} = \underline{J} + \frac{\partial \underline{D}}{\partial t}, \quad \nabla \cdot \underline{B} = 0, \quad \underline{B} = \mu_0 \underline{H}, \tag{6}$$

with charge densities and motions expressed by ρ and $\underline{J}$, and an equation of charged particle motion expressing the Lorentz force law in a form suitable for the problem in hand. Three special cases may be distinguished,

$$m \frac{d\underline{v}}{dt} = q(\underline{E} + \underline{v} \times \underline{B}) \quad \text{(single-particle or cold plasma)}, \tag{7}$$

$$\frac{\partial f}{\partial t} + \underline{v} \cdot \frac{\partial f}{\partial \underline{r}} + \frac{q}{m}(\underline{E} + \underline{v} \times \underline{B}) \cdot \frac{\partial f}{\partial \underline{v}} = 0 \quad \text{(microscopic)}, \tag{8}$$

$$mn \frac{d\underline{v}_D}{dt} + \nabla \cdot \overleftrightarrow{\underline{P}} + qn(\underline{E} + \underline{v}_D \times \underline{B})$$

$$= \int m\underline{v} \left[\frac{\partial f}{\partial t} \right]_C d\underline{v} \quad \text{(macroscopic)}, \tag{9}$$

and we shall discuss the Lagrangian densities corresponding to these equations of motion in Section II.

If it is to be valuable, the integral formulation should be applicable to linear waves and resonances; nonlinear wave-wave and wave-particle interactions, both in the random-phase approximation and involving a small number of coherent waves; modulational instabilities, and solitons. These phenomena typically involve effects occurring on widely different space/time scales, e.g., one high-frequency plasma wave decaying slowly in space and time into two high-frequency waves. As we shall see in Section IV, it is in precisely this situation that the averaged-Lagrangian method is most useful.

This chapter is not intended to constitute a thorough, balanced literature review: such a review will be presented elsewhere [3]. For convenience to the author, many of the references are consequently to Stanford work. They are simply to be taken as illustrative, and contain numerous references to work carried out elsewhere.

II. LAGRANGIAN DENSITIES

II.1. Single-Particle

The means by which Lagrangian densities can be constructed will be appreciated by presenting first a Lagrangian density, L_{pf}, capable of giving the particle motion when the electromagnetic field is prescribed. We have,

$$L_{pf} = \int L_{pf}\, dt, \quad L_{pf} = -q(\phi - \underline{v} \cdot \underline{A}) + \frac{1}{2} m\underline{v}^2 . \tag{10}$$

Note that it is most convenient to use the scalar and vector potentials ϕ, $\underline{A}$ in terms of which the Lorentz force law is

$$\frac{d\underline{p}}{dt} = \frac{d}{dt}(m\underline{v} + q\underline{A}) = -q\frac{\partial}{\partial \underline{r}}(\phi - \underline{v} \cdot \underline{A}), \tag{11}$$

where

$$\underline{E} = -\left[\nabla\phi + \frac{\partial \underline{A}}{\partial t}\right], \quad \underline{B} = \nabla \times \underline{A}. \tag{12}$$

The Hamiltonian density corresponding to L_{pf} is

$$H_{pf} = \underline{v} \cdot \frac{\partial L_{pf}}{\partial \underline{v}} - L_{pf}$$

$$= \frac{1}{2m}(\underline{p} - q\underline{A})^2 + q\phi . \tag{13}$$

We may combine (10) with the well known Lagrangian density, L_f, yielding electromagnetic field equations for free space,

$$L_f = \int L_f \, d\underline{r} \, dt \ , \ L_f = \frac{\varepsilon_0}{2}\left[\nabla\phi + \frac{\partial \underline{A}}{\partial t}\right]^2 - \frac{(\nabla \times \underline{A})^2}{2\mu_0} , \tag{14}$$

to obtain the combined field-particle Lagrangian density, L, and the corresponding Hamiltonian density, H,

$$L = \int\left[\frac{\varepsilon_0 \underline{E}^2}{2} - \frac{\underline{B}^2}{2\mu_0}\right] d\underline{r} - q\left[\phi_p - \underline{v}_p \cdot \underline{A}_p\right] + \frac{m\underline{v}_p^2}{2} ,$$

$$H = \int\left[\frac{\varepsilon_0 \underline{E}^2}{2} + \frac{\underline{B}^2}{2\mu_0}\right] d\underline{r} + \frac{m\underline{v}_p^2}{2} . \tag{15}$$

The subscript p denotes that ϕ and $\underline{A}$ are to be evaluated at the particle.

II.2. Cold Plasma

It will be evident from the single-particle Lagrangian density that the variable $\underline{r}$ plays a dual role: in the variation with respect to $\underline{r}$ (leading to the particle equation of motion),

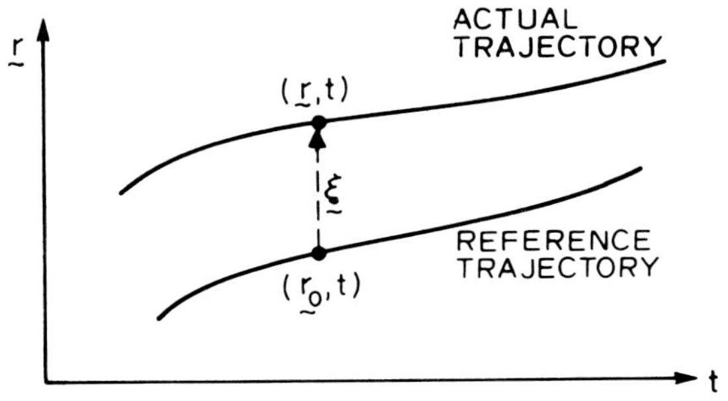

FIGURE 2. Polarization variables.

$\underline{r}$ is a *dependent* variable, and t is the independent variable; in the variations of $\underline{A}$ and ϕ (leading to the field equations), both $\underline{r}$ and t are *independent* variables. In the many-particle models, where charges are effectively pulverized, it is possible to carry out a full Eulerian treatment by introducing the polarization variable, $\underline{\xi}$, defined in Figure 2. The Lagrangian density is then of the form $L(\underline{A},\phi,\underline{\xi}\ ;\ \underline{r}_0,t)$, where $\underline{r}_0$ and t are independent variables along a prescribed reference trajectory. Variations with respect to $\phi,\underline{A}$ and $\underline{\xi}$ now give the field equations and the correct particle equation of motion.

The cold plasma Lagrangian density is a straightforward extension of (15) to

$$L = \left[\frac{\varepsilon_0\underline{E}^2}{2} - \frac{\underline{B}^2}{2\mu_0}\right] - n(q(\phi - \underline{v}\cdot\underline{A}) - m\underline{v}^2), \tag{16}$$

where n is the charged particle number density. Although there is no mathematical requirement that $\underline{\xi}$ be small, in practice we shall be interested in perturbation analyses where this is assumed to be the case. Evaluation of ϕ and $\underline{A}$ at the particle in (16) then proceeds by expansion

$$\underline{E} \to -\left(\nabla\phi_0 + \nabla\phi_1 + \frac{\partial \underline{A}_0}{\partial t} + \frac{\partial \underline{A}_1}{\partial t}\right),$$

$$\underline{B} \to \nabla \times \underline{A}_0 + \nabla \times \underline{A}_1 \;, \quad \underline{v} \to \underline{v}_0 + \dot{\underline{\xi}} \;,$$

$$\text{ZERO ORDER:} \quad \begin{cases} L_{f0} = \dfrac{\varepsilon_0}{2}\left(\nabla\phi_0 + \dfrac{\partial \underline{A}_0}{\partial t}\right)^2 - \dfrac{1}{2\mu_0}(\nabla \times \underline{A}_0)^2 \;, \\[2ex] L_{p0} = \dfrac{1}{2} m\underline{v}_0^2 - q(\phi_0 - \underline{v}_0 \cdot \underline{A}_0) \;. \end{cases} \tag{17}$$

$$\text{SECOND ORDER:} \quad L_{f2} = \frac{\varepsilon_0}{2}\left(\nabla\phi_1 + \frac{\partial \underline{A}_1}{\partial t}\right)^2 - \frac{1}{2\mu_0}(\nabla \times \underline{A}_1)^2,$$

$$L_{p2} = \frac{1}{2} m\, \dot{\underline{\xi}}^2 - q\left((\dot{\underline{\xi}} \cdot \nabla)\phi_1 + \frac{1}{2}(\underline{\xi} \cdot \nabla)^2\phi_0 - \dot{\underline{\xi}} \cdot (\underline{A}_1 + (\underline{\xi} \cdot \nabla)\underline{A}_0)\right.$$

$$\left. - \underline{v}_0 \cdot \left((\dot{\underline{\xi}} \cdot \nabla)\underline{A}_1 + \frac{(\underline{\xi} \cdot \nabla)^2}{2!}\right)\underline{A}_0\right) . \tag{18}$$

The field terms are zero at the third and higher orders ($L_{fn} = 0$, $n > 2$), while the plasma contributions continue indefinitely ($L_{pn} \neq 0$).

II.3. *Microscopic*

If the plasma is described by the Vlasov equation, the appropriate Lagrangian density is [4]

$$L = \left[\frac{\varepsilon_0 \underline{E}^2}{2} - \frac{\underline{B}^2}{2\mu_0}\right] - \int f(q(\phi - \underline{v} \cdot \underline{A}) - m\underline{v}^2)\, d\underline{v} \;, \tag{19}$$

where $f(\underline{v},\underline{r},t)$ is the charged particle velocity distribution.

II.4. *Macroscopic*

If the treatment is to be macroscopic, i.e., employ moments of the velocity distribution, there are two distinctly different ways to proceed. In the first, the microscopic Lagrangian of

(19) is expanded, and then velocity integrations are carried out [5]. In the second, a macroscopic Lagrangian density is defined, and then expanded. A suitable Lagrangian density is [6,7]

$$L = \left[\frac{\varepsilon_0 \underline{E}^2}{2} - \frac{\underline{B}^2}{2\mu_0} \right] - n(q(\phi - \underline{v} \cdot \underline{A}) - m\underline{v}^2) - \frac{1}{2}\, Tr\, \overleftrightarrow{\underline{P}}, \tag{20}$$

where $\overleftrightarrow{\underline{P}}$ in this equation and in (9) is the pressure tensor.

In practice, it is found that expansion of the microscopic Lagrangian, followed by velocity integration, is easier and more flexible than direct use of (20), since the latter must be varied subject to constraints set by the continuity and heat flow equations [7].

III. AVERAGED-LAGRANGIAN METHOD

III.1. Expansions

We need not describe the averaged-Lagrangian method in detail here. It is treated generally in [1], and in its applications to plasma physics in [4-20]. The sequence of the steps to be followed is discussed in [17]. It will suffice to say that the Lagrangian density is first expanded in terms of a suitable set of field components, ψ_i, and $\underline{\xi}$, according to the perturbation scheme described in Section II.2, to obtain

$$L = L_0 + L_1 + L_2 + L_3 + L_4 \;\ldots . \tag{21}$$

If the problem concerned coherent wave-wave interaction, for example, L_0 would embody information on the plasma background. The next term, L_1, would be of no interest, since its variation provides no information on the component waves, while L_2 could be used to give information on uncoupled propagation of the individual waves, and L_3, L_4, ... on three-wave, four-wave, ..., nonlinear interactions, respectively.

This information is obtained by introducing the variables

$$\psi_{i1} = \Psi_{i1} \sin(\theta + \alpha) \;, \qquad \theta(\underline{r},t) = \omega t - \underline{k} \cdot \underline{r} \;, \tag{22}$$

where ψ_i, ω, $\underline{k}$ all vary slowly in time and space, in general. Now consider

$$L_2 \equiv L_2\left(\psi_{i1}, \psi_{i1}, \frac{\partial \psi_{i0}}{\partial \underline{r}_0}, \frac{\partial \psi_{i1}}{\partial \underline{r}_0}, \frac{\partial \psi_{i0}}{\partial t}, \frac{\partial \psi_{i1}}{\partial t}, \frac{\partial \underline{\xi}}{\partial \underline{r}_0}, \frac{\partial \underline{\xi}}{\partial t} \;;\; \underline{r}_0, t\right),$$

in which $\partial\psi_{i0}/\partial\underline{r}_0$, $\partial\psi_{i0}/\partial t$ are zero only for a homogeneous, time-invariant background plasma. Introduction of the new variables, followed by averaging over one period of the phase to remove rapidly-varying terms, provides an averaged Lagrangian, $\overline{L}_2$, in which only slow temporal and spatial variations survive.

Variation with respect to each of the perturbation quantities leads to relations between them, which can in turn be used to express $\overline{L}_2$, in terms of a single quantity, Ψ, and the small-signal dispersion relation $D(\omega,\underline{k})$,

$$\overline{L}_2 = D(\omega,\underline{k})\Psi^2 = 0. \tag{23}$$

Higher-order terms are simplified similarly. Since the manipulations are carried out on the Lagrangian, rather than on the corresponding (Euler-Lagrange) differential equations, there is a great saving of effort.

III.2. Conservation Relations

It is easy to establish small-signal energy, momentum and action conservation relations. For example, for a single wave, we have the energy conservation relation,

$$\frac{d\varepsilon_2}{dt} + \nabla \cdot (\underline{v}_g \varepsilon_2) = - \frac{\partial \overline{L}_2}{\partial t} = - \frac{\partial D}{\partial t}\psi^2 \tag{24}$$

where ε_2 is the second-order perturbation term in the expansion of the corresponding averaged-Hamiltonian density.

The momentum conservation relation is

$$\frac{dP_j}{dt} + \nabla \cdot (\underline{v}_g P_j) = \frac{\partial \overline{L}_2}{\partial x_j} = \frac{\partial D}{\partial x_j} \psi^2 \ , \ P_j = \left(\frac{\varepsilon_2}{(\omega / k_j)} \right) , \tag{25}$$

where $\underline{v}_g$ is the group velocity.

If the dispersion relation, $D(\omega, \underline{k}, \underline{r}, t)$, is time-invariant, energy is conserved; if it is space-invariant, momentum is conserved. Neither condition is required for action conservation according to the relation,

$$\frac{d}{dt}\left(\frac{\varepsilon_2}{\omega} \right) + \nabla \cdot \left(\underline{v}_g \frac{\varepsilon_2}{\omega} \right) = 0. \tag{26}$$

which is easily derived by use of the averaged Lagrangian.

It may be noted further that inclusion of other waves in the analysis provides driving terms in (26) that establish the Manley-Rowe relations [4]. For three waves, we have

$$\frac{1}{\omega_1}\left(\frac{\partial \varepsilon_1}{\partial t} + \nabla \cdot (\underline{v}_{g1} \varepsilon_1) \right) = \frac{1}{\omega_2}\left(\frac{\partial \varepsilon_2}{\partial t} + \nabla \cdot (\underline{v}_{g2} \varepsilon_2) \right)$$

$$= - \frac{1}{\omega_3}\left(\frac{\partial \varepsilon_3}{\partial t} + \nabla \cdot (\underline{v}_{g3} \varepsilon_3) \right) = \varepsilon_0 , \tag{27}$$

where the synchronism conditions

$$\omega_1 + \omega_2 = \omega_3 \ , \ \underline{k}_1 + \underline{k}_2 = \underline{k}_3 \ , \tag{28}$$

are satisfied.

It may be remarked that derivations of this type carry a high degree of symmetry in subscripts, and are consequently easier to carry out than the commonly used iterative method. It is the author's experience with that method that the complexity of the algebra may preclude a clear demonstration that the Manley-Rowe relations are satisfied [20].

IV. APPLICATIONS

We shall not attempt to give a detailed catalogue here of problems that have been treated by averaged-Lagrangian (or Hamiltonian) methods; that will be done elsewhere. We shall simply indicate the range of problems covered by [4-20], thus effectively restricting ourselves to those involving many charged particles, and ignoring applications to single-particle motion and its associated adiabatic invariants (but see, for example, [22-24]).

Linear wave propagation and three-wave interactions have been treated in the cold plasma, microscopic and macroscopic models for homogeneous, time-invariant plasmas, and in some cases for weak inhomogeneity, i.e., in the eikonal approximation. Four-wave interactions have been studied, taking into account both synchronous combinations and virtual waves, and pursued to special cases of sideband instability and envelope solitons. The effects of background plasma variation have been worked out in the microscopic approximation to describe wave-induced, quasilinear, velocity-space diffusion.

A reasonable overall conclusion to be drawn from this work is that, apart from the possible exception of linear wave propagation, the averaged-Lagrangian method offers relative ease of manipulation, and leads to results in highly symmetrical and physically transparent form. This is not to say that the manipulations are easy, and indeed many nonlinear problems are at the limits of tractability for human agents. However, further exploitation of the power of the well known MACSYMA symbolic computation program might well extend those limits substantially [25].

V. FURTHER WORK

So far, the analogies with quantum-mechanical relations, which emerge most strongly when a spectrum of waves is studied in the random-phase approximation [26], have not been fully demonstrated by use of averaged-Lagrangian methods, and merit further work. So, also, does the theory of plasma solitons [27], since Lagrangian densities are known for many of the equations whose soliton solutions have been studied [28,29].

Precisely how the foregoing problems may best be treated depends on developments in an improved perturbation technique, distinguishable from the averaged-Lagrangian method: it employs an averaged-Hamiltonian density expressed in canonical variables whose choice is determined from Lie algebra considerations. This technique was introduced into celestial mechanics nearly fifteen years ago [2], and into plasma physics more recently, particularly by Dewar and Kaufman, and their co-workers [30-40], in the course of their developments of oscillation-center theory, and studies of turbulence.

The new technique offers the advantages that it may be even more economical in manipulation than the averaged-Lagrangian method, and that a mathematical prescription is available for making canonical transformations using a generating function in which the old and new coordinates are not mixed. Further, Poisson brackets are strongly in evidence, contributing to physical insight, and extensions of the technique in non-perturbation form are possible.

ACKNOWLEDGMENT

This work was supported by the National Aeronautics and Space Administration.

REFERENCES

1. Whitham, G.B., "Linear and Nonlinear Waves" Chapters 14-16. John Wiley, New York, (1974).
2. Nayfeh, A., "Perturbation Methods" Chapter 5. John Wiley, New York, (1973).
3. Crawford, F.W., *Rev. Mod. Phys.* (in preparation).

4. Galloway, J.J. and Kim, H., *J. Plasma Phys. 6*, 53 (1971).
5. Kim, H. and Crawford, F.W., Stanford University Institute for Plasma Research Report No. 763, October (1978).
6. Peng, Y.-K. and Crawford, F.W., Stanford University Institute for Plasma Research Report No. 764, October (1978).
7. Kim, H. and Crawfore, F.W., Stanford University Institute for Plasma Research Report No. 765, October (1978).
8. Galloway, J.J. and Crawford, F.W., Proc. Fourth European Conference on Controlled Fusion and Plasma Physics, Rome, Italy (CNEN, Rome 1970), p. 161 (1970).
9. Dougherty, J.P., *J. Plasma Phys. 4*, 761 (1970).
10. Dougherty, J.P., *J. Plasma Phys. 11*, 331 (1974).
11. Dewar, R.L., *J. Plasma Phys. 7*, 267 (1972).
12. Boyd, T.J.M. and Turner, J.G., *J. Phys. A, 5*, 881 (1972).
13. Boyd, T.J.M. and Turner, J.G., *J. Phys. A, 6*, 272 (1973).
14. Boyd, T.J.M. and Turner, J.G., *J. Math. Phys. 19*, 1403 (1978).
15. Dysthe, K.B., *J. Plasma Phys. 11*, 63 (1974).
16. Kim, H. and Crawford, F.W., *Radio Sci. 12*, 941 (1977).
17. Kim, H. and Crawford, F.W., *Radio Sci. 12*, 953 (1977).
18. Galloway, J.J. and Crawford, F.W., *Radio Sci. 12*, 965 (1977).
19. Larsen, J.-M. and Crawford, F.W., *Int. J. Electron. 46*, 577 (1979).
20. Larsen, J.-M. and Crawford, F.W., *Int. J. Electron. 47*, 317 (1979).
21. Harker, K.J. and Crawford, F.W., *J. Appl. Phys. 39*, 5959 (1968).
22. Gardner, C.S., *Phys. Rev. 115*, 791 (1959).
23. Lichtenberg, A.J., "Phase-space Dynamics of Particles" John Wiley, New York, (1969).
24. Dragt, A. and Finn, J., *J. Math. Phys. 17*, 2215 (1976).
25. Bers, A., Kulp, J.L. and Karney, C.F.F., *Comput. Phys. Comm. 12*, 81 (1976).
26. Harris, E.G., *in* "Advances in Plasma Physics" (A. Simon and W.B. Thompson, eds.), p. 157. John Wiley, New York, (1969).
27. Gibbons, J., Thornhill, S.G., Wardrop, M.J., and ter Haar, D., *J. Plasma Phys. 17*, 153 (1977).
28. Scott, A.C., Chu, F.Y.F. and McLaughlin, *Proc. IEEE 61*, 1443 (1973).
29. "Solitons in Action" (K. Lonngren and A. Scott, eds.) Academic Press, New York, (1978).
30. Dewar, R.L., *Phys. Fluids 13*, 2710 (1970).
31. Dewar, R.L., *J. Plasma Phys. 1*, 267 (1972).
32. Dewar, R.L., *Phys. Fluids 16*, 1102 (1973).
33. Dewar, R.L., *J. Phys. A, 9*, 2043 (1976).
34. Johnston, S., *Phys. Fluids 19*, 93 (1976).
35. Johnston, S. and Kaufman, A., *in* "Plasma Physics" (H. Wilhelmsson, ed.), Plenum, New York, (1977).
36. Kaufman, A.N., Cary, J.R. and Pereira, N.R., Proc. Third Topical Conference on RF Plasma Heating, Pasadena,

California, January (1978).
37. Kaufman, A.N., *in* "Topics in Nonlinear Dynamics" (S. Jorna, ed.), American Institute of Physics, New York, (1978).
38. Johnston, S. and Kaufman, A.N., *Phys. Rev. Letters 40*, 1266 (1978).
39. Johnston, S. and Kaufman, A.N., *J. Plasma Phys. 22*, 105 (1979).

ELECTROMAGNETIC PROBLEMS IN COMPOSITE MATERIALS IN LINEAR AND NONLINEAR REGIMES[1]

George C. Papanicolaou

Courant Institute
New York University
New York, New York

I. INTRODUCTION

We shall analyze electromagnetic problems in materials that can be described macroscopically by linear or nonlinear constitutive laws that arise by a suitable averaging process from microscopic laws. We focus attention on this averaging process and articulate it by the usual multiscaling arguments. This gives us the form of the macroscopic equations and the identification of the effective constitutive laws. Rarely can one relate the effective parameters to the microscopic laws and the structure of the material in an explicit manner. It is necessary to introduce additional simplifications in order to obtain explicit results.

In section II we review the multiscaling argument for the identification of the effective dielectric constant of a composite medium with periodic structure (cf. [1-6]). In section III we consider a nonlinear version of the same problem which was considered first by Babuška. In section IV we continue with the computation of the effective dielectric constant of a composite

[1] *Research supported by the Air Force Office of Scientific Research under Grant No. AFOSR-78-3668.*

ISBN 0-12-709660-4

medium with random structure following [7] and [8]. We find here the possibility of showing that effective parameters exist (much like one proves existence of averages in ergodic theory) but, contrary to the periodic case, their explicit computation is in general impossible, even by numerical methods.

The explicit determination of the linear or nonlinear effective dielectric constant can only be done under additional conditions (not just stationarity) such as small volume fraction in a model of dielectric spheres randomly dispersed. We do not discuss this here, limiting ourselves to the ergodic theory. We refer however to [9], [10] and [11] for further discussion, expecially on the use of the so-called coherent potential approximation.

II. DIELECTRIC MEDIUM WITH SPATIALLY PERIODIC STRUCTURE (LINEAR)

The time-harmonic Maxwell's equations are

$$\begin{aligned} i\omega\mu H &= \nabla \times E \quad , \quad \nabla \cdot (\mu H) = 0 \\ -i\omega\varepsilon E &= \nabla \times H - J \quad , \quad \nabla \cdot (\varepsilon E) = \rho \ . \end{aligned} \tag{2.1}$$

Here μ and ε are the magnetic permeability and dielectric constant, ω is the radian frequency ($e^{-i\omega t}$ is the time factor) and ρ and J are the charge and current density, respectively. In this section we shall assume that

(i) ρ and J are given functions satisfying the continuity equation $-i\omega\rho + \nabla \cdot J = 0$.

(ii) μ is a constant.

(iii) ε is a periodic function of x with period ℓ in each direction.

We shall analyze (2.1) in the asymptotic limit $\ell \to 0$. This means that we want to find the behavior of the electric and magnetic fields excited by some external currents or sources when the dielectric medium has periodic structure with small spatial period. We want, in particular, to find an *effective dielectric constant* ε^* which describes the $\ell \to 0$ asymptotic limit.

Since $\varepsilon(x)$ is ℓ-periodic we may write it in the form $\varepsilon\left(\frac{x}{\ell}\right)$ with $\varepsilon(y)$ now a unit periodic function. We also look for E and H in the form

$$\begin{aligned} E(x) &= E_0\left(x, \frac{x}{\ell}\right) + \ell E_1\left(x, \frac{x}{\ell}\right) + \cdots \\ H(x) &= H_0\left(x, \frac{x}{\ell}\right) + \ell H_1\left(x, \frac{x}{\ell}\right) + \cdots, \end{aligned} \tag{2.2}$$

where $E_0(x,y)$, $E_1(x,y)$, etc. are periodic vector functions of the variable y. Using this form in (2.1) and collecting terms in powers of ℓ we obtain the following equations.

$$\nabla_y \times E_0 = 0 \, , \quad \nabla_y \times H_0 = 0 \tag{2.3}$$

$$\nabla_y \times E_1 + \nabla_x \times E_0 = i\omega\mu H_0 \tag{2.4}$$

$$\nabla_y \times H_1 + \nabla_x \times H_0 = -i\omega\varepsilon(y) E_0 + J, \text{ etc.}$$

From (2.3) we see that we can write E_0 in the form

$$E_0 = (I + \nabla\chi(y)) E_{00}(x), \quad H_0 = H_{00}(x) \tag{2.5}$$

where $\chi(y) = (\chi^{(1)}(y), \chi^{(2)}(y), \chi^{(3)}(y))$ is a vector periodic function to be determined and I is the identity (δ_{ij}). Inserting (2.5) into (2.4) and averaging y over a unit cell yields

$$\nabla_x \times E_{00} = i\omega\mu H_{00} \tag{2.6}$$

$$\nabla_x \times H_{00} = -i\omega \, \overline{\varepsilon(I + \nabla\chi)} \, E_{00} + J, \quad \left[\bar{a} \equiv \int_{\text{unit cell}} a(y)\, dy \right].$$

Using (2.5) in (2.4) again and taking the y-divergence leads to the following equation for χ

$$\nabla_y \cdot [\varepsilon(y)(I + \nabla\chi(y))] = 0. \tag{2.7}$$

If we assume that ε is strictly positive and bounded and we normalize χ so that $\bar{\chi} = 0$ then (2.7) uniquely determines χ.

It is clear that the *effective dielectric constant*, which will be a tensor in general, is given by

$$\varepsilon^* = \overline{\varepsilon(I + \nabla\chi)} \ . \tag{2.8}$$

The macroscopic or effective fields $E_{00}(x)$ and $H_{00}(x)$ will satisfy Maxwell's equation in the form (2.1) with ε replaced by ε^*. The interpretation of this result is this: if macroscopic fields E_{00} and H_{00} are to be established, the local field observed in each cell has the form $E_{00}(x) + \nabla\chi\left(\frac{x}{\ell}\right)E_{00}(x)$ and $H_{00}(x)$ (to first order in ℓ) and the dielectric displacement is given by $\varepsilon E_{00} + \varepsilon\nabla\chi E_{00}$ so that $\varepsilon\nabla\chi$ is the induced polarization per unit applied field. It follows that ε^* is the average dielectric constant $\bar{\varepsilon}$ plus the averaged induced polarization $\overline{\varepsilon\nabla\chi}$ as (2.8) shows.

For a more detailed discussion of the above we refer to [6] (Chapter 4). We should point out that even in the case of a cubic lattice of homogeneous dielectric spheres in vacuum, $\overline{\varepsilon\nabla\chi}$ cannot be computed explicitly for arbitrary volume fraction (fraction of volume occupied by the dielectric sphere per unit period cell). In the periodic case, however, the solution of (2.7) can be done numerically in a straightforward way. The random case is considered in section IV.

III. DIELECTRIC MEDIUM WITH SPATIALLY PERIODIC STRUCTURE (NONLINEAR)

The time-harmonic Maxwell's equations are again given by (2.1) but we shall now assume that ε depends on the field strength $|E|^2$ and is spatially periodic with period ℓ

$$\varepsilon = \varepsilon\left(\frac{x}{\ell}\right)\left[1 + \gamma|E|^2\right]. \tag{3.1}$$

We assume that γ is a small number measuring the strength of the nonlinearity.

We repeat the expansion process (2.2). We obtain the following equations

$$\nabla_y \times E_0 = 0 \, , \quad \nabla_y \times H_0 = 0 \tag{3.2}$$

$$\nabla_y \times E_1 + \nabla_x \times E_0 = i\omega\mu H_0 \tag{3.3}$$

$$\nabla_y \times H_1 + \nabla_x \times H_0 = -i\omega\varepsilon(y)\left[1 + \gamma|E_0|^2\right]E_0 + J, \text{ etc.}$$

We write $E_0(x,y)$ $(y = x/\ell)$ in the form

$$E_0 = (I + \nabla\chi)E_{00}(x) \, , \quad H_0 = H_{00}(x) \tag{3.4}$$

where $\chi(y) = (\chi^{(1)}(y), \chi^{(2)}(y), \chi^{(3)}(y))$ is a vector periodic function and $I = (\delta_{ij})$ as before.

Inserting (3.4) into (3.3) and averaging y over a unit cell yields

$$\nabla_x \times E_{00} = i\omega\mu H_{00} \tag{3.5}$$

$$\nabla_x \times H_{00} = -i\omega \, \overline{\varepsilon\big(I + \nabla\chi)(1 + \gamma|I + \nabla\chi|^2|E_{00}|^2\big)} \; E_{00} + J \, .$$

Let U be a fixed constant vector. Taking the y-divergence of (3.3) and replacing E_{00} with U (since E_{00} is only a function of the slow scale x) we obtain

$$\nabla_y \cdot \left[\varepsilon(y)(I + \nabla\chi(y))\big(1 + \gamma|U|^2|I + \nabla\chi(y)|^2\big)\right] = 0. \tag{3.6}$$

If we assume that ε is strictly positive and bounded and $\gamma|U|^2$ is small then (3.6) has a unique solution $\chi = \chi(y;|U|^2)$ such that $\bar{\chi} = 0$.

It is now clear that *effective nonlinear dielectric* constant ε^* will have the form

$$\varepsilon^*\big(|E|^2\big) = \overline{\varepsilon(I + \nabla\chi(\cdot,|E|^2))(1 + \gamma|E|^2|I + \nabla\chi(\cdot,|E|^2)|^2)} \, . \tag{3.7}$$

For $|E|^2$ small this expression can be simplified by expanding and keeping only the first two terms.

$$\varepsilon^*(|E|^2) = \overline{\varepsilon(I+\nabla\chi_0)} \qquad (3.8)$$

$$+ \overline{\left[\gamma\varepsilon(I+\nabla\chi_0)\,|I+\nabla\chi_0|^2 + \varepsilon\gamma\nabla\chi_1\right]}\;|E|^2 \;.$$

Here $\chi_0 = \chi(y;0)$ and $\chi_1 = \dfrac{\partial\chi(y;\lambda)}{\partial\lambda}$ at $\lambda = 0$. Thus, the first term on the right side of (3.8) is the *linear* effective dielectric constant (2.8) while the second term is the nonlinear correction to it.

The nonlinear correction is composed of two terms where $|E|^2$ is small. The first term is obtained by averaging a nonlinear function of the linear cell functions χ_0. The second term takes into account nonlinearity within a single cell since it involves χ_1. Of course χ_1 can be expressed simply in terms of the solution of a suitable cell problem (by perturbation from (3.6) when χ is small).

IV. DIELECTRIC MEDIUM WITH RANDOM STRUCTURE

Let us return to Maxwell's equations (2.1) and assume (i) and (ii) of section II but replace (iii) now by the hypothesis:

(iv) The microscopic dielectric constant ε is a stationary random process of the spatial variable x. Moreover, it has the form $\varepsilon(x/\ell)$ where ℓ is a spatial scale parameter that is small.

At this point one must explain what the limitations of assumption (iv) are and why it is a convenient starting point for understanding how an effective dielectric constant emerges in the limit $\ell \to 0$.

The assumed form $\varepsilon = \varepsilon(x/\ell)$ of the microscopic dielectric constant, with $\varepsilon(y)$ a stationary random process, implies that there is only one relevant length scale associated with the microstructure. In many physical problems (for example in models of rock, porous media, soils, etc.) there are many scales that enter the description and the microstructure cannot be expressed even roughly by the form $\varepsilon(x/\ell)$.

The principal reason for considering the model $\varepsilon = \varepsilon(x/\ell)$ is because it can be analyzed mathematically completely [7,8] and the result turns out to be *identical* in structure to the one obtained for the periodic model.

We now explain in more detail how the stochastic case is analyzed and what the result is.

We begin with expansion form (2.2) where now dependence on $y = x/\ell$ is not periodic but random and, one hopes, stationary. Stationarity is what one expects will replace periodicity.

We continue as in section II up to (2.7) and now look at that problem carefully. In fact we rewrite it here in a slightly different form as follows:

$$\nabla_y \cdot [\varepsilon(y)(e_k + G^{(k)}(y))] = 0 \ , \quad \nabla_y \times G^{(k)}(y) = 0, \tag{4.1}$$

$$k = 1,\ 2,\ 3, \qquad y \in \mathbb{R}^3,$$

where e_1, e_2 and e_3 are the column vectors with components (1,0,0), (0,1,0) and (0,0,1), respectively. The physical meaning of $G^{(k)}(y)$ is clearly this: when unit electric field is impressed across the kth coordinate direction, $e_k + G^{(k)}(y)$ is the resulting total field in the randomly inhomogeneous, stationary, dielectric medium.

The difficulty with (4.1) comes from the fact that even though $\nabla_y \times G^{(k)} = 0$, the field $G^{(k)}(y)$ cannot be written as the gradient of a *stationary* random potential $\chi^{(k)}$ (as in the periodic case $\nabla\chi^{(k)} = G^{(k)}$). A potential $\chi^{(k)}(y)$ exists but it is not stationary. The total potential $y_k + \chi^{(k)}(y)$ (whose gradient is $e_k + G^{(k)}(y)$) must be such, however, that $\chi^{(k)}(y)$ grows slower than $|y|$ as $|y| \to \infty$. Otherwise the ansatz of the multiple scale expansion (2.2) does not make sense because the second term on the right side of (2.2) is as big as the first. It turns out that $\chi^{(k)}(y)$ can be constructed with the desired properties even though *stationary* ones cannot be found except in very special cases such as the periodic ones. The details are given in [8].

The effective dielectric constant ε^* is again given by (2.8) (and is a tensor in general).

$$\varepsilon^*_{kj} = \overline{\varepsilon(\delta_{kj} + G_j^{(k)})} \quad , \text{(bar is ensemble average now)}. \tag{4.2}$$

Since ε^* depends on the solution $G^{(k)}$ of (4.1) which is a full stochastic partial differential equation there is little hope of computing it explicitly, even numerically as one can handle the periodic case. Therefore, as we mentioned in the introduction, the results of [7] and [8] should be considered, as general ergodic theorems for stoachastic PDE's that prove the existence of effective parameters in a very general context.

In the nonlinear case one must solve, in addition to (4.1) the stochastic PDE that corresponds to $\nabla\chi_1$. If we denote $\nabla\chi_1^{(k)}$ by $F^{(k)}$ we find from (3.6) that $F^{(k)}$ must satisfy the problem

$$\nabla_y \cdot \left[\varepsilon(y)((e_k + G^{(k)}(y))|e_k + G^{(k)'}(y)|^2 + F^{(k)}(y))\right] = 0, \tag{4.3}$$

$$\nabla_y \times F^{(k)} = 0, \quad k = 1,\ 2,\ 2, \quad y \in \mathbb{R}^3.$$

The nonlinear effective dielectric constant for small fields is given by

$$\varepsilon^*_{kj}(|E|^2) = \overline{\varepsilon(\delta_{kj} + G_j^{(k)})} \tag{4.4}$$
$$+ \overline{\gamma\left[\varepsilon(\delta_{kj} + G_j^{(k)})|e_k + G^{(k)}|^2 + F_j^{(k)}\right]}\ |E|^2$$

which is another way of writing (3.8) -- but with stochastic ε now.

REFERENCES

1. Palencia, Sanchez E., "Comportements local et macroscopique d'un type de milieux physique heterogenes," *Int. J. Eng. Sci. 12*, 331-351 (1974).
2. DeGiorgi, E. and Spagnolo, S., "Sulla convergenza degli integrali dell'energia per operatori ellittici del secondo ordine," *Bull. U.M.I. 8*, 391-411 (1973).
3. Babuška, I., "Solution of interface problems by homogenization. I, II, III," *SIAM J. Math. Analysis 7*, 603-634, 635-645 (1976) and *8*, 923-937 (1977).
4. Bensoussan, A., Lions, J.L. and Papanicolaou, G.C., "Homogenization in deterministic and stochastic problems," *in* "Stochastic Problems in Dynamics" (B.L. Clarkson, ed.), pp. 106-115. Pitman, London, (1977).
5. Keller, J.B., "Effective behavior of heterogeneous media," *in* "Statistical Mechanics and Statistical Methods in Theory and Applications, (U. Landman, ed.), Plenum Press, (1977).
6. Bensoussan, A., Lions, J.L. and Papanicolaou, G.C., "Asymptotic Analysis for Periodic Structures" North Holland, (1978).
7. Kozlov, C.M., *Doklady Adad. Nauk 236* (#5), (1979).
8. Papanicolaou, G.C. and Varadhan, S.R.S., "Boundary value problems with rapidly oscillating coefficients," Proceedings of Conference in Statistical Mechanics, Ezstergom, Hungary, 1979, North Holland, (to appear).
9. Lax, M., "Wave propagation and conductivity in random media," *SIAM-AMS Proceedings, VI*, Providence, R.I., 35-96 (1973).
10. Elliott, R.J., Kramhansl, J.A. and Leath, P.L., "Theory and properties of randomly disordered crystals and related physical systems," *Rev. Mod. Phys. 46*, 465-543 (1974).
11. Brown, W.F., Jr., "Dielectrics," Handbuch der Physik *XVII*, Springer, Berlin, (1956).

STATIONARY REGIMES IN PASSIVE NONLINEAR NETWORKS[1]

A.G. Ramm

Department of Mathematics
The University of Michigan
Ann Arbor, Michigan

I. INTRODUCTION

Nonlinear network theory is of great interest to applications. There are many questions in the theory which are treated by different methods. There is no possibility here to go into any details. Very schematically the main trends in the theory of one-loop systems can be described as follows. Let us consider a one-loop network consisting of an electromotive force $e(t)$ in series with a linear two-port L and a nonlinear two-port N (see Fig. 1). Many practical problems can be reduced to calculation of stationary regimes in such a network, to the qualitative study of these regimes and especially to the study of stability of these regimes towards small (in some sense) perturbations of $e(t)$ or/and arbitrary initial data on the linear two-port. In the theory of differential equations in analytical mechanics these questions are known as stability under constantly acting disturbances and stability in the large, respectively. Of course, the first questions are about existence and uniqueness of stationary regimes. The second question is about

[1] *This work was partly supported by AFOSR F4962079C0128.*

ISBN 0-12-709660-4

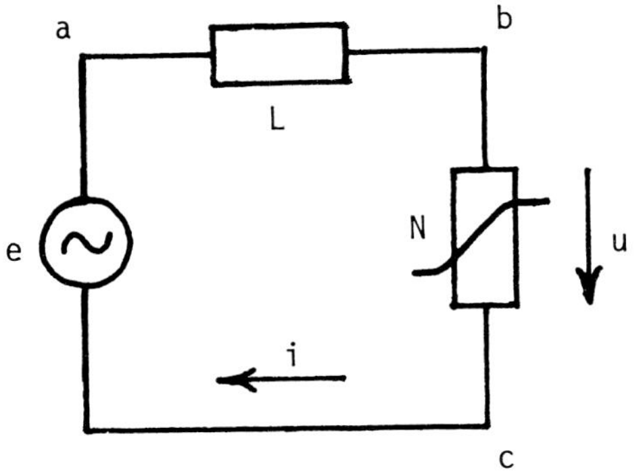

FIGURE 1. A one-loop nonlinear network.

stability, and the third is about calculation of the stationary regimes. In what follows, by stationary regime we mean a periodic, almost periodic, or just uniformly bounded for $t \in I = (-\infty, \infty)$, solution of the fundamental equation governing the network. This equation can be written as

$$u + Z\,F(u) = e. \tag{1}$$

Here $i = F(u)$ is the characteristic of the nonlinear two-port, $u = u_{bc}$ (see Fig. 1) and Z is the linear operator describing the linear two-port L. It means that the current i through L and the voltage u on L are connected by the relation $Zi = u$. In the literature mostly the case of time invariant L was treated. In this case

$$Zi = \int_{-\infty}^{t} g(t-\tau)\,i(\tau)\,d\tau, \tag{2}$$

$$g(t) = 0 \text{ for } t < 0. \tag{3}$$

In the general case

$$Zi = \int_{-\infty}^{t} g(t,\tau)\,i(\tau)\,d\tau\ , \tag{4}$$

$$g(t,\tau) = 0 \quad \text{for} \quad t < \tau . \tag{5}$$

Relations (3), (5) follow from the causality principle: $g(t,\tau)$ is the response of L on the impulse $i = \delta(t-\tau)$, where $\delta(t)$ is the Dirac function. Effect $g(t,\tau)$ can only follow the cause $\delta(t-\tau)$. Thus (5) is valid. For time invariant L the concept of impedance is useful. Impedance $Z(p)$ is defined by the formula

$$Z(p) = \int_0^\infty g(t)\ exp(-pt)dt,\ p = \sigma + i\lambda. \tag{6}$$

A linear time invariant two-port is called stable if

$$\int_0^\infty |g(t)|dt = G < \infty . \tag{7}$$

From (7) it follows that for any bounded input signal $|i(t)| \leq \mu$ the response will be bounded:

$$|u(t)| = |\int_{-\infty}^t g(t-\tau)\ i(\tau)d\tau| \leq \mu G.$$

There are other definitions of stability. For example, if the definition of stability of L requires that for any input signal with finite energy $(\int_{-\infty}^\infty |i(t)|^2dt)^{\frac{1}{2}} \leq \mu$ the response $u = \int_{-\infty}^t g(t-\tau)i(\tau)d\tau$ must be founded, then the two-port is stable if

$$\int_0^\infty |g(t)|^2dt < \infty . \tag{8}$$

In both cases (7) and (8) the impedance (6) is analytic in the half-plane $\sigma > 0$. In case (7), $Z(i\lambda) = \lim_{\sigma\to 0} Z(\sigma + i\lambda)$ exists and is continuous for $-\infty < \lambda < \infty$. In case (8), $Z(i\lambda) \in L^2(-\infty, \infty)$ and $Z(i\lambda) = \lim_{\sigma\to 0} Z(\sigma + i\lambda)$. The linear two-port L is called exponentially stable if

$$\int_0^\infty exp(vt)|g(t)|dt < \infty ,\ v > 0. \tag{9}$$

In this case $Z(p)$ is analytic in the half-plane $\sigma > -\gamma$ and

$$Re\ Z(i\lambda) \geq \frac{\gamma}{(\gamma^2 + \lambda^2)^{\frac{1}{2}}} \left|Z(i\lambda)\right| , \tag{10}$$

$$-\lim_{\lambda\to+\infty} \lambda\ Im\ Z(i\lambda) = \frac{2}{\pi}\int_0^\infty Re\ Z(i\lambda)d\lambda . \tag{11}$$

Properties of impedance and concepts of stability are widely discussed in the literature [1,2]. The fundamental equation (1) can be rewritten as

$$Au + Fu = J , \tag{12}$$

where

$$A = Z^{-1},\ J = Ae. \tag{13}$$

The operator A is called admittance operator of L and the passage from (1) to (12) in network theory is known as the theorem about equivalent generator (Thevenin theorem). The function J in (12), (13) is the equivalent current generator. A two-port is called passive if the energy consumed by the two-port during a sufficiently long period of time is positive. The typical questions of interest both from the mathematical and practical points of view are the following: 1) If $e(t) = e(t+T)$ does there exist a periodic regime in the network of Fig. 1?; 2) When is this regime stable under small periodic perturbations of $e(t)$?; 3) Will any transient regime in the network for an arbitrary initial data on L tend to the stationary regime as $t \to +\infty$? (if this is the case, the network is called convergent); 4) How to calculate the stationary regime? 5) How to answer similar questions when $e(t)$ is almost periodic? 6) How to answer similar questions in case $e(t)$ is a uniformly bounded measurable function, $e(t) \in L^\infty(I)$, $I = (-\infty, \infty)$, e.g., $e(t)$ is a sequence of random impulses with a fixed amplitude?

In the litarature there are three major trends in the study of nonlinear networks. The first trend deals with the systems with small (in some sense) nonlinearity and includes methods of small parameter, averaging procedures and their variants. There is extensive literature on the subject, but we cannot go into any detail [3,4]; mostly the case when the nonlinear network can be described by a system of differential equations is under study in these books.

The second trend deals with the systems with a linear two-port satisfying the filter property. It means that the frequency amplitude characteristic has a narrow peak near some frequency ω_0 and is small outside a small neighborhood of ω_0. In this case, methods of harmonic linearization are popular in the engineering literature [2].

The third trend deals with the questions of stability in the large and absolute stability of control systems. In [2] one can find references and main papers on this topic.

In what follows we are going to describe a general method for studying the network in Fig. 1. We give answers to questions 1 - 6 above. No assumptions concerning smallness of nonlinearity or filter property of linear two-port are made. From the engineering point of view the theory developed here can possibly be used for designing some new systems with broad-band frequency characteristic of the linear two-port and steep characteristic of the nonlinear two-port. Our main physical assumption is the following: the two-port *L-N* in Fig. 1 is passive. This assumption will be formulated mathematically and discussed in detail. Thus, from the engineering point of view we consider systems which are not generators. We expect that there exists a unique stationary regime which is stable under the constantly acting perturbations of $e(t)$, which preserves period, and has the same period as $e(t)$. Any transient regime caused by an arbitrary initial data on L will tend to the stationary regime as $t \to +\infty$. Iterative methods for calculating the stationary regime will be given. The problem is treated globally: there is no need to start with some special zero approximation (e.g., with approximation lying in some neighborhood of the stationary regime). Also some apriori estimates of solutions of integral equations for transient processes in the network of Fig. 1 will be given. From these estimates, stability in the large of the transient regime will follow. The method developed and results obtained are due to the author [5-11].

In what follows (6.1) means formula (6) in Section I. We use only one number in references to formulas in the same section.

II. INVESTIGATION OF OPERATOR EQUATIONS GOVERNING PASSIVE NONLINEAR NETWORKS

II.1. INTRODUCTION

In this section we study the cases of periodic and almost periodic exterior force $e(t)$. Equation (12.1)

$$Bu = Au + Fu = J \tag{1}$$

is of principal interest for us. In what follows equation (1) will be studied in a Hilbert space H. If $J(t+T) = J(t)$ (this is the case of periodic exterior force) then H is the $L^2[0,T]$ space of T periodic functions. If $J(t)$ is almost periodic we take $H = B_2$, where B_2 is the Besicovich space of almost periodic functions which is the completion of the set of trigonometric polynomials in the norm, generated by the scalar product $(u,v)_{B_2}$ $- \lim_{T\to} \frac{1}{2T}\int_{-T}^{T} u(t)v^*(t)dt$, where the asterisk denotes complex conjugation.

II.2. Main Assumptions

Let us assume that:

i) The operator F: $H \to H$ is bounded, defined on all H and semicontinuous, i.e., the function $(F(f+\lambda g),L)$ is continuous in λ, $-\infty < \lambda < \infty$, for all $f,g,h \in H$.

ii) The operator A: $H \to H$ is linear, closed, densely defined, $D(B) = D(A)$

$$Re\ (Bu - Bv,\ u-v) = 0 \Longrightarrow u = v. \tag{2}$$

There exists a sequence of linear bounded operators A_n such that

$$A_n u \to Au\ \forall u \in D(A);\ A_n^* u \to A^* u\ \forall u \in D(A^*), \tag{3}$$

$$Re(B_n u - B_n v,\ u-v) \geq 0\ \forall u,v \in H,\ B_n \equiv A_n + F,\ n > n_o, \tag{4}$$

$$Re(B_n u,\ u) \geq \gamma(\|u\|)\ \|u\|, \tag{5}$$

where $\|u\|$ is the norm in H, n_o is an arbitrary large fixed number, $0 \leq \gamma(t) \to +\infty$ as $t \to +\infty$, the arrow $\to$ denotes strong convergence in H (while by $\rightharpoonup$ we will denote weak convergence in H).

iii) $$Re(Bu-Bv,\ u-v) \geq \nu_R(\|u-v\|)\ \|u-v\| \tag{6}$$

for $\|u\|$, $\|v\| \leq R$, where $\nu_R(t)$ is continuous in t, $\nu(0) = 0$, $\nu(t) > 0$ for $t > 0$.

Of course (6) implies (2), but we are going to use assumptions i), ii) and iii) separately.

Instead of these assumptions, the following set of assumptions can be useful:

j) $$Re(Au,u) \geq \delta\ \|u\|^2,\ \forall u \in D(A);\ R(A) = \text{range of } A = H, \tag{7}$$

jj) $$\|Fu\| \leq \varepsilon\ \|u\| + C(\varepsilon),\ \forall \varepsilon > 0,\ \forall u \in H, \tag{8}$$

where $C(\varepsilon) = const$,

$$Re(Fu - Fv,\ u-v) \geq 0,\ \forall u,\ v \in H. \tag{9}$$

jjj) $$\|Fu-Fv\| \leq C(\rho)\ \|u-v\|,\ \|u\| \leq \rho,\ \|v\| \leq \rho, \tag{10}$$

$0 < \rho < \infty$, $C(\rho) = const.$

Remark 1. If A is a generator of a strongly continuous semigroup in H, then the sequence A_n with property (3) exists (e.g.,

Yosida approximation of A) (see [12]).

Let us denote by H_A the Hilbert space which is the completion of $D(A)$ with respect to metric generated by the form $Re(Au,u)$. Sesquilinear form $[u,v] = \frac{1}{2}\{(Au,v) + (u,Av)\}$ is the inner product in H_A, $[u,u] = Re(Au,u)$. By R_λ we denote the operator $(A+\lambda I)^{-1}$, $\lambda > 0$; I is the identity in H.

II.3. Discussion of Assumptions

Before we formulate the results let us discuss the physical meaning of the assumptions.

Assumption i) usually is satisfied and the volt-amper characteristic of nonlinear elements is often a smooth monotone and bounded function $i = f(u)$, $|f(-\infty)| < \infty$, $|f(+\infty)| < \infty$, so that (8) is satisfied. In network theory the behavior of $f(u)$ as $|u| \to \infty$ is of no importance as long as we are interested in stationary regimes which are uniformly bounded for $t \in I = (-\infty,\infty)$. Assumption (3) in case when linear two-port is time invariant and $J(t)$ is periodic can be interpreted as follows. In this case the operator A is an integral operator of the type

$$Au = \int_0^T \phi(t-\tau)\, u(\tau)d\tau, \tag{11}$$

where

$$\phi(t) = T^{-1} \sum_{m=-\infty}^{\infty} exp(im\omega t)\, Z^{-1}(im\omega), \tag{12}$$

where the impedance $Z(p)$ is defined by formula (6.1), $\omega = 2\pi T^{-1}$. The sequence A_n can be constructed by truncating the series in (12):

$$A_n u = \int_0^T \phi_n(t-\tau)\, u(\tau)d\tau, \tag{13}$$

where

$$\phi_n(t) = T^{-1} \sum_{m=-n}^{n} exp(im\omega t) Z^{-1}(im\omega). \tag{14}$$

Assumption (4) means that the operator B_n describes a passive two-port $L_n - N$. Assumption (7) means that the two-port L is passive and δ can be considered as some measure of its passiveness. The bigger $\delta > 0$ is, the less assumptions about F we need. For example, if F is not monotone but $F + aI$ is monotone, then for $\delta > a$ the two-port $L - N$ can be transformed into an equivalent two-port $L_a - N_a$, where N_a is described by the monotone operator $F + aI$ and L_a is described by the operator $A - aI$ which satisfies the inequality (7) with $\delta_a = \delta - a > 0$.

Stationary regime in the system in Fig. 1 is a uniformly bounded (on I) solution of the equation

$$u(t) = e(t) - \int_{-\infty}^{t} g(t,\tau)\, f(\tau,u(\tau))d\tau, \quad t \in I \tag{15}$$

where $i = f(t,u(t))$ is the characteristic of N. But if $e(t)$ is T-periodic and $f(t+T,u) = f(t,u)$, $\forall u$, we look for a T-periodic stationary regime. This regime satisfies equation

$$u(t) = e(t) - \int_{0}^{T} \Phi(t,\tau)\, f(\tau,u(\tau))d\tau, \tag{16}$$

where

$$\Phi(t,\tau) = T^{-1} \sum_{m=-\infty}^{\infty} exp\{im\omega(t-\tau)\}\, Z(im\omega;\, t), \quad \omega = \frac{2\pi}{T} \tag{17}$$

and

$$Z(i\lambda;t) = Z(i\lambda;t+T) = \int_{0}^{\infty} g(t;t-s) exp(-i\lambda s) ds. \tag{18}$$

Equation (16) can be rewritten in the form (1) with

$$Au = \int_{0}^{T} \psi(t,\tau) u(\tau) d\tau \tag{19}$$

and

$$\psi(t,\tau) = T^{-1} \sum_{m=-\infty}^{\infty} exp\{im\omega(t-\tau)\} \; Y(im\omega;t). \tag{20}$$

Here $Y(i\lambda;t)$ plays the role of operator admittance and can be defined by

$$Y(i\lambda;t) = \int_0^{\infty} h(t,t-s)exp(-i\lambda s)ds, \tag{21}$$

where $h(t,\tau)$ is defined by formula

$$i(t) = \int_{-\infty}^{t} h(t,\tau)u(\tau)d\tau. \tag{22}$$

Let us explain formula (16). The current and voltage on L satisfy the relation

$$u(t) = \int_{-\infty}^{t} g(t,\tau)i(\tau)d\tau.$$

If $i = exp(i\lambda t)$, then

$$u(t) = \int_{-\infty}^{t} g(t,\tau)exp(i\lambda\tau)d\tau = exp(i\lambda t)\int_0^{\infty} g(t,t-s) \times exp(-i\lambda s)ds.$$

Expanding $i(\tau) = f(\tau,u(\tau))$ in Fourier series we get equation (16) from (15) with Φ defined by (17) and $Z(i\lambda;t)$ defined by (18). Equation for almost periodic regimes takes the form

$$u(t) = e(t) - \int_{-\infty}^{\infty} h(t,\tau)f(\tau,u(\tau))d\tau, \tag{23}$$

where

$$h(t,\tau) = \frac{1}{2\pi}\sum_m exp\{i\lambda_m(t-\tau)\} \; Z(i\lambda_m;t). \tag{24}$$

Transient regimes in the network in Fig. 1 can be described

by the equation

$$u(t) = e(t) + m(t) - \int_0^t g(t,\tau)f(\tau,u(\tau))d\tau, \quad t \geq 0; \qquad (25)$$

here $m(t)$ is the reaction to initial conditions on the linear two-port.

The following questions will be discussed:

(1) Does equation (16) have a unique periodic solution in $H = L^2[0,T]$? Is this solution stable in the large, i.e., with respect to arbitrary perturbations of initial conditions? Is it stable with respect to small periodic perturbations of $e(t)$? Is the network convergent? A network is called convergent if there exists only one stationary regime in the network (i.e., a uniformly bounded (on I) solution of equation (15)) and if for any initial data the transient regime converges to the stationary regime as $t \to +\infty$ (i.e., for any admissible $m(t)$ the solution of equation (25) tends to the solution of equation (15) as $t \to +\infty$).

(2) Similar questions are of interest when the exterior force is almost periodic or belongs to $L^\infty(I)$, where $L^\infty(I)$ is the space of measurable functions with the finite norm $\|u\|_{L^\infty} = \operatorname{ess\,sup}_{t \in I} |u(t)|$.

Though all of the abstract results can be formulated for operators acting from a Banach space X into X^* provided that X is reflexive, we present the theory for operators acting in a Hilbert space H. This is done for simplicity of the presentation and also because the results in this case will be sufficient to treat the problems of interest in network theory.

II.4. Main Results

Theorem 1. Let conditions i), ii) be satisfied. Then equation (1) is uniquely solvable in H. If conditions i), ii), iii) hold then the map B^{-1} is continuous in H.

Theorem 2. Let conditions j), jj), jjj) be satisfied. Then for any $u_0 \in H$ and sufficiently large $\lambda > 0$ the sequence

$$u_{n+1} = \lambda R_\lambda u_n - R_\lambda Fu_n + R_\lambda J, \; u_0 \in H, \; R_\lambda \equiv (A + \lambda I)^{-1} \tag{26}$$

converges in H_A to the solution of equation (1) at the rate of geometrical progression. Equation (1) is uniquely solvable in H and the map B^{-1}: $H \to H_A$ is continuous. (Remark: In the proof of Theorem 2 it will be shown how to choose $\lambda \to 0$ in order that the sequence u_n converges at the maximal rate.)

Surjectivity in Theorem 2 is known from monotonicity theory, uniqueness is obvious. Conditions (5), (6) are used for construction of the solution. The map T^{-1} from H to $D(A)$ equipped with the norm $(\|u\|^2 + \|Au\|^2)^{\frac{1}{2}}$ is continuous.

Theorem 3. Let F be a Frechet differentiable operator on a Hilbert space H, T a linear injective operator, $D(T) \supset R(F'(u))$, $D(T) \supset R(F(u))$, $ReTF'(u) \geq a > 0$, $\|TF'(u)\| \leq b$, $\forall u \in H$, $b > a$. Here $D(T)$ denotes the domain of A and $R(A)$ denotes the range of A. Let $\gamma = ab^{-2}$, $q = (1 - a^2 b^{-2})^{\frac{1}{2}}$. Then the equation $Fu = 0$ is uniquely solvable in H, its solution can be obtained by the iterative process

$$u_{n+1} = u_n - \gamma \, TFu_n, \; u_0 \in H, \tag{26a}$$

and the following estimate of the rate of convergence is valid

$$\| u-u_n \| = 0(q^n). \tag{27}$$

Theorem 4. Let

$$Au + Fu = 0. \tag{28}$$

$$A \geq d > 0; \; 0 \leq F'(u) \leq \mu \quad \forall \, u \in H. \tag{29}$$

Let $T \equiv A^{-\frac{1}{2}}$, $F_1(v) \equiv A^{\frac{1}{2}} v + F(Tv)$, $b = 1 + \mu d^{-1}$, $v = A^{\frac{1}{2}} u$. Then equation $F_1(v) = 0$ is uniquely solvable in H, equation (28) is uniquely solvable in H, $u = Tv$ and the iterative process

$$v_{n+1} = v_n - b^{-2} TF_1(v_n), \; v_0 \in H \tag{30}$$

converges to the solution v of equation $F_1(v) = 0$, $\| v_n - v \| = 0(b^{-2n})$.

It is often convenient to investigate the dependence of the stationary regimes on the exterior force by using the following theorem.

Theorem 5. Let A be a continuous compact map of a Banach space X into itself, $T = I + A$, I be the identity in X. If T is injective and T^{-1} is bounded then T is a homeomorphism of X onto X.

Theorem 6. If Q is a bounded linear operator on H, QF is compact, F is a bounded nonlinear operator and

$$(Qu,u) \geq 0, \ (Fu,u) \geq \gamma \| F_u \| - C(\gamma), \forall \gamma > 0, \ C(\gamma) > 0 \tag{31}$$

then equation

$$U + QFu = e \tag{32}$$

has a solution in H.

If QF is compact, Q is a linear bounded operator and

$$\| Fu \| \leq A \| u \|^a + B, \ A > 0, \ B = const, \ 0 \leq a < 1 \tag{33}$$

then equation (32) has a solution in H.

Remark. It is essential that no compactness assumptions are made in theorems 1-4 since in the almost periodic case the operator in the main equation (e.g., equation (23)) is not compact.

II.5. Proofs

Proofs of Theorem 1. Consider the sequence of equations

$$B_n u_n = A_n u_n + Fu_n = J. \tag{34}$$

According to conditions (4), (5) B_n is a monotone, bounded, defined on all H coercive mapping. Thus B_n is surjective (see [13]) and equation (34) is solvable. From (34) we get

$$\gamma(\|u_n\|)\,\|u_n\| \le Re[(A_n u_n, u_n) + (Fu_n, u_n)]$$
$$= Re(J, u_n) \le \|J\|\,\|u_n\|. \quad (35)$$

Since $\gamma(t) \to +\infty$ as $t \to +\infty$ it follows from (35) that $\|u_n\| \le C$. By C we denote here and below various constants which do not depend on n. Since F is bounded $\|Fu_n\| \le c$. This and (34) imply that $\|A_n u_n\| \le C$. Since H is weakly compact the sequences $\{u_n\}$, $\{Fu_n\}$, $\{A_n u_n\}$ converge weakly. Let $u_n \rightharpoonup u$, $Fu_n \rightharpoonup v$, $A_n u_n \rightharpoonup w$. We prove that $u \in D(A)$ and $Au + Fu = J$. Indeed

$$(u, A^*y) = (u - u_n, A^*y) + (u_n, A^*y - A_n^*y) + (A_n u_n, y), \forall y \in D(A^*). \quad (36)$$

Taking into account that $u_n \rightharpoonup u$, $A_n y_n \rightharpoonup w$ and $A^*y_n \rightharpoonup A^*y$ we get

$$(u, A^*y) = (w, y), \forall y \in D(A^*).$$

This implies $u \in D(A)$, $Au = w$. Since B_n is monotone we have $0 \le Re(x - u_n, B_n x - B_n u_n)$, $\forall x \in D(A)$. Taking into account that $B_n u_n = J$ and $u_n \rightharpoonup u$, $B_n x \rightharpoonup Bx$ we get $0 \le Re(x - u, Bx - J)$, $\forall x \in D(A)$. Since A is linear and F is semicontinuous $B = A + F$ is semicontinuous. For $x - u = \lambda y \in D(A)$, $\lambda > 0$, we have $0 \le Re(y, B(u + \lambda y) - J)$. If $\lambda \to 0$ it follows that $0 \le Re(y, Bu - J)$, $\forall y \in D(A)$. Thus $B_u = J$. Condition (2) guarantees uniqueness of the solution of equation (1). It remains to prove that B^{-1} is continuous. Let $Bu_n = J_n$, $Bu = J$, $J_n \to J$. From (6) we get $\nu_R(\|u_n - u_m\|)\,\|u_n - u_m\| \le Re(Bu_n - Bu_m, u_n - u_m) = Re(J_n - J_m, u_n - u_m) \le \|J_n - J_m\|\,\|u_n - u_m\|$. Thus $\|u_n - u_m\| \to 0$ as $n, m \to \infty$, $\lim u_n = u_0$ exists. Below we prove that B is closed. Hence $u_0 \in D(B)$, $Bu_0 = J$ and since the solution of equation (1) is unique we conclude that $u_0 = u$. It means that B^{-1} is continuous. To end the proof it remains to verify that B is closed. Let $Bu_n = Au_n + Fu_n \to J$, $u_n \to u$. Then $\|Fu_n\| \le C$, $\|Au_n\| \le C$. As above we conclude that $Fu_n \rightharpoonup v$, $Au_n \rightharpoonup w$, $w = Au$. Thus $u \in D(A) = D(B)$, $Bu = J$.

Proof of Theorem 2. The proof can be described as follows.

First we prove that the sequence (26) is bounded. Then we prove that this sequence converges. It is easy to prove that its limit is the solution of equation (1).

Step 1. Denote by $a_n = \|u_n\|$, $\eta_n = \|u_n - u_{n-1}\|$, $b_n = \|\eta_n\|$, $\psi_n = Fu_n - Fu_{n-1}$, $y_n = (A + \lambda I)\eta$, $\|y_n\| = h_n$.

From (26) we get

$$a_{n+1} \leq \lambda \|R_\lambda\| a_n + \|R_\lambda\| (\varepsilon a_n + C(\varepsilon)) + \|R_\lambda\| \|J\|. \tag{37}$$

From (7) it follows that $\|R_\lambda\| \leq (\lambda + \delta)^{-1}$. Thus

$$a_{n+1} \leq \gamma a_n + C,\ 0 < \gamma = \frac{\lambda + \varepsilon}{\lambda + \delta} < 1 \quad \text{if} \quad 0 < \varepsilon < \delta. \tag{38}$$

From (38) it follows that

$$a_n \leq R,\ R = const. \tag{39}$$

Step 2. From (26) it follows that

$$\eta_{n+1} = \lambda R_\lambda \eta_n - R_\lambda \psi_n,\ y_{n+1} = \lambda\eta_n - \psi_n. \tag{40}$$

Thus

$$\begin{aligned}\|y_{n+1}\|^2 &= \lambda^2\|\eta_n\|^2 + \|\psi_n\|^2 - 2\lambda Re(\eta_n,\psi_n)\\ &\leq \lambda^2\|\eta_n\|^2 + \|\psi_n\|^2.\end{aligned} \tag{41}$$

Here we used monotonicity of F:

$$Re(\eta_n,\psi_n) = Re(u_n - u_{n-1},\ Fu_n - Fu_{n-1}) \geq 0. \tag{42}$$

From (10) it follows that $\|\psi_n\| \leq C(R)b_n$. Thus $h^2_{n+1} \leq b^2_n(\lambda^2 + C^2(R))$. But $b_n \leq (\lambda + \delta)^{-1} h_n$. Hence $h_{n+1} \leq \tilde{q}h_n$, $\tilde{q} = \dfrac{(\lambda^2 + C(R))^{\frac{1}{2}}}{\lambda+\delta}$. Minimizing in λ we get $\lambda_{min} = C^2(R)\delta^{-1}$,

$$q = q_{min} = C(R)(C^2(R) + \delta^2)^{-\frac{1}{2}}. \tag{43}$$

It is clear that $h_n = O(q^n)$, i.e., $\|(A + \lambda)\eta_n\| = O(q^n)$. Thus

$\|\eta_n\| \leq (\lambda+\delta)^{-1}O(q^n)$. Since $\eta_n = u_n - u_{n-1}$ it means that the sequence u_n converges. If $u = lim\ u_n$ then $u = \lambda R_\lambda u - R_\lambda Fu + R_\lambda J$. Applying $A + \lambda I$ one gets equation (1).

Step 3. Uniqueness of the solution of equation (1) follows from (7) and (9). Indeed, if $Bu = J$, $Bv = J$ then $0 = Re(Bu-Bv, u-v) \geq \delta\|u-v\|^2$. Thus $u = v$. It remains to prove that $B^{-1}: H \to H_A$ is continuous. Let $Bu_n = J_n$. We have

$$\delta\|u_n - u_m\|^2 \leq Re(Au_n - Au_m, u_n - u_m)$$

$$\leq Re(Bu_n - Bu_m, u_n - u_m) < \|J_n - J_m\|\ \|u_n - u_m\|.$$

Thus $\|u_n - u_m\| \leq \delta^{-1}\|Bu_n - Bu_m\|$, $\|u_n - u_m\|_{H_A} \leq \delta^{-\frac{1}{2}}\|Bu_n - Bu_m\|$.

Proof of Theorem 3. Let $Q \equiv BF'(v)$, $ReQ \geq a$, $\|Q\| \leq b$, $v \in H$ is arbitrary. We have $\|I-\alpha Q\|^2 = \|I-2\alpha ReQ + \alpha^2 QQ^*\| \leq 1-2\alpha a+b^2\alpha^2 \equiv \kappa^2(\alpha)$, $\kappa^2_{min} = \kappa^2(\gamma) \equiv q^2 = 1 - a^2b^{-2}$, $\gamma = ab^{-2}$. Equation $Fu = 0$ is equivalent to equation $u = u - \gamma BFu \equiv Tu$, where $\|Tu - Tv\| \leq \sup_{w\in H}\|I - \gamma BF'(w)\|\ \|u-v\| \leq q\|u-v\|$. Thus from the contractive maps principle the statement of theorem 3 follows.

Proof of Theorem 4. Equation (28) is equivalent to equation $F_1(v) = 0$, where $Tv = u$. We have $TF_1'(w)h - h + TF'(Tw)Th$, $\forall w \in H$. Thus $TF_1'(w) \geq 1$, $\|TF_1'(w)\| \leq 1 + \mu d^{-1} \equiv b$. From Theorem 3 it follows that the iterative process (30) converges to the unique solution of equation $F_1(v) = 0$ and $\|v - v_n\| = O(b^{-2n})$. Therefore $u = Tv$ is the solution of (28) and the sequence $u_n = Tv_n$ converges to this solution, $\|u - u_n\| \leq d^{-\frac{1}{2}}\|v_n - v\| \leq d^{-\frac{1}{2}}O(b^{-2n})$.

Proof of Theorem 5. First we prove that $R(T)$ is closed. Then we prove that $R(T)$ is open. From this it follows that $R(T) = X$. In the course of the proof it will be shown that T^{-1} is continuous on $R(T)$. It means that T is a homeomorphism X onto X. Step 1. Let $Tu_n = f_n \to f$. Since T^{-1} is bounded we have $\|u_n\| \leq C$. Since A is compact we get $Au_{n_k} \to v$. Hence by u_{n_k} we denote some convergent subsequence of $\{u_n\}$. Thus

$u_{n_k} = f_{n_k} - Au_{n_k} \to u$. Passing to the limit we get $u = f - Au$. Since T is injective $u = T^{-1}f$. Thus $u_n \to T^{-1}f$. Hence $R(T)$ is closed in X and T^{-1} is continuous on $R(T)$.

Step 3. $R(T)$ is open according to a variant of the invariance of domain theorem ([14], p. 161).

Proof of Theorem 6. If (33) holds, then $(u,u) \geq (e-QFu,u)$ on the sphere $||u|| = R$ provided that $R > 0$ is large enough. This implies existence of a solution of equation (32). To see this one can use the following known proposition ([14], p. 339). If T is compact and $(Tu,u) < ||u||^2$ for $||u|| = R$ then equation $u = Tu$ has a solution in the ball $||U|| < R$. In our case $T = e - QF$.

To prove the first statement of Theorem 6 let us consider equation $u = \lambda(e - QFu)$, $0 \leq \lambda \leq 1$. From this we get $(u,Fu) - \lambda(QFu,Fu) = \lambda(e,Fu)$. From (31) it follows that

$$\gamma\ ||Fu|| - C(\gamma) \leq \lambda\ ||e||\ ||Fu||\ .$$

Hence, taking $\gamma > ||e||$ we get

$$||Fu|| \leq C,\ ||u|| \leq ||e|| + ||Q||C \leq C_1.$$

By Leray-Schauder principle ([14], p. 298) we conclude that equation (32) has a solution in H.

II.6. Discussion of Applications

First let us show that in some sense the results given in theorems 1,2 are close to best possible for the network in Fig. 1. Namely, consider the network in Fig. 1 and assume that $e(t) = e(t+T)$, N has a smooth bounded monotone increasing characteristic $i = f(u)$, $0 < f \leq \mu$ and L is a capacitance C. Conditions (8), (9), (10) are satisfied, but $(Au,u) = \int_0^T C \frac{du}{dt} u^* dt = 0$ if u is periodic. Thus δ in (7) is equal to zero. It is easy to show that there do not exist periodic solutions in the network. Indeed, $f(e-u) = c\ du/dt$. Thus $0 < \int_0^T f(e-u)dt =$

$c\int_0^T \frac{du}{dt}\,dt = 0$. This contradiction proves the statement. In this example all conditions of Theorem 1 hold except for the coercivity condition. If L is a time invariant two-port then (7) is fulfilled if $\inf_{-\infty<n<\infty} Re\, Z^{-1}(in\omega) \geq \delta$. Indeed, in the periodical case

$$Re(Au,u) = Re\int_0^T Au \cdot u^* dt = Re \sum_{n=-\infty}^{\infty} Z^{-1}(in\omega) u_n u_n^*$$

$$\geq \min_{-\infty<n<\infty} Re Z^{-1}(in\omega) \sum_{n=-\infty}^{\infty} |u_n|^2$$

Here $Z(i\lambda)$ is the impedance of L defined by formula (6.1), $Z^{-1}(i\lambda) = Y(i\lambda)$ is the admittance of L. A similar argument is valid for the almost periodic case. Conditions (7), (9) mean that L and N are passive. From the discussion in Section II.3 it follows that neither L nor N individually should be passive but the L-N tow-port should be passive in order that the main results on existence, uniqueness and stability of the stationary regime be true.

III. INTEGRAL EQUATIONS, TRANSIENT AND STATIONARY REGIMES

III.1. Main Result

If the exterior force is in L^∞, but neither periodic nor almost periodic, then the space where the solution lies is not a Hilbert space. In this case instead of concepts of nonlinear functional analysis we shall use analytical methods. In what follows $\dot{e} = de/dt$, $|u| = |u|_{L^\infty}$, $I = (-\infty, \infty)$, $I_+ = (0,\infty)$ and all functions are real-valued. Our main result is the following theorem.

Theorem 1. Let $|e(t)| + |\dot{e}| < \infty$, $f(t,u)$ is measurable in t, uniformly bounded in u for $|u| \leq R$, $\forall R > 0$ and uniformly continuous in u for $t \in I$. If

$$\sup_{t \in I} \int_I |g(t,\tau)|d\tau \equiv G < \infty \tag{1}$$

$$|f(t,u)| \leq \varepsilon|u| + c(\varepsilon), \forall \varepsilon > 0, \ c(\varepsilon) = const \tag{2}$$

then all uniformly bounded solutions of equation (23.2) are apriori bounded on I. If moreover $0 \leq \frac{\Delta f(t,u)}{\Delta u} \leq \mu$ and for sufficiently large numbers ℓ the inequality holds

$$\varepsilon \int_{-\ell}^{\ell} u^2 dt \leq \mu^{-1} \int_{-\ell}^{\ell} u^2 dt + \int_{-\ell}^{\ell} dt \ u(t) \int_{-\infty}^{t} g(t,\tau)u(\tau)d\tau \tag{3}$$

where $\varepsilon > 0$ and $u(t)$ is an arbitrary bounded measurable function, then equation (23.2) has no more than one uniformly bounded on I solution $U(t)$. If moreover the function $g(t,\tau)$ is continuously differentiable in t for $t \geq \tau$, $\tau \in I$ and

$$\sup_{t \in I} \left(|g(t,t)| + \int_{-\infty}^{t} |\dot{g}(t,\tau)|d\tau \right) < \infty \tag{4}$$

then there exists and is unique the uniformly bounded on I solution $U(t)$ of equation (23.2). If moreover the following inequalities hold

$$\int_0^{\infty} dt \left(\int_{-\infty}^{0} |g(t,\tau)|^2 d\tau \right)^2 < \infty, \ m(t) \in L^2(I_+), \ \underset{t \to +\infty}{m(t) \to 0}, \tag{5}$$

$$\sup_{t \geq 0} |m(t)| < \infty,$$

$$\sup_{s \geq 0} \int_0^{\infty} \int_0^{\infty} |g(t,\tau)g(t,s)|dtd\tau < \infty, \tag{6}$$

$$\sup_{t \geq 0} \int_0^{\infty} |g(t,\tau)|^2 d\tau < \infty, \quad |g(t,\tau)| \xrightarrow[|t-\tau| \to \infty]{} 0 \tag{7}$$

$$\int_{-\infty}^{0} |g(t,\tau)|d\tau \to 0 \quad \text{as} \quad t \to +\infty , \tag{8}$$

then every solution of equation (25.2) satisfies the equality

$$\lim_{t\to+\infty} |U(t)-u(t)| = 0, \tag{9}$$

i.e., the network in Fig. 1 is convergent.

Remark 1. If $g(t,\tau) = g(t-\tau)$ and $|g(t)| + |\dot{g}(t)| \le c\ exp(-at)$, $a > 0$, and $\mu^{-1} + ReZ(i\lambda) \ge \varepsilon > 0$ for $\lambda \in I$, then all the assumptions of theorem 1 are fulfilled.

Remark 2. Theorem 1 remains valid if $g(t,\tau) = R\delta(t-\tau) + g_1(t,\tau)$, where $R = const > 0$, $\delta(t)$ is the delta-function and $g_1(t,\tau)$ satisfies assumptions of theorem 1.

Remark 3. Uniqueness of solution of equation (23.2) can be proved under the assumption

$$0 < \mu^{-1}\int_{-\ell}^{\ell} u^2 dt + \int_{-\ell}^{\ell} dt u(t)\int_{-\infty}^{t} g(t,\tau)u(\tau)d\tau, \forall \ell > T_o \tag{10}$$

where T_o is an arbitrary large fixed number and $u(t)$ is an arbitrary bounded measurable function.

III.2. Proofs

Proof of Theorem 1. Step 1. From (1), (2) and (23.2) we get $|u| \le \varepsilon G|u| + G\ c(\varepsilon) + |e|$. Taking $\varepsilon\ G = 0.5$ we have

$$|u| \le 2\bigl(|e| + c\ G\bigr). \tag{11}$$

This proves that all the solutions of equation (23.2) are uniformly bounded on I.

Step 2. Let us show that (3) implies uniqueness of the uniformly bounded solution of equation (23.2). If u,v are such solutions, $w = u - v$, $\Phi = Fu - Fv$, $Fu = f(t,u(t))$, then

$$w + \int_{-\infty}^{t} g(t,\tau)\ \Phi(\tau)d\tau; \tag{12}$$

multiplying (12) by $\Phi(t)$, integrating over $(-\ell,\ell)$ and taking into account that $\mu^{-1}\Phi^2 \le \Phi w$ one gets:

$$0 \ge \mu^{-1}\int_{-\ell}^{\ell} \Phi^2 dt + \int_{-\ell}^{\ell} dt\Phi(t) \int_{-\infty}^{t} g(t,\tau)\ \Phi(\tau)d\tau$$

$$\ge \varepsilon \int_{-\ell}^{\ell} \Phi^2 dt. \tag{13}$$

Thus, $\Phi \equiv 0$, $u \equiv v$. Actually we proved also Remark 3.

Step 3. Let us prove existence of the uniformly bounded solution. Let

$$u_n(t) + \int_{-n}^{t} g(t,\tau)f(\tau,u_n(\tau))d\tau = e(t),\ t \ge -n. \tag{14}$$

This equation has a solution for any $n = 1, 2, \ldots$ and the solution is unique. It follows from the fact that (14) is a Volterra equation with a nonlinearity, which satisfies the inequality $|f(t,u)| \le A + B|u|$. For u_n inequality (11) holds. Therefore $|u_n| \le c$, where c does not depend on n. From (4) and (14) it follows that

$$\dot{u}_n + g(t,t)f(t,u_n(t)) + \int_{-n}^{t} \dot{g}(t,\tau)f(\tau,u_n(\tau))d\tau = \dot{e} \tag{15}$$

From (15) we conclude that $|\dot{u}_n| \le c$. Thus

$$|u_n| + |\dot{u}_n| < c. \tag{16}$$

Let us choose a subsequence, denoted again u_n, which converges to a limit $U(t)$ on any finite segment. This can be done by using the diagonal process. Using Lebesque theorem one can pass to the limit in (14). As a result one gets equation (23.2) for $U(t)$. Because of uniqueness of the uniformly bounded solution of equation (23.2) all the subsequences of the sequence $\{u_n\}$ converge to the same limit $U(t)$. Thus $\lim u_n(t) = U(t)$.

Step 4. Let us prove (9). We have

$$U(t) = e(t) + n(t) - \int_0^t g(t,\tau)f(\tau,U(\tau))d\tau, \tag{17}$$

$$n(t) \equiv -\int_{-\infty}^0 g(t,\tau)f(\tau,U(\tau))d\tau. \tag{18}$$

Let

$$v = U - u, \quad q(t) = n(t) - m(t) \tag{19}$$

where $u(t)$ is the solution of equation (25.2) and $m(t)$ is the function in that equation. We have

$$v = q(t) - \int_0^t g(t,\tau)\psi(\tau)d\tau. \tag{20}$$

Multiplying (20) by ψ, integrating over $(0,\infty)$ and applying (3) one gets

$$\varepsilon\int_0^\infty \psi^2 dt \le \mu^{-1}\int_0^\infty \psi^2 dt + \int_0^\infty dt\psi(t)\int_0^t g(t,\tau)\psi(\tau)d\tau$$

$$\le \left(\int_0^\infty q^2 dt\right)^{\frac{1}{2}} \left(\int_0^\infty \psi^2 dt\right)^{\frac{1}{2}} . \tag{21}$$

Hence

$$\int_0^\infty \psi^2 dt \le c. \tag{22}$$

From (22), (20), (6), (5) it follows that

$$\int_0^\infty v^2 dt \le 2\int_0^\infty q^2 dt + 2\int_0^\infty \left(\int_0^t g(t,\tau)\psi(\tau)d\tau\right)^2 dt$$

$$\le 4\int_0^\infty m^2 dt + 4\int_0^\infty n^2 dt + c_1 \int_0^\infty \psi^2 dt \le C. \tag{23}$$

Here we took into account the fact that (6) implies boundedness in $L^2(I_+)$ of the integral operator $T\psi \equiv \int_0^t g(t,s)\psi(s)ds$. It is not difficult to verify that

$$\|T\|_{L^2(I_+)\to L^2(I_+)} \le \sup_{s\ge 0} \int_0^\infty \int_0^\infty |g(t,s)g(t,\tau)|dtd\tau. \tag{24}$$

From (20) and (8) we conclude that

$$v(t) \to 0 \quad \text{as} \quad t \to +\infty. \tag{25}$$

To prove it, let us show that

$$q(t) \to 0, \int_0^t g(t,\tau)\psi(\tau)d\tau \to 0 \quad \text{as} \quad t \to +\infty. \tag{26}$$

We have

$$|q(t)| \le \int_{-\infty}^0 |g(t,\tau)f(\tau,U(\tau))|d\tau + |m(t)|$$

$$\le C\int_{\infty}^0 |g(t,\tau)|d\tau + |m(t)| \to 0 \quad \text{as} \quad t \to +\infty. \tag{27}$$

Here we used (5) and (8) and took into account that $|U(t)| \le c$ implies $|f(t,U(t))| \le c_1$. Further we have

$$\left|\int_0^t g(t,\tau)\psi d\tau\right| \le \int_0^N |g(t,\tau)|\ |\psi|d\tau$$

$$+ \left(\int_N^t |g(t,\tau)|^2 d\tau\right)^{1/2}\left(\int_N^t |\psi|^2 dt\right)^{1/2}. \tag{28}$$

We can prove that all solutions of equation (25.2) are uniformly bounded on I_+ provided that $|e| + |m| \leq c$ and conditions (1), (2) hold. The proof is similar to the proof given in Step 1. Thus $|\psi| \leq c$. Let us take N so large that

$$\left(\int_N^t |g(t,\tau)|d\tau\right)^{\frac{1}{2}}\left(\int_N^t \psi^2 dt\right)^{\frac{1}{2}} < \alpha \tag{29}$$

for any given fixed $\alpha > 0$; this is possible because of (22) and (1). Then let us fix N and take t so large that

$$\int_0^N |g(t,\tau)|\ |\psi| d\tau < \alpha, \tag{30}$$

which is possible because $|\psi| \leq c$ and $|g(t,\tau)| \to 0$ as $|t-\tau| \to \infty$ (see (7)). Since $\alpha > 0$ is arbitrary we have proved that

$$\left|\int_0^t g(t,\tau)\psi d\tau\right| \to 0 \quad \text{as} \quad t \to +\infty . \tag{31}$$

Formula (9) follows from (27) and (31).

Proof of Remark 1. Under the assumptions of this remark conditions (1), (4)-(8) are obviously satisfied. It remains to prove (3). Let

$$u_\ell = \begin{cases} 0, & |H| > \ell \\ u(t), & |t| \leq \ell . \end{cases}$$

The right-hand side of (3) can be rewritten, by applying Parseval quality, as:

$$J_1 + J_2 \equiv \frac{1}{2\pi}\int_{-\infty}^{\infty} [Re\ Z(i\lambda) + \mu^{-1}]\ |\tilde{u}_\ell|^2 d\lambda$$
$$+ \int_{-\ell}^{\ell} dt\ u(t) \int_{-\infty}^{-\ell} g(t,\tau)u(\tau)d\tau. \tag{32}$$

Here $\tilde{u}$ is the Fourier transform of $u(t)$, $\tilde{u} = \int_0^\infty exp(-i\lambda t)u(t)dt.$

Since $Re\ Z(i\lambda) + \mu^{-1} \geq \varepsilon > 0$ we have

$$J_1 \geq \varepsilon \int_{-\ell}^{\ell} |u|^2 dt. \tag{33}$$

Let us prove that

$$|J_2| \leq \delta \int_{-\ell}^{\ell} |u|^2 dt \tag{34}$$

where $\delta > 0$ is arbitrarily small if ℓ is sufficiently large. If $u \in L^2(I)$ then $J_2 \to 0$ as $\ell \to \infty$. Indeed

$$|J_2| \leq \int_{-\ell}^{\ell} dt\ u(t) \int_{-\infty}^{-\ell} exp\{-a(t-\tau)\} |u(\tau)| d\tau$$

$$\leq \int_{-\ell}^{\ell} dt |u(t)|\ exp(-at) \left(\int_{-\infty}^{-\ell} u^2 dt \right)^{\frac{1}{2}} \frac{exp(-a\ell)}{\sqrt{2a}}$$

$$\leq \left(\int_{-\ell}^{\ell} u^2 dt \right)^{\frac{1}{2}} \frac{exp(a\ell)}{\sqrt{2a}} \left(\int_{-\infty}^{-\ell} u^2 dt \right)^{\frac{1}{2}}$$

$$\times \frac{exp(-a\ell)}{\sqrt{2a}} \longrightarrow 0 \quad \text{as} \quad \ell \to \infty .$$

If $u \notin L^2(I)$ then $\int_{-\ell}^{\ell} u^2 dt \to \infty$ as $\ell \to \infty$. Since $|u| \leq c$ we have:

$$|J_2| \leq \int_{-\ell}^{\ell} dt |u(t)| exp(-at) a^{-1}\ exp(-a\ell)$$

$$\leq c\ a^{-1}\ exp(-a\ell) \left(\int_{-\ell}^{\ell} u^2 dt \right)^{\frac{1}{2}} \frac{exp(a\ell)}{\sqrt{2a}}$$

$$\leq c \left(\int_{-\ell}^{\ell} u^2 dt \right)^{\frac{1}{2}} = o \left(\int_{-\ell}^{\ell} u^2 dt \right) \text{ as } \ell \to \infty .$$

This completes the proof.

Proof of Remark 2. Let us rewrite equations (16.2) and (25.2) as

$$u + Rf(t,u) + \int_0^T \Phi_1(t,\tau) f(\tau, u(\tau)) d\tau = e(t), \tag{35}$$

and

$$u + Rf(t,u) = e(t) + m_1(t) - \int_0^t g_1(t,\tau) f(\tau, u(\tau)) d\tau. \tag{36}$$

Let us set $v = u + Rf(t,u)$ and note that the function $u = u(t,v)$ inverse to $v = v(t,u)$ increases monotonically. If we set $F_1 v = f(t,u(t,v)) \equiv f_1(t,v)$ then (35) and (36) take the following forms

$$v + \int_0^T \Phi_1(t,\tau) f_1(\tau,v) d\tau = e(t), \tag{37}$$

$$v = e + m_1(t) - \int_0^t g_1(t,\tau) f_1(\tau,v) d\tau. \tag{38}$$

Since $0 \le \frac{\Delta f}{\Delta u} \le \mu$, $1 \le \frac{\Delta v}{\Delta u} < 1 + R\mu$, we have

$$1 \ge \frac{\Delta u}{\Delta v} \ge (1 + R\mu)^{-1}, \quad 0 \le \frac{\Delta f_1}{\Delta f} = \frac{\Delta f_1}{\Delta u} \cdot \frac{\Delta u}{\Delta v} \le \mu. \tag{39}$$

Further we have:

$$|f_1(t,v)| \le \varepsilon |u(t,v)| + c(\varepsilon)$$

$$|u| - \varepsilon R|u| - Rc(\varepsilon) \le |v| \le |u| + \varepsilon R|u| + Rc(\varepsilon).$$

Thus

$$|u(t,v)| \le (1 - \varepsilon R)^{-1} |v| + c_1(\varepsilon).$$

Therefore

$$|f_1(t,v)| \le \varepsilon |v| + c(\varepsilon), \tag{40}$$

if $\varepsilon > 0$ is sufficiently small. Now we can apply theorem 3. This completes the proof.

IV. ESTIMATES OF TRANSIENT REGIMES

IV.1. *Main Assumptions*

Consider the equation

$$u(t) = e(t) - \int_0^t g(t-\tau)f(\tau,u(\tau))d\tau, \quad t \geq 0 \tag{1}$$

and assume that

$$uf(t,u) \geq 0, \; \forall u, \; t \in I. \tag{2}$$

We derive estimates of the solution of (1) in terms of $e(t)$ for any nonlinearity which satisfies (2). About $g(t)$ we assume the following:

$$\int_0^\infty exp(at)|g(t)|dt < \infty \; , \; a > 0. \tag{3}$$

If $Z(p)$ is defined by (6.1) then from (3) it follows that

$$Z(p) \text{ is analytic for } \sigma > -a, \; p = \sigma + i\lambda. \tag{4}$$

We also assume that

$$Re\; Z(i\lambda) \geq \frac{a}{\sqrt{\lambda^2 + a^2}} |Z(i\lambda)|, \; -\infty < \lambda < \infty \; . \tag{5}$$

This inequality is true provided that the linear two-port described by the function $g(t)$ is passive [16]. It is known [16] that

$$-\lim_{\lambda \to +\infty} \lambda Im Z(i\lambda) = \frac{\zeta}{\pi} \int_0^\infty Re\; Z(i\lambda)d\lambda. \tag{6}$$

Let us introduce the following notations:

$$\tilde{u}(\lambda) = \int_0^\infty u(t)exp(-i\lambda t)dt \tag{7}$$

$$N_1(u) = \left\{ a\int_{-\infty}^{\infty} \frac{|\tilde{u}|^2 d\lambda}{|Z(i\lambda)| \sqrt{\lambda^2+a^2}} \right\}^{\frac{1}{2}} \tag{8}$$

$$N_2(u) = N_1(u) + \left\{ a^{-1}\int_{-\infty}^{\infty} \frac{|\tilde{u}|^2 \sqrt{\lambda^2+a^2}}{|Z(i\lambda)|} d\lambda \right\}^{\frac{1}{2}} . \tag{9}$$

IV.2. Main Results

Theorem 1. Let conditions (2), (5) be valid and $N_2(e) < \infty$. Then the solution of equation (1) satisfies the inequality

$$N_1(u) \leq N_2(e). \tag{10}$$

Remark 1. If $0 < C_1 \leq \sqrt{\lambda^2+a^2}\ |Z(i\lambda)| \leq C_2 < \infty$ then the norm (8) is equivalent to the norm of $L^2(I_+)$ and the norm (9) is equivalent to the norm $W_2^1(I_+)$, $||u||^2_{W_2^1(I_+)} = \int_0^\infty \{|u|^2 + |u'|^2\}dt$.

Remark 2. Let $e_d(t) = exp(dt)e(t)$. If $d < a$, $N_2(e_d) < \infty$ then $\tilde{N}_1(u_d) \leq \tilde{N}_2(e_d)$, where the norms with the twiddle are defined by formulas (8), (9) in which the following substitutions should be made: $a \to a - d$, $i\lambda \to i\lambda - d$.

Remark 3. Condition (5) means that $Z(p)$ is a positive real function, i.e., $Z(p)$ is analytic in the half-plane $Re\ p > 0$, $Z(\sigma)$ is real if σ is real and $Re\ Z(p) \geq 0$ if $\sigma \geq 0$. There is an extensive literature on positive real functions [17].

IV.3. Proofs

Proof of Theorem 1. Let $\tilde{\psi} = \tilde{f}(t,u(t))$. From (1) it follows that

$$\tilde{u} = \tilde{e} - Z(i\lambda)\tilde{\psi}\ , \tag{11}$$

multiplying (11) by $\tilde{\psi}^*$ and integrating over I we get:

$$\int_I \tilde{u}\,\tilde{\psi}^* d\lambda = \int_I \tilde{e}\,\tilde{\psi}^* d\lambda - \int_{-\infty}^{\infty} Z(i\lambda)\,|\tilde{\psi}|^2 d\lambda \,. \tag{12}$$

From (2) using Parseval's equality we get:

$$\int_I \tilde{u}\,\tilde{\psi}^* d\lambda = 2\pi \int_I u f(t,u(t))\,dt \geq 0. \tag{13}$$

Thus (5) and (12) imply that

$$a \int_I \frac{|Z(i\lambda)|}{(\lambda^2 + a^2)^{\frac{1}{2}}} |\tilde{\psi}|\, d\lambda \leq \left(\int_I \frac{|\tilde{e}|^2 (a^2 + \lambda^2)^{\frac{1}{2}}}{a|Z(i\lambda)|} d\lambda \right)^{\frac{1}{2}}$$

$$\times \left(\int_I |\tilde{\psi}|^2 \frac{|Z(i\lambda)|a}{(a^2 + \lambda^2)^{\frac{1}{2}}} d\lambda \right)^{\frac{1}{2}} .$$

From here it follows that

$$a \int_I \frac{|Z(i\lambda)|}{(a^2 + \lambda^2)^{\frac{1}{2}}} |\tilde{\psi}|^2 d\lambda \leq \int_I |\tilde{e}| \frac{(a^2 + \lambda^2)^{\frac{1}{2}}}{a|Z(i\lambda)|} d\lambda \,. \tag{14}$$

Let us multiply the identity

$$|u|^2 = \tilde{e}\,\tilde{u}^* - Z(i\lambda)\tilde{\psi}\,\tilde{u}^*$$

by the function $A(\lambda) = a(a^2 + \lambda^2)^{-\frac{1}{2}} |Z^{-1}(i\lambda)|$; we get:

$$N_1^2(u) \leq \int_I A(\lambda)|\tilde{e}\tilde{u}^*|d\lambda + \int_I A(\lambda)|Z(i\lambda)|\tilde{\psi}\,\tilde{u}^* d\lambda$$

$$\leq N_1(u) \left[\left(\int_I A(\lambda)|\tilde{e}|^2 d\lambda \right)^{\frac{1}{2}} \right.$$

$$\left. + \left(\int_I A(\lambda)|Z(i\lambda)|^2|\tilde{\psi}|^2 d\lambda \right)^{\frac{1}{2}} \right]$$

$$= N_1(u) N_2(e) . \tag{15}$$

In the last step we used (14). From (15) inequality (10) follows immediately.

The statements of the remarks are obvious.

IV.4. Additional Remarks

Remark 4. From (4), (5) and (6) it follows that a number $q > 0$ exists such that

$$Re(1 + i\lambda q)\ Z(i\lambda) > 0, -\infty < \lambda < \infty. \tag{16}$$

Proof. For $|\lambda| \leq R$, where R is an arbitrarily large fixed number, and $q > 0$ sufficiently small (16) holds. If $|\lambda| > R$ and q is fixed (16) holds because (5) implies that $Re\ Z(i\lambda) > 0$ and (6) implies that $Re\ i\lambda q Z(i\lambda) = -\lambda q\ Im\ Z(i\lambda) > 0$ for sufficiently large R.

Remark 5. Let $f(t,u)$ be monotonically increasing and (2.3) hold. Let $u_j(t)$ be the solution of equation (1) with $e(t)$ replaced by $e(t) + e_j(t)$, j = 1, 2, where $N_2(e_j) < \infty$. Then

$$N_1(u_1 - u_2) \leq N_2(e_1 - e_2). \tag{17}$$

Proof. Let $v = u_1 - u_2$, $\phi = f(t,u_1) - f(t,u_2)$. By assumption $\phi \cdot (u_1 - u_2) \geq 0$. Let $h = e_1 - e_2$. Then $\tilde{v} = \tilde{h} - Z(i\lambda)\tilde{\phi}$ and the arguments given in the proof of theorem 1 lead to (17).

Remark 6. If $Z(p)$ is rational, then equation (1) can be reduced to a system of differential equations but such a reduction is often not convenient. To show this, let us consider an example. Let $Z(p) = \dfrac{p+c}{p^2 + ap + b}$ where a,b,c are constants, $c > 0$ and a,b are real and chosen so that $Z(p)$ is analytic in the half-plane $Re\ p > -d$, $d > 0$. In this case the corresponding two-port is exponentially stable and passive. Taking the Laplace transform of (1) we get

$$(p^2 + ap + b)\bar{u} = (p^2 + ap + b)\bar{e} - (p + c)\ f(t,u) \tag{18}$$

where

$$\bar{u} = \int_0^\infty exp(-pt)u(t)dt. \tag{19}$$

If we let $x_1 = u$, $x_2 = \dot{u} = \frac{du}{dt}$, we can rewrite (18) as

$$\begin{cases} \dot{x}_1 = x_2 \\ \dot{x}_2 = -ax_2 - bx_1 - \frac{d}{dt} f(t,x_1) - cf(t,x_1) - \ddot{e} - a\dot{e} - be \end{cases},$$

or

$$\dot{x} = Px + \Phi(t,x) + F. \tag{20}$$

Here

$$P = \begin{pmatrix} 0 & 1 \\ -b & -a \end{pmatrix}, \tag{21}$$

$$\Phi = \begin{pmatrix} 0 \\ -\frac{df(t,x)}{dt} - cf \end{pmatrix}, \quad F = \begin{pmatrix} 0 \\ -\ddot{e} - a\dot{e} - be \end{pmatrix}. \tag{22}$$

Matrix P possesses the Hurwitz property, i.e., its eigenvalues lie in the half-plane $Re\ p \le -d$, $d > 0$. But the nonlinearity Φ is not necessarily monotone or sign-preserving (i.e., satisfying (1)), even if $f(t,u)$ is monotone in u or sign-preserving. That is why it is not advisable to reduce integral equation (1) to the system (20).

V. NONLINEARITIES WITHOUT GROWTH RESTRICTIONS

V.1. Main Assumption and Result

It is interesting to know whether equation (12.1) is solvable if the nonlinearity can grow arbitrarily at infinity.

Suppose that $Fu = f(u)$ where

$$f \in C_{loc}, \quad uf(u) \ge 0 \quad \text{for} \quad |u| \ge R. \tag{1}$$

Here R is an arbitrarily large fixed number, $f \in C_{loc}$ means that $f(u)$ is continuous in every bounded domain. We assume that

$$J \in L^{\infty}, \quad L^{\infty} \equiv L^{\infty}(I), \quad I = (-\infty, \infty) \tag{2}$$

and

$$Au = -u'' + \kappa^2 u, \quad \kappa^2 > 0 \tag{3}$$

though A can be of more general type. We discuss some possible generalizations at the end of this section.

Theorem 1. Equation

$$-u'' + \kappa^2 u + f(u) = J(x), \quad x \in I, \quad \kappa^2 > 0 \tag{4}$$

has a solution in L^{∞} provided that (1), (2) hold. There exists $J(x) \in L^{\infty} \cap L^2$ for which a solution $u \in L^{\infty}$ of equation (4) does not belong to L^2.

Remark 1. Equation (4) is of interest in various applications; in particular, it is used as a model equation in quantum field theory. The technique given in proof of theorem 1 can be used without any alterations to study the following equation:

$$Lu + f(u) = J(x), \quad x \in R^d \tag{5}$$

where L is a self-adjoint positive definite elliptic operator of the second order, $Lu = -\sum_{i,j=1}^{d} \partial_i (a_{ij}(x) \partial_j u) + c(x)$, $c(x) \geq d > 0$, $a_{ij}(x) = \delta_{ij}$ for $|x| > R$, $0 < a_1 |t|^2 \leq \sum_{i,j=1}^{d} a_{ij}(x) t_i \bar{t}_j \leq a_2 |t|^2$, $0 < a_1 \leq a_2$, $a_{ij}(x) \in c_1(R^d)$, $c(x) \in L^{\infty}$.

V.2. *Proof.*

The outline of the proof is the following. First we prove that for any uniformly bounded nonlinearity $f \in C_{loc}$ the integral equation

$$u = -\int_{-\infty}^{\infty} \frac{\exp(-\kappa|x-y|)}{2\kappa} f(u(y))dy + h(x), \quad x \in I, \tag{6}$$

where $h(x) \equiv \int_{-\infty}^{\infty} \frac{exp(-\kappa|x-y|)}{2\kappa} J(y)dy$, which is equivalent in L^{∞} to equation (4), is solvable in L^{∞}. Actually solutions of equation (6) belong to $C(R)$. Secondly, we prove that any solution of (4) satisfies the estimate

$$|u|_{L^{\infty}} \leq |J|_{L^{\infty}} \kappa^{-2} + R \leq c. \tag{7}$$

If both above statements are proved then we can complete the proof as follows. Let us take $f_c = f(u)$ if $|u| \leq c$, $f_c = \psi(u)$ if $|u| > c$, where $\psi(u)$ is a uniformly bounded function so chosen that $f_c \in C_{loc}$. For $f = f_c$ equation (6) is solvable in L^{∞}. But because of (7) this solution will be a solution of equation (6) with the original nonlinearity. It remains to prove the above statements.

Lemma 1. If $|f(u)| \leq \mu$, $\forall u \in I$ and $f \in C_{loc}$ then equation (6) is solvable in $C(R)$. Proof. Let

$$\eta_n(x) \equiv \begin{cases} 1, & |x| \leq n \\ 0, & |x| \geq n+1 \end{cases} \quad \text{and} \quad |\eta_n'| \leq 2. \tag{8}$$

Consider the equation

$$u_n = -\eta_n(x) \int_{-\infty}^{\infty} \frac{exp(-\kappa|x-y|)}{2\kappa} f(u_n(y))dy + \eta_n h(x), \; x \in I. \tag{9}$$

This equation is solvable in $C(R)$. This follows from the estimate

$$|u_n| \leq \mu_1,$$

where μ_1 is some constant which does not depend on n and x, and from Schauder's fixed point theorem, if we take into account that the operator on the right hand side of (9) is compact. From (8) and (9) it follows that the family $\{u_n(x)\}$ is equicontinuous. Thus we can choose a subsequence, called again $u_n(t)$, such that

$$u_n(x) \to u(x) \text{ as } n \to \infty ,$$

where the arrow means uniform convergence on any bounded interval. It is clear that $|u(x)| \le \mu_1$, $u(x) \in C_{loc}$. We complete the proof by passing to the limit in equation (9) and using Lebesgue's theorem.

Lemma 2. Any solution of equation (4) in L^∞ satisfies inequality (7).

Proof. Let $u^+ = \begin{cases} u, & u \ge 0 \\ 0, & u < 0 \end{cases}$. Let $N = R + |J|_{L^\infty} \kappa^{-2}$. We have

$$-u''(u-N)^+ + \kappa^2(u-N)^+(u-N) + f(u)(u-N)^+ = (J-\kappa^2 N)(u-N)^+.$$

If $u > N$ then $f(u)u \ge 0$, $f(u)(u-N)^+ = f(u)u - f(u)N \ge 0$. Suppose that $u \ge N$ on some interval (a,b), $u(a) = u(b) = N$. Then

$$\kappa^2(u-N)^2 - u''(u-N)^+ \le 0 \quad a < x < b. \tag{10}$$

Thus

$$\kappa^2 \int_a^b (u-N)^2 dx + \int_a^b |u'|^2 dx \le 0 . \tag{11}$$

From (10) it follows that $u \le N$ on (a,b). Here we took into account that any solution $u \in L^\infty$ of equation (4) belongs actually to $C_{1\ell oc}$. If $u > N$ for $a < x < \infty$ (or $u > N$ on I) then u' either change sign on $a < x < \infty$, or $\lim_{x \to +\infty} u'(x) = 0$. Thus a number $b > a$ exists such that either $u'(b) = 0$ or $u'(b) - u'(a) \le 0$. In both cases, integrating (10) over (a,b) we get (11) and conclude that $u \le N$. The inequality $u \ge -N$ can be proved similarly (or one can use the above proof for $v = -u$).

Lemma 3. There exists $J \in L^\infty \cap L^2$ for which a solution $u \in L^\infty$ of equation (4) does not belong to L^2.

Proof. First we note that if $uf(u) \ge 0$ for all $u \in I$ then $J \in L^\infty \cap L^2$ implies $u \in L^\infty \cap L^2$. Thus a new property of solutions of equation (4) is caused by the nonmonotonicity of $f(u)$. Consider $f(u) = -\kappa^2 u$ for $|u| \le c$, where c is the constant in (7), $f(u) = au$, $a > 0$ for $|u| \geq c+1$ and $f(u) \in C_{loc}$. Then for $|u|_{L^\infty} \leq c$ equation (4) takes the following form:

$$-u'' = J(x). \tag{12}$$

We have the formula for the general solution of (12):

$$u(x) = C_0 + C_1 x + \int_0^x (x-t)J(t)dt = C_0 + C_1 x + x\int_0^x J(t)dt$$

$$- \int_0^x t\, J(t)dt. \tag{13}$$

Let us take $J(t)$ such that

$$\int_0^\infty J(t)dt = A < \infty, \int_0^{-\infty} J(t)dt = A, \tag{14}$$

$$\left|x\int_x^\infty J(t)dt\right| \leq B, \left|\int_0^{\pm\infty} tJ(t)dt\right| \leq B. \tag{15}$$

Then, taking $C_1 = -A$ in (13), we get by formula (13) a solution $u \in L^\infty$ of equation (4) which does not belong to L^2. For example, we can take $J = \frac{\sin^5 t}{t^4}$ in order to satisfy (14) and (15).

From Lemmas 1-3 the conclusion of theorem 1 follows.

VI. DISCONTINUOUS NONLINEARITIES

If the nonlinearity Fu is not continuous the results given in theorems 1,2 Section II are not valid. We describe another iterative process for solution of equation (12.1). Let us assume that $(Au,u) \geq d(u,u)$, Fu is bounded and monotone and rewrite equation (12.1) as

$$Tu = u + A^{-1} Fu - e(t) = 0. \tag{1}$$

The operator T is defined on all H_A (H_A is the Hilbert space with the inner product

$$[u,v] = (A^{\frac{1}{2}}u, A^{\frac{1}{2}}v))$$

and

$$[Tu-Tv,u-v] = (ATu-ATv,u-v) = [u-v,u-v] + (Fu-Fv,u-v)$$
$$\geq [u-v,u-v]. \tag{2}$$

We can use now the following theorem (see [18]).

Theorem. Suppose that $T:H \to H$ satisfies inequality $(Tu-Tv,u-v) \geq \gamma \, ||u-v||^2$. Then the iterative process

$$u_{n+1} = u_n - \beta_n h_n, \quad h_n = \frac{Tu_n}{||Tu_n||}, \quad \beta_n = min\left(\rho_n, \frac{d_n}{2\rho_n}\right),$$

$$d_{n+1} = (1 - q_n)d_n, \quad q_n = (1 - \beta_n\rho_n d_n^{-1})\beta_n\rho_n^{-1}, \tag{3}$$

$$d_1 = \frac{||Tu_1||}{\gamma}, \quad \rho_n = \frac{||Tu_n||}{2\gamma}, \quad n > 1$$

converges in H to an element u, $||u-u_n|| \leq d_n^{\frac{1}{2}} = 0(n^{-\frac{1}{2}})$ and u is the unique generalized solution of equation $Tu = 0$. Element u is called a generalized solution of equation $Tu = 0$ if $(Tv,v-u) \geq \gamma \, ||v-u||^2$, $\forall v \in H$.

We omit the proof of this theorem and explain how to calculate by formulas (3). We take an arbitrary $u_1 \in H$, find ρ_1, d_1 then β_1, q_1, d_2, h_1 and u_2. This is the first step of calculations. Then we find $\rho_2, \beta_2, q_2, h_2, d_3, u_3$ and so on.

VII. AN EXAMPLE: STATIONARY REGIME IN A NONLINEAR FEEDBACK AMPLIFIER

Consider the amplifier in Fig. 2. Using the theorem about equivalent generator (Thevenin theorem) we pass over to the circuit in Fig. 3, where e_e is the equivalent electromotive force, and Z_e is the equivalent impedance. The theory presented in the previous sections can be applied to the circuit in Fig. 3.

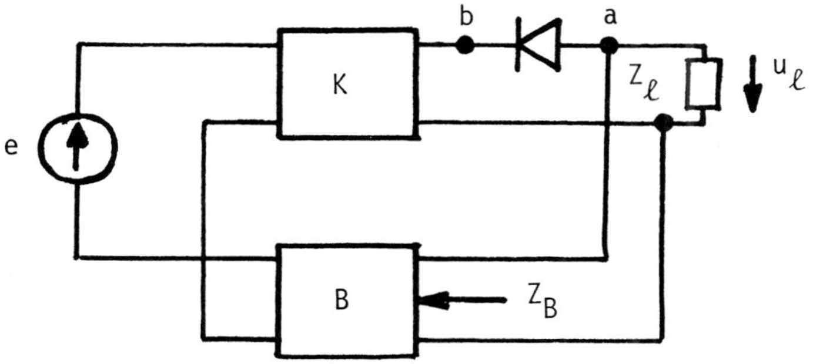

FIGURE 2. Nonlinear feedback amplifier.

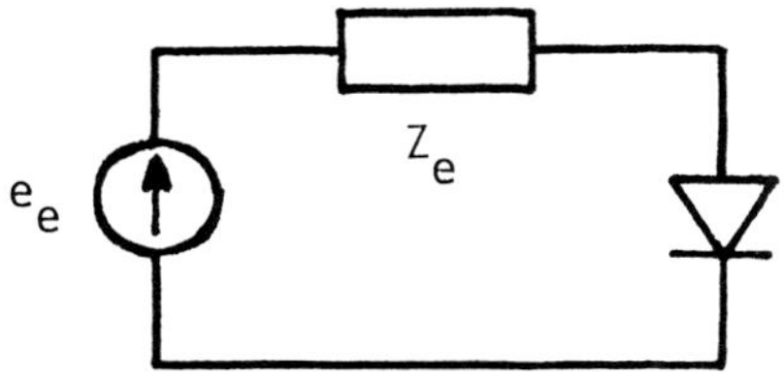

FIGURE 3. Equivalent circuit.

If the characteristic of the nonlinearity satisfies conditions (8.2), (9.2), (10.2), then in the network in Fig. 2, to insure stability of the periodic regime in the large and stability towards constantly acting periodic perturbations it is sufficient that $Re\ Z_e > 0$. This condition is easy to verify in practice.

VIII. BIBLIOGRAPHICAL NOTE

Good guides to the literature are references [1,2]. Linear network analysis is given in references [16,17]. Mathematical theory of nonlinear operator equations and applications is given in references [13,14]. Oscillations in nonlinear systems are studied in [3,4]. The theory presented in this chapter is due to the author [5-11]. Although the list of references is incomplete, the reader can find many further references in [1,2,13,14]. The example given in Section VII will be discussed in detail in a forthcoming paper by G.S. Ramm and the author.

REFERENCES

1. Aggarwal, J. and Vidyasager, U., "Nonlinear Systems" Dowden, Pennsylvania, (1977).
2. Hsu, J. and Meyer, A., "Modern Control Principles and Applications" McGraw-Hill, New York, (]968).
3. Hale, J., "Oscillations in Nonlinear Systems" McGraw-Hill, New York, (1963).
4. Bogoljubov, N. and Mitropolsky, Yu, "Asymptotic Methods in Nonlinear Oscillations Theory" Gordon and Breach, New York, (1967).
5. Ramm, A.G., "Existence of periodic solutions to some non-linear problems," *Diff. Eqs. 13*, 1186-1191 (1977).
6. Ramm, A.G., "Stability of control systems," *Diff. Eqs. 14*, 1188-1193 (1978).
7. Ramm, A.G., "An iterative process for calculation of periodic and almost periodic oscillations in some nonlinear systems," *Radio. Engr. Electr. Phys. 21*, 137-140 (1976); *24*, 190-191 (1979).
8. Ramm, A.G., "Investigation of some classes of integral equations and their applications," *in* "Abel Inversion and its Generalization" (N. Preobrazhensky, ed.), pp. 120-179. *Sib. Dept. of Acad. Sci. USSR*, Novosibirsk (1978).
9. Ramm, A.G., "Existence uniqueness and stability of solutions to some nonlinear problems," *Proc. Int. Congr. on Appl. Math. in Engineering*, Weimar, 345-351 (1978).
10. Ramm, A.G., "Existence uniqueness and stability of periodic regimes in nonlinear networks," *Proc. 3rd Int. Symp. on Network Theory*, Split, 623-628 (1975).
11. Ramm, A.G., "Investigation of Some Classes of Integral Equations and their Applications" Springer, New York, (1980).

12. Kato, T., "Perturbation Theory for Linear Operators" Springer-Verlag, New York, (1966).
13. Lions, J., "Quelques Methodes de Resolution des Problemes aux Limites Non Lineares" Dunod, Paris, (1959).
14. Krasnoselsky, M. and Zabreiko, P., "Geometrical Methods in Nonlinear Analysis" (in Russian), Nauka, Moscow, (1975).
15. Lefschetz, S., "Stability of Nonlinear Control Systems" Academic Press, New York, (1965).
16. Kontorovich, M., "Operational Calculus and Processes in Electrical Circuits" (in Russian), Nauka, Moscow, (1964).
17. Seshu, S. and Balabanian, N., "Linear Network Analysis" Wiley, New York, (1964).
18. Perov, A. and Jurgelas, J., "On convergence of an iterative process," *J. Vycisl. Math. Phys. 17*, 859-870 (1977).

NONLINEARLY LOADED ANTENNAS[1]

Giorgio Franceschetti[2]
Innocenzo Pinto[3]

Istituto Elettrotecnico
Università di Napoli
Italy

Nonlinearly loaded antennas, e.g., wire, loop or spiral antennas connected to a nonlinear load, e.g., a diode, are considered. The nonlinear model of the load is analyzed, and it is then shown that, under suitable assumptions, computation of the voltage across the load is reduced to the solution of a Volterra-type nonlinear integral equation. The well known Volterra series solution is reviewed and, as a new contribution, convergence properties and truncation error are (heuristically) discussed. Furthermore, new and known numerical methods are presented, together with some computational examples.

I. INTRODUCTION

The problem of nonlinearly loaded antennas, (*NLA*), e.g., wire, loop or spiral antennas whose terminals are connected to a diode, exhibits both a practical and an academic interest.

[1] *Work supported in part by the Italian National Research Council (CNR), and in part by the U.S. Air Force Office of Scientific Research under grant AFOSR-77-3253.*

[2] *Also; Communications Laboratory, University of Illinois at Chicago Circle, Chicago, Illinois, USA.*

[3] *Also; Istituto Universitario Navale, Napoli, Italy.*

ISBN 0-12-709660-4

Diode loaded small probes implanted in the human body to provide non invasive electrical stimulation [1,2], harmonic radar applications, e.g., for lost people rescuing and avalanche victims search [3,4], are some examples of practical motivations. The same nonlinear problem arises in connection with spurious harmonic emission from contacts between different conductors, e.g., in antenna reflectors with weldings, rivets or nails, after aging and/or oxidation [5,6].

The academic interest stems out from the fact that the nonlinearly loaded antenna is probably the simplest E.M. boundary value nonlinear problem, wherein the nonlinearity is locally concentrated. The next problem, in increasing difficulty, would be that of an antenna whose boundary conditions are nonlinear, so that a distributed nonlinearity, instead of a pointwise one, should be considered [7].

Before proceeding to examine techniques for studying *NLA*, a remark seems in order. Let us consider the simple case of a dipole antenna whose terminals are connected to a diode (fig. 1). Being the nonlinearity squeeze across the gap, it seems reasonable and convenient to cast the problem in a circuit form, therefore replacing the antenna with its equivalent circuit. This requires, however, that the equivalent circuit be properly defined, which postulates, in turn, that the boundary-value problem associated with the gap of the antenna is rigorously solved. The solution is known, for some idealized geometries, under the assumption that the gap is filled with some linear material [8]. Accordingly, in order to replace the antenna with its equivalent circuit, one should remove the nonlinear load from the gap region, e.g., inserting a transmission line section between the antenna and the load itself. (Obviously we are tacitly assuming that the equivalent circuits of the line terminations, the one containing the diode mount, and the other joining to the antenna terminas, are known [9,10].)

FIGURE 1. Diode loaded dipole antenna.

Another limitation of the equivalent circuit approach is that it gives, in general, incomplete information about scattered power and fields. In the simple but relevant case of a sinusoidal incident field, however, the total power radiated on the harmonics may be correctly identified with the power dissipated into the equivalent circuit antenna resistance.

With this in mind, the problem in fig. 1 is changed into the equivalent circuit problems depicted in fig. 2, upon use of Thévenin's or Norton's theorem at the load terminals; $\hat{I}_g$ and $\hat{V}_g$ in the figure are the short circuit current and open circuit voltage, respectively, of the equivalent generator; $\hat{Y}(\omega)$ and $\hat{Z}(\omega)$ are its frequency domain input admittance and impedance, and the nonlinear load is described by some known nonlinear functional, say:

$$i_d = F(v_d) \; ; \; v_d = \Phi(i_d) \tag{1.1}$$

respectively.

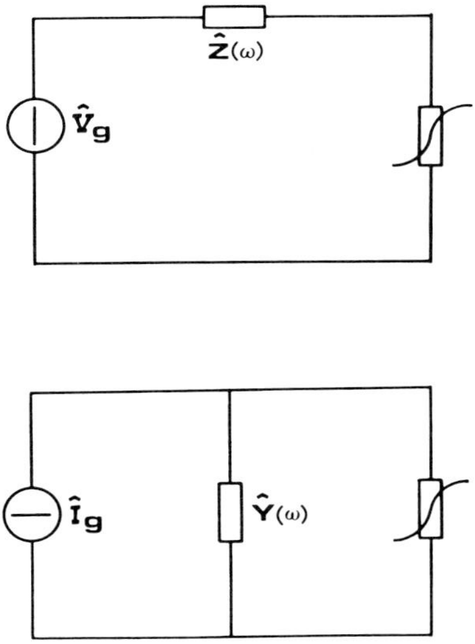

FIGURE 2. Equivalent circuit of nonlinearly loaded antenna (Thévenin and Norton forms).

As shown in the following Section II, the circuit model of the load is composed of nonlinear and linear elements as well; it is convenient to incorporate the latter within $\hat{Y}$ or $\hat{Z}$, so that the functional relationships (1.1) are representative of the nonlinear part only of the equivalent circuit.

A number of techniques are available for tackling the *NLA* problem. They can be broadly grouped in two categories: analytical and numerical methods.

The former are mostly based on a Volterra series solution [11-16] of the circuit equations:

$$v_g(t) = v_d(t) + \int_{-\infty}^{\infty} z(t-\tau)F(v_d(\tau))\,d\tau \tag{1.2}$$

$$v_g(t) = \Phi(i_d(t)) + \int_{-\infty}^{\infty} z(t-\tau)\, i_d(\tau)\, d\tau \qquad (1.3)$$

(Thévenin's equivalent circuit),

$$i_g(t) = F(v_d(t)) + \int_{-\infty}^{\infty} y(t-\tau)\, v_d(\tau)\, d\tau \qquad (1.4)$$

$$i_g(t) = i_d(t) + \int_{-\infty}^{\infty} y(t-\tau)\, \Phi(i_d(\tau))\, d\tau \qquad (1.5)$$

(Norton's equivalent circuit),

wherein $y(t)$ and $z(t)$ are the responses of the linear part of the circuit to a unit-pulse voltage or current, i.e., the inverse Fourier transforms of $\hat{Y}(\omega)$ and $\hat{Z}(\omega)$, respectively.

Several numerical techniques exist, on the other hand, for solving the *NLA* problem, both in its field (method of moments [17-20]) and circuit formation [21-23]. More recently, a piece-wise harmonic balance technique has been applied, which seems particularly suitable for the large signal steady-state case [24,25].

After the nonlinear load modelling problem has been discussed in Section II, all these techniques will be reviewed and compared.

II. NONLINEARITY MODELLING

It is quite important to find a good compromise between accuracy and simplicity in modelling real nonlinear loads, in order to solve *NLA* problems successfully.

Some authors do assume [12,14,15]:

$$i_d = g_1\, v_d + g_3\, v_d^3 \qquad (2.1)$$

as the i - v relationship of a back-to-back diode pair, thus neglecting nonlinear capacitive effects. This is justified, however, only at low frequencies (< 1 MHz). Then some value is given to the constants g_1 and g_3 so as to match computed and measured results.

Other authors define some ideal nonlinear element, usually memoryless, without reference to any specific existing device.

The $i-v$ relationship of metal-oxide-metal contacts (occurring, e.g., in aged rivets or nails) has been thoroughly investigated [26,27]. For $Al - Al_2O_3 - Al$ contacts at small voltages, eq. (2.1) holds, with [5,28]:[4]

$$\left.\begin{aligned} g_1 &= G_o\left(\frac{a\phi^{1/2}}{2} - 1\right) \\ g_3 &= G_o\left(\frac{q^2a^3\phi^{-1/2}}{192} - \frac{q^2a^2\phi^{-1}}{64} - \frac{q^2a\phi^{-3/2}}{64}\right) \\ a &= \frac{2\delta}{h}(2m)^{1/2} \end{aligned}\right\} \qquad (2.2)$$

wherein ϕ is the work function of the metal, q and m are the electron charge and mass, respectively, h is Planck's constant, δ is the oxide layer thickness, and G_o is a constant conductance, whose value depends on contact's geometry. Experimental verifications of (2.2) are reported in [29].

A commonly accepted equivalent circuit for semiconductor diodes is the one depicted in fig. 3, wherein C_p is the package capacitance (typically, some tenths of pF); L_s is the lead inductance (typically, some tenths of nH); R_s is the overall lead plus bulk resistance (typically, some fractions of Ω); R_n and C_n are the nonlinear resistance and capacitance of the junction, respectively [30]. Note that carrier inertia and skin effect result in a complex, frequency dependent Z_s in place of R_s at small wavelengths [31-32].

For usual *pn*, point contact and Schottky barrier hot-carrier (*HC*) diodes, the nonlinear resistance is (implicitly) described

[4] *Note the error in eqs. (18) and (22) of [5].*

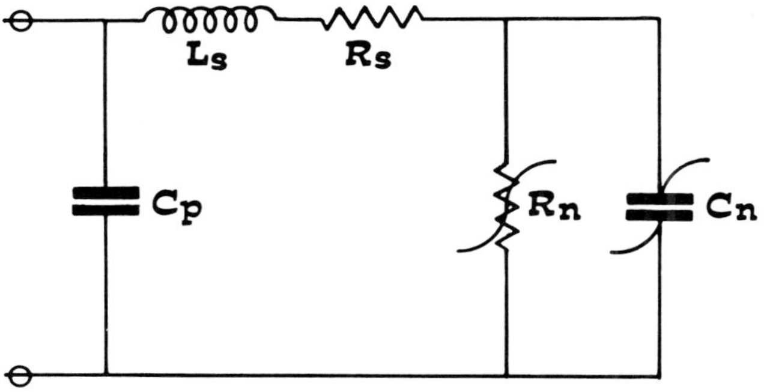

FIGURE 3. Semiconductor diode small signal equivalent circuit.

by the well known (static) diode law [33-34]:

$$i = I_S(exp\ (v/nV_T) - 1) \qquad (2.3)$$

wherein I_S is the reverse saturation current (ranging from some nA to some μA, depending on technology), which roughly doubles for every 10°K rise in temperature, $V_T \simeq T(°K)/11{,}600$ volt, and n is an ideality factor, accounting semi-empirically for several effects ($n \simeq 1$ for HC diodes; n in the range $1 \div 2$ for pn junctions). As we depart from the static condition, the junction conductance is still described by (2.3), wherein I_S turns out to be fairly independent of frequency in HC diodes, and nearly inversely proportional to the square root of frequency in pn junctions [41].

At large reverse voltages, where Zener and/or avalanche breakdown occurs, an additional Miller factor [35] should be inserted in (2.3) yielding:

$$i_d \simeq I_s(1 - (v/V_B)^m)^{-1} \qquad (2.4)$$

wherein V_B is the junction breakdown voltage (typically in the range -1 to -100 volt, depending on technology [36] and temperature), and m is an *ad hoc* Miller exponent (usually in the range 2 - 5). A more rigorous expression than (2.4), avoiding the empirical determination of m, has been given in [37].

The junction nonlinear capacitance consists of two main contributions: depletion layer capacitance and minority carrier diffusion capacitance. The former is modelled by:

$$C(v) = C_o(1 - v/V_\phi)^{-q} \tag{2.5}$$

wherein C_o is typically some pF, V_ϕ is the built-in junction barrier potential (0.4 - 0.9 volt, typically) and $q = 0.2 - 7$, depending on technology [38], slightly on the applied voltage [39], little or nothing on frequency [40]. Minority carrier diffusion capacitance, on the other hand, is nearly inversely proportional to the square-root of frequency [41], and may be safely neglected, of course, in majority carrier devices (e.g., *HC* and point contact diodes).

At moderate voltage excursions, a series expansion of (2.3) and (2.5) is appropriate, hence:

$$G_n(v) = R_n^{-1}(v) = G_o + G_1\left(\frac{v}{V_T}\right) + G_2\left(\frac{v}{V_T}\right)^2 + \dots \tag{2.6}$$

$$C_n(v) = C_o + C_1\left(\frac{v}{V_\phi}\right) + C_2\left(\frac{v}{V_\phi}\right)^2 + \dots \tag{2.7}$$

wherein:

$$G_o = \frac{I_s}{nV_T} \; ; \; G_1 = \frac{I_s}{2n^2 V_T} \; ; \; G_2 = \frac{I_s}{6n^3 V_T} \; ; \; \dots \tag{2.8}$$

$$C_1 = -qC_o \; ; \; C_2 = \frac{q(q+1)}{2} C_o \; ; \; \dots \tag{2.9}$$

The linear terms G_1, C_1 can be absorbed in the linear part of the circuit. Then, the current in the (pure) nonlinear part of the diode is given by:

$$i_d = \sum_{2}^{\infty}{}_m \gamma_m v_d^m + \sum_{2}^{\infty}{}_m \xi_m \frac{d}{dt} v_d^m \qquad (2.10)$$

wherein v_d is the voltage across the diode, and:

$$\gamma_m = \frac{G_m}{V_T^{m-1}} \quad ; \quad \xi_m = \frac{C_m}{V_\phi^{m-1}} \quad . \qquad (2.11)$$

Note that diffusion capacitance has been ignored in writing (2.10), since only majority carrier devices will be considered hereafter.

It should be stressed that the nonlinear capacitive current cannot in general be disregarded. In a typical diode ($I_S = 10\, nA$; $n = 1$; $V_T = 52\ mV$; $C_o = 1\ pF$; $V_\phi = 0.5V$; $q = 0.5$), the first terms ($m = 2$) of the two series in (2.10) are equal, for a sinusoidal applied voltage, at a frequency $f \simeq 150$ KHz. Accordingly, a pure memoryless diode model, as adopted in [12,14,15,19,20,22,23] turns out to be a too crude approximation.

Equations (2.3) and (2.5) are not adequate for describing special devices as, e.g., tunnel [42], backward [43] p-i-n [44] and MIS (45) diodes.

The typical $i - v$ and $C - v$ relationships of backward diodes and MIS varactors, respectively, are shown in figs. 4 and 5. Note that strong nonlinear behaviour should be expected even in the small signal range, rendering these devices very attractive, whenever use is made of nonlinearity. Though no simple theory exists relating their observed characteristics to physical processes, the former are easily modelled, both in the large (piecewise linear approximations) and in the small signal range (polynomial approximations).

Possibly, distinct nonlinear (quadratic) behaviour in the extremely low voltage range could be obtained by space charge limited diodes (*SCLD*) [47].

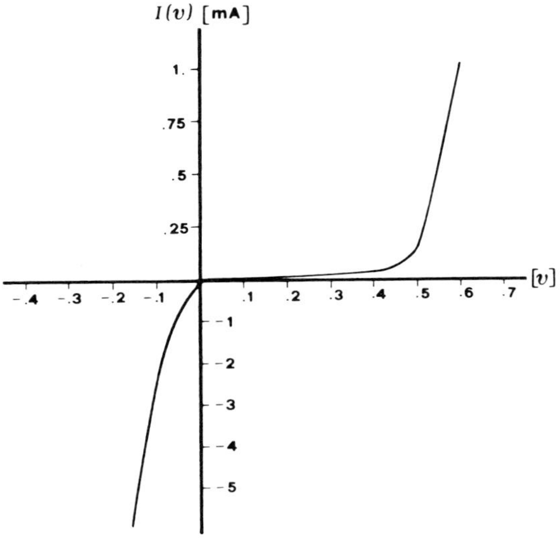

FIGURE 4. Current vs. voltage characteristic of typical backward diode.

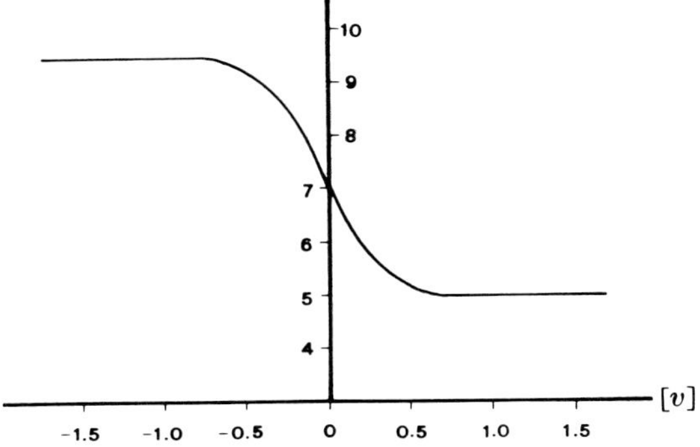

FIGURE 5. Capacitance vs. voltage characteristic of typical MIS varactor diode.

II. VOLTERRA SERIES APPROACH

Let us consider an antenna connected to a nonlinear load, e.g., the one depicted in fig. 1. As discussed in Section I, and with limitations therein pointed out, the problem may be reduced to the study of one of the equivalent circuits of fig. 2. Since the explicit expression of the nonlinear functional relationship $i_d = F(v_d)$ is known (see eq. 2.10), it is convenient to utilize eqs. (1.2 or 4), wherein $v_g(t)$ and $i_g(t)$ are the open circuit voltage and short circuit current of the equivalent generator, respectively; $z(t)$ and $y(t)$ are the inverse Fourier transforms of its (linear) impedance $\hat{Z}(\omega)$ and admittance $\hat{Y}(\omega)$.

Assuming $v_d(t)$ to be a continuous stationary nonlinear functional of $i_g(t)$, the following convergent Volterra series expansion holds [47]:

$$v_d(t) = \sum_1^\infty {}_k \quad v_k(t) \tag{3.1}$$

$$v_k(t) = \int_{-\infty}^{\infty} d\tau_1 \ldots \int_{-\infty}^{\infty} d\tau_k \, h_k(\tau_1, \ldots, \tau_k) \prod_1^k {}_i \; i_g(t-\tau_i) \tag{3.2}$$

The k-fold convolutional nature of (3.2) suggests the convenient k-fold Fourier transform representation:

$$\hat{V}_k(\omega_1, \ldots, \omega_k) = \hat{H}_k(\omega_1, \ldots, \omega_k) \prod_1^k {}_i \; \hat{I}_g(\omega_i) \tag{3.3}$$

wherein $\hat{I}_g(\omega)$ is the Fourier transform of $i_g(t)$, and the Volterra kernel:

$$\hat{H}_k(\omega_1, \ldots, \omega_k) = \int_{-\infty}^{\infty} dt_1 \ldots \int_{-\infty}^{\infty} dt_k \; h_k(t_1, \ldots, t_k)$$

$$\exp(-j\omega_1 t_1 - j\omega_2 t_2 + \ldots - j\omega_k t_k) \tag{3.4}$$

is the k-fold Fourier transform of the generalized (in the Wiener sense [48]) k-th order impulse response function, $h_k(t_1,\ldots,t_k)$.

Before presenting a procedure for the computation of the general kernel $h_k(\cdot)$ appearing in (3.2), or, equivalently, its Fourier transform (3.4), let us remark that the series (3.1) can be obtained as a convenient rearrangement of the iterative solution of eq. (1.2) -- Picard's method [49] -- obtained by collecting all and only terms of k-degree with respect to the peak amplitude of $i_g(t)$ from the given iterated solution, to form $v_k(t)$. Accordingly, it is instructive to derive the first two kernels, $h_1(t)$, $h_2(t_1,t_2)$, for the simple case of a quadratic nonlinear operator:

$$F(v_d(t)) = \gamma\, v_d^2 + \xi \frac{d}{dt} v_d^2 \tag{3.5}$$

modelling, e.g., a diode load, under the assumption of very small signals (see eqs. 2.10,11, where higher order terms have been neglected). Note that the nonlinear capacitance of the diode is taken into accoutn, so that a "memory-type" of nonlinearity is being considered.

Equation (1.2) is written as follows, with formal changes of symbols:

$$v_d(t) = v_g(t) - L\{N(v_d(t))\} \tag{3.6}$$

wherein L is a linear operator:

$$L(x(t)) = z(t) * x(t) = \int_{-\infty}^{\infty} z(t-\tau)x(\tau)d\tau\,, \tag{3.7}$$

$$z(t) = \frac{1}{2\pi} \int_{-\infty}^{\infty} \hat{z}(\omega)\; exp\; j\omega t\; d\omega\,, \tag{3.8}$$

$\hat{z}(\omega)$ being the (linear) impedance of the equivalent generator; and N is the nonlinear operator (3.5):

$$N(x(t)) = F(x(t))\,. \tag{3.9}$$

The first approximation to the unknown load voltage $v_d(t)$, say $v_d^{(1)}(t)$, is obtained just neglecting the nonlinear part of the load, i.e.:

$$v_d^{(1)}(t) = v_g(t) = \int_{-\infty}^{\infty} z(t-\tau) i_g(\tau) d\tau , \tag{3.10}$$

whence:

$$h_1(t) = z(t) . \tag{3.11}$$

The second approximation, say $v_d^{(2)}(t)$, is obtained by taking $v_d^{(1)}(t)$ as the voltage across the nonlinear load, hence:

$$v_d^{(2)}(t) = v_g(t) - L\{N(v_d^{(1)}(t))\} \tag{3.12}$$

wherein:

$$N(v_d^{(1)}(t)) = \left(\gamma + \xi \frac{d}{dt}\right) \int_{-\infty}^{\infty} z(t-\tau_1) i_g(\tau_1) d\tau_1 \times \int_{-\infty}^{\infty} z(t-\tau_2) i_g(\tau_2) d\tau_2 . \tag{3.13}$$

Hence:

$$L\{N(v_d^{(1)}(t))\} = \int_{-\infty}^{\infty} d\theta \; z(t-\theta) \int_{-\infty}^{\infty} \int_{-\infty}^{\infty} \left(\gamma + \xi \frac{d}{d\theta}\right) \times \{z(\theta-\tau_1) z(\theta-\tau_2) i_g(\tau_1) i_g(\tau_2)\} d\tau_1 \, d\tau_2 . \tag{3.14}$$

Interchanging the order of integration, and using the new variables:

$$t - \theta = \alpha \; ; \; \theta - \tau_1 = t - \tau_1 - \alpha \; ; \; \theta - \tau_2 = t - \tau_2 - \alpha$$

we get:

$$L\{N(v_d^{(1)}(t))\} = -\int_{-\infty}^{\infty}\int_{-\infty}^{\infty} h_2(t-\tau_1, t-\tau_2) \times i_g(\tau_1) i_g(\tau_2) d\tau_1 d\tau_2 \tag{3.15}$$

wherein:

$$h_2(t_1, t_2) = -\int_{-\infty}^{\infty} d\alpha\{z(\alpha)(\gamma\, z(t_1 - \alpha) z(t_2 - \alpha) - \xi \frac{d}{d\alpha} z(t_1 - \alpha) z(t_2 - \alpha))\} \tag{3.16}$$

Then, integrating by parts,

$$\int_{-\infty}^{\infty} d\alpha\, z(\alpha) \frac{d}{d\alpha} z(t_1 - \alpha) z(t_2 - \alpha) = z(\alpha) z(t_1 - \alpha) z(t_2 - \alpha) \Big|_{-\infty}^{\infty} + -\int_{-\infty}^{\infty} \frac{dz}{d\alpha} z(t_1 - \alpha) z(t_2 - \alpha) d\alpha \tag{3.17}$$

Since $z(\pm\infty) = 0$, we have finally:

$$h_2(t_1, t_2) = -\int_{-\infty}^{\infty} z(t_1 - \alpha) z(t_2 - \alpha) \left(\gamma z(\alpha) + \xi \frac{dz}{d\alpha}\right) d\alpha \tag{3.18}$$

or, by double Fourier transformation,

$$\begin{aligned} \hat{H}_2(\omega_1, \omega_2) &= \int_{-\infty}^{\infty}\int_{-\infty}^{\infty} h_2(\tau_1, \tau_2)\, exp(-j\omega_1\tau_1 - j\omega_2\tau_2) d\tau_1 d\tau_2 \\ &= -\hat{Z}(\omega_1)\hat{Z}(\omega_2) \int_{-\infty}^{\infty} \{\gamma\, z(\alpha) + \xi \frac{dz}{d\alpha}\} exp(-j(\omega_1 + \omega_2)\alpha) d\alpha \\ &= -(\gamma + j(\omega_1 + \omega_2)\xi)\hat{Z}(\omega_1)\hat{Z}(\omega_2)\hat{Z}(\omega_1 + \omega_2) \end{aligned} \tag{3.19}$$

The iterative procedure can be continued, and then the terms rearranged to obtain the Volterra series (3.1) and kernels (3.4).

Note that N and only N iterations are required to get the complete expression of all the series terms $v_k(t)$, $k = 1,2,\ldots,N$.

However, it is more convenient to compute directly each kernel as follows. The method will be illustrated taking $\hat{H}_3(\omega_1,\omega_2,\omega_3)$ as an example.

First of all we will assume that each kernel is a completely symmetric function of its arguments ω_j, i.e.:

$$\hat{H}_3(\omega_1,\omega_2,\omega_3) = \hat{H}_3(\omega_2,\omega_1,\omega_3) = \hat{H}_3(\omega_3,\omega_1,\omega_2) =$$

$$\hat{H}_3(\omega_1,\omega_3,\omega_2) = \hat{H}_3(\omega_2,\omega_3,\omega_1) = \hat{H}_3(\omega_3,\omega_2,\omega_1) \tag{3.20}$$

which turns out to be a self-consistent *ansatz* [50]. Then we will assume an analytic input:

$$i_g(t) = exp(j\omega_1 t) + exp(j\omega_2 t) + exp(j\omega_3 t) , \tag{3.21}$$

$$\hat{I}_g(\Omega) = 2\pi(\delta(\Omega - \omega_1) + \delta(\Omega - \omega_2) + \delta(\Omega - \omega_3)) , \tag{3.22}$$

wherein the angular frequencies, ω_i, are assumed strictly positive and linearly independent. Accordingly, the k-order term of the response $v_k(t)$ is given by:

$$v_k(t) = (2\pi)^{-k}\int_{-\infty}^{\infty} d\Omega_1 \ldots \int_{-\infty}^{\infty} d\Omega_k \, \hat{I}_g(\Omega_1) \ldots \hat{I}_g(\Omega_k)$$

$$\times \hat{H}_k(\Omega_1,\ldots,\Omega_k) exp\left[j(\Omega_1 + \ldots + \Omega_k)t\right] \tag{3.23}$$

and the total response, $v_d(t)$, may be written as:

$$v_d(t) = \sum_1^{\infty}{}_k \, v_k(t) = \sum_1^{3}{}_i \, \hat{Z}(\omega_i) exp\ j\omega_i t$$

$$+ 2!\{\hat{H}_2(\omega_1,\omega_2) exp\ j(\omega_1 + \omega_2)t + \hat{H}_2(\omega_1,\omega_3) exp\ j(\omega_1 + \omega_3)t$$

$$+ \hat{H}_2(\omega_2,\omega_3) exp\ j(\omega_2 + \omega_3)t\} + 3!\ \hat{H}_3(\omega_1,\omega_2,\omega_3)$$

$$exp\ j(\omega_1 + \omega_2 + \omega_3)t + \text{"extra terms"} \tag{3.24}$$

wherein the "*extra terms*" are those having at least two equal arguments, ω_i, each. Note that factorials account for terms differing but for arguments permutation, according to (3.20).

Expansion (3.24) and (3.21) are now inserted into eq. (1.4), wherein the nonlinear operator $N(v_d(t))$ may be specified in terms of a shunt connection of nonlinear conductance and capacitance:

$$i_g(t) = \int_{-\infty}^{\infty} y(t-\tau)v_d(\tau)d\tau + \sum_2^{\infty}{}_n \left(\gamma_n + \xi_n \frac{d}{dt}\right)v_d^n(t) \; . \tag{3.25}$$

Next, we single out and equate those terms whose time dependence is of type *exp* $j(\omega_1 + \omega_2 + \omega_3)t$, on both sides of (3.25).

The left hand side contributes with no such a term. The linear term at the right hand side generates:

$$3!\; \hat{H}_3(\omega_1,\omega_2,\omega_3)\; \hat{Y}(\omega_1 + \omega_2 + \omega_3) \tag{3.26}$$

whereas the nonlinear operator produces several terms containing the *exp* $j(\omega_1 + \omega_2 + \omega_3)t$ factor. These terms, due to the assumed strict positivity and linear independence of the angular frequencies, ω_i, may be easily laid down. The resulting equation:

$$\begin{aligned} 0 = 3!\; &\hat{H}_3(\omega_1,\omega_2,\omega_3)\hat{Y}(\omega_1 + \omega_2 + \omega_3) + 2(\gamma_2 + j(\omega_1 + \omega_2 + \omega_3)\xi_2) \\ &\times 2!(\hat{Z}(\omega_1)\hat{H}_2(\omega_2,\omega_3) + \hat{Z}(\omega_2)\hat{H}_2(\omega_1,\omega_3) + \hat{Z}(\omega_3)\hat{H}_2(\omega_1,\omega_2)) \\ &+ 3!(\gamma_3 + j(\omega_1 + \omega_2 + \omega_3)\xi_3)\hat{Z}(\omega_1)\hat{Z}(\omega_2)\hat{Z}(\omega_3) \end{aligned} \tag{3.27}$$

may then be solved for the unknown kernel $\hat{H}_3(\omega_1,\omega_2,\omega_3)$.

Explicit expressions for kernels up to fifth order are given in Appendix A, wherein a general *recipe* for their formal computation is also given.

IV. CONVERGENCE PROPERTIES OF VOLTERRA SERIES

The complexity of successive Volterra kernels is rapidly increasing with their order. Accordingly, it is important to ascertain, even heuristically, the convergence properties of (3.1).

A first point worth of discussion is the comparison between successive iterative solutions of (3.6). Let then $v_d^{(m)}(t)$ be the m-th order iterated solution of (3.6), as discussed in Section III, and consider any two approximate solutions of (3.6), say $v_d^{(k)}(t)$ and $v_d^{(h)}(t)$. Iteration of (3.6) gives the new voltages:

$$v_d^{(k+1)}(t) = v_g(t) - \int_{-\infty}^{\infty} z(t-\tau)F(v_d^{(k)}(\tau))d\tau, \tag{4.1}$$

$$v_d^{(h+1)}(t) = v_g(t) - \int_{-\infty}^{\infty} z(t-\tau)F(v_d^{(h)}(\tau))d\tau, \tag{4.2}$$

wherein $v_g(t)$ is the equivalent generator open circuit voltage, and $z(t)$ is the inverse Fourier transform of its (linear) impedance $\hat{Z}(\omega)$.

The mapping provided by (3.6) is a contraction, and the iterative solution with its Volterra reordering (3.1) is convergent, if:

$$\frac{\| v_d^{(k+1)}(t) - v_d^{(h+1)}(t) \|}{\| v_d^{(k)}(t) - v_d^{(h)}(t) \|} < 1 \tag{4.3}$$

in a suitable norm. We assume as such the maximum absolute value of the difference in the time interval in which the solution is considered, and use the shorthand notation for (4.3):

$$\frac{\Delta v'}{\Delta v} < 1 \tag{4.4}$$

wherein:

$$\Delta v' = \left\| \int_{-\infty}^{\infty} z(t-\tau)\{F(v_d^{(k)}(\tau)) - F(v_d^{(h)}(\tau))\}d\tau \right\| . \tag{4.5}$$

If the linear part of the circuit is purely resistive, then $z(t) = R\delta(t)$, R being the series resistance of the circuit of fig. 2. Then (4.5) simplifies as follows:

$$\Delta v' = R\,\Delta F \tag{4.6}$$

and inequality (4.4) becomes:

$$R\,\frac{\Delta F}{\Delta v} < 1. \tag{4.7}$$

Disequation (4.7) suggests a very simple conclusion: if the linear resistance of the circuit times the maximum value of the differential conductance of the load does not exceed unity, in the range of the excitation, a convergent Volterra series solution may be obtained.

When the hypothesis of a purely resistive memoryless circuit is relaxed, we get, instead of (4.7):

$$\frac{\Delta F}{\Delta v} \int_{-\infty}^{\infty} |z(t)|dt < 1 , \tag{4.8}$$

yielding a similar interpretation.

Truncation of Volterra series in the general case is by no means a simple problem, and is among the less studied topics in literature. It has been almost ignored, hitherto, in the *NLA* field.

A few heuristic remarks on the subject will be drawn here from previous discussion.

Under the assumption of mild nonlinearity and/or excitation, in view of the very definition (3.2) of the n-term of Volterra series (3.1), we have to the lowest order:

$$v_d^{(k+1)}(t) - v_d^{(k)}(t) \sim v_{k+1}(t) \propto V_g^{k+1} \tag{4.9}$$

wherein V_g is the peak amplitude of the excitation, $v_g(t)$.

On the other hand, letting

$$\left.\frac{\Delta F}{\Delta v}\right|_{max} \int_{-\infty}^{\infty} |z(t)|dt \triangleq a < 1 , \tag{4.10}$$

we get from (3.3):

$$|v_d^{(k+1)}(t) - v_d^{(k)}(t)| < a|v_d^{(k)}(t) - v_d^{(k-1)}(t)| \tag{4.11}$$

whence, iterating and taking into account that $v_d^{(1)}(t) - v_d^{(0)}(t) = v_g(t) < V_g$,

$$|v_d^{(k+1)}(t) - v_d^{(k)}(t)| < a^k \; V_g . \tag{4.12}$$

From (4.9 and 12) an upper bound is obtained for the truncation error of (3.1):

$$\sum_{n+1}^{\infty}{}_k \; v_k(t) \sim \sum_{n+1}^{\infty}{}_k \left(v_d^{(k)}(t) - v_d^{(k-1)}(t)\right) < \frac{a^n \; V_g}{1-a} . \tag{4.13}$$

As a last remark, we finally note that inclusion in (3.1) of terms of order higher than *n* may not be consistent, as can be seen from (4.9), with an *n*-degree truncated polynomical representation of the nonlinear load characteristic (2.10).

V. NUMERICAL TECHNIQUES

In this section we briefly present the numerical techniques used in the *NLA* field, and, with somehow greater detail, introduce a new one.

The method of moments has been successfully applied for studying transmitting and receiving *NLA* [18-20] as an extension of the same well known technique in the time-domain [51-53].

Let us consider a wire antenna, lying along the *z*-axis, and let $\underline{E}_i$ be the incident electric field. We have for the total (incident plus scattered) tangential field on the surface of the antenna:

$$E_z(z,t) = E_{zi}(z,t) - \frac{\partial A_z}{\partial t} - \frac{\partial \phi}{\partial z} \tag{5.1}$$

wherein the vector potential $\underline{A}$, and the scalar potential ϕ, related by the Lorentz gauge:

$$- \nabla \cdot \underline{A} = \frac{\partial \phi}{\partial t} \tag{5.2}$$

can be expressed in terms of the (unknown) current $I(z,t)$ and (linear) charge density $\rho_s(z,t)$ distributions along the wire, which are related, in turn by the continuity equation:

$$\frac{\partial I(z,t)}{\partial z} + \frac{\partial \rho_s}{\partial t} = 0. \tag{5.3}$$

The total tangential field $E_z(z,t)$, however, depends on the properties of the wire. It is zero for a perfectly conducting wire. For a wire with nonlinear loads, it will be a nonlinear functional of the local current, hence:

$$E(z,t) = \overline{N}(I(z,t)). \tag{5.4}$$

Accordingly, eqs. (5.1) and (5.3) make up a couple of functional equations in the unknown current and charge density distributions. Then, by introducing convenient expansion basis, e.g., rectangular pulses of width Δz and time duration $\Delta t = \Delta z/c$, and test basis, e.g., Dirac pulses (point-matching), eqs. (5.1 and 3) are converted into two coupled sets of equations in the unknown expansion coefficients. Note that the method can handle, in principle, distributed nonlinear loadings as well.

An interesting numerical technique for solving *NLA* problems using the circuit approach, has been proposed by Liu and Tesche [18-20]. Eq. (1.4):

$$F\{v_d(t)\} = i_g(t) - \int_0^t y(t-\tau) v_d(\tau) d\tau, \tag{5.5}$$

wherein it is assumed that the antenna is excited at $t = 0$, is solved step-by-step as follows. At time $t = n\Delta t$, $n = 0,1,\ldots,$ we have:

$$F\{v_d(n\Delta t)\} = i_g(n\Delta t) - \int_0^{(n-1)\Delta t} y(n\Delta t-\tau)v_d(\tau)d\tau + - \int_{(n-1)\Delta t}^{n\Delta t} y(n\Delta t-\tau)v_d(\tau)d\tau. \tag{5.6}$$

If $\Delta t << L/c$, L being the length of the wire, the second integral in (5.6) may be written as:

$$\int_{(n-1)\Delta t}^{n\Delta t} y(n\Delta t-\tau)v_d(\tau)d\tau = G_\infty\, v_d(n\Delta t) + C_0 \frac{v_d(n\Delta t) - v_d((n-1)\Delta t)}{\Delta t}, \tag{5.7}$$

wherein G_∞ is the limit as $\omega \to \infty$ of the input conductance of an infinitely long antenna [54], and C_0 is the static gap capacitance [55].

Furthermore:

$$F\{v_d(n\Delta t)\} = g(v_d(n\Delta t))v_d(n\Delta t) + c(v_d(n\Delta t)) \times \frac{v_d(n\Delta t) - v_d((n-1)\Delta t)}{\Delta t}, \tag{5.8}$$

wherein $g(v)$ and $c(v)$ are the voltage dependent nonlinear load conductance and capacitance, respectively. Using (5.7 and 8), eq. (5.6) can be written finally as follows:

$$i_g(n\Delta t) - \int_0^{(n-1)\Delta t} y(n\Delta t-\tau)v_d(\tau)d\tau$$

$$+ \frac{C_0 + c\{v_d((n-1)\Delta t)\}}{\Delta t} v_d((n-1)\Delta t)$$

$$= \left\{ G_\infty + g(v_d(n\Delta t)) + \frac{C_0 + c(v_d(n\Delta t))}{\Delta t} \right\} v_d(n\Delta t) \ , \tag{5.9}$$

which provides the value of the load voltage at time $n\Delta t$ as a function of all previous values (step-by-step algorithm).

A further numerical method, the (piecewise) harmonic balance technique [56-60], has been recently applied by the authors for the first time to *NLA* problems. The method makes direct use of frequency domain data for the antenna equivalent circuit, and is very well suited to handle the large signal case [24-25].

Let us assume a sinusoidal excitation, $i_g(t)$, with angular frequency $\omega = 2\pi/T$,[5] and let us expand the (unknown) load voltage $v_d(t)$ into a Fourier series:[6]

$$v_d(t) = \frac{a_0}{2} + \sum_{1}^{N}{}_{n} (a_n \cos n\omega t + b_n \sin n\omega t) \ . \tag{5.10}$$

Equation (1.4) may be written:

$$i_g(t) = i_1(t) - i_n(t) \tag{5.11}$$

wherein:

$$i_n(t) = F(v_d(t)) \tag{5.12}$$

is the specified nonlinear load characteristic, and

$$i_1(t) = y(t) * v_d(t) = \int_{-\infty}^{\infty} y(t-\tau) v_d(\tau) d\tau \ , \tag{5.13}$$

$$y(t) = \frac{1}{2\pi} \int_{-\infty}^{\infty} \hat{Y}(\omega) \exp j\omega t \ d\omega \ , \tag{5.14}$$

[5]*The method can also be applied in the case of time-limited signals, by use of the sampling theorem.*

[6]*The lack of subharmonics in the response is realted to uniqueness of solution [61], which depends, in turn, on the one-to-one nature of the load characteristic (5.12).*

$\hat{Y}(\omega)$ being the (linear) admittance of the equivalent circuit of fig. 2. Accordingly,

$$i_1(t) = Y(0)\frac{a_o}{2} + \sum_{1}^{N}{}_n |\hat{Y}(n\omega)|\{a_n \cos(n\omega t + \psi_n) + b_n \sin(n\omega t + \psi_n)\} \quad (5.15)$$

wherein $\psi_n = \arg(\hat{Y}(n\omega))$.
Should (5.10) be the exact solution, eq. (5.11) would be fulfilled. It is therefore reasonable to introduce a quadratic mean error:

$$\bar{E}^2 = \int_0^T (i_g(t) - i_1(t) - i_n(t))^2 dt \quad (5.16)$$

to be minimized by a proper choice of the $(2N+1)$ expansion coefficients, a_n, b_n. Accordingly, (5.16) may be evaluated by introducing (at least) $2N+1$ voltage samples and using, e.g., the trapezoidal rule to get:

$$\bar{E}^2 = \frac{1}{2N+1} \sum_{1}^{2N+1}{}_n (i_g(t_n) - i_1(t_n) - i_n(t_n))^2 . \quad (5.17)$$

The problem is now to search for the minimum of (5.17). This can be accomplished by starting from a suitable approximate solution, and then repeatedly applying some suitable gradient method [62]. An alternative procedure, which avoids computation of derivatives [63], is the following. Let us consider a_n, b_n as the coordinates x_k of a point $\underline{X}$ in $2N+1$ dimensional space. Then let us change one coordinate at a time, say x_n, the others being kept fixed, to a new value, $x_n + \Delta x_n$ in order to minimize the quadratic mean error (5.17). Let us repeat this procedure for all n. It is reasonable that the vector $\Delta\underline{X}$ of components Δx_n has the direction of the gradient of the error function (5.17), and opposite sense. Accordingly, the minimum of (5.17) will be presumably located somewhere between $\underline{X}$ and $\underline{X} + \Delta\underline{X}$.

Minimization of (5.17) is thus obtained by $2N+2$ sequential one-dimensional minimizations, each of which can be conveniently implemented by using the Golden-Section or Fibonacci algorithm [64] (see Appendix B), provided the quadratic mean error (5.17) is a (locally) unimodal function of its arguments, a_n, b_n.

The last condition is reasonably met if the minimization algorithm is implemented as follows:

a) an initial choice for $v_d(t)$ is made, with a few harmonics, using, e.g., the Volterra series;

b) the excitation, $i_g(t)$, is slightly increased, and the minimization procedure is repeatedly applied, until the error (5.17) is made lower than a fixed threshold, $\bar{E}_t^2$;

c) point b) is iterated, until the error (5.17) cannot be made smaller than $\bar{E}_t^2$. This occurs because, as the excitation is increased, higher order harmonics in the response v_d cannot be neglected anymore. A successive harmonic is therefore added, by introducting a new "a" and a new "b", whose starting values may be taken equal to zero. Point b) is now again iterated.

Previous procedure is flexible, and reasonably fast.

Attractive perspectives are those based on the random search of the (global) minimum of (5.17) [65-67]. The actual feasibility of such an approach is being actively investigated.

VI. COMPUTED EXAMPLES

Some of the techniques described in previous sections have been applied to the study of antennas with realistic nonlinear loads, whose assumed characteristics have been collected in Table I.

Figure 6 illustrates a parasitic effect. The antenna is a half-wavelength transmitting wire (thickness $\Omega = 10$) with oxidated terminals (Table I,1) fed by a matched generator at a

TABLE I. Models of nonlinear loads. ([i] = [ampere]; [v] = [volt]; [C] = [picofarad]; [R] = [ohm]; [L] = [nanohenry].)

1.	*Oxidated contacts* $i = 0.018\bar{3}v + 0.00375v^3$	
2.	*Backward diode* $i = 0.0114v - 0.1521v^2 + 0.4375v^3 + 0.8334v^4 - 1.667v^5 \; , \; \lvert v \rvert < 0.3$ $C(v) \simeq 1$	$C_p = 0.1$ $L_s = 1$ $R_s = 1$
3.	*MIS varactor diode* $C(v) = \begin{cases} 9.5 - 4.5(v+0.5) \; , & \lvert v \rvert < 0.5 \\ 9.5 \; , & v < -0.5 \\ 4.5 \; , & v > 0.5 \end{cases}$ $i \simeq 0$	
4.	*Hyper-abrupt Schottky barrier diode* $i = 10^{-8}(exp(\frac{v}{0.052}) - 1)$ $C(v) = (1 - v/0.45)^{-2}$	

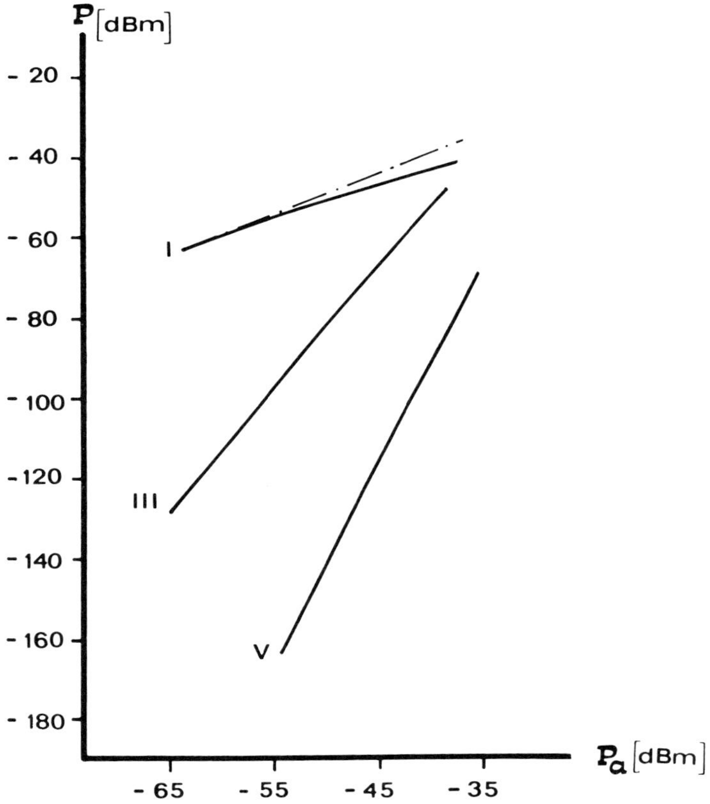

FIGURE 6. Half-wave transmitting dipole with oxidated terminals. Total radiated vs. available power (matched generator).

frequency f = 300 MHz. The total radiated power on fundamental, 3rd and 5th harmonic is displayed as a function of the available power. The total delivered power in the absence of oxidation is also shown (dashed line).

Figures 7 and 8 refer, respectively, to the same half-wavelength wire loaded by a backward diode (Table I,2), and to a plane-sheet self-complementary spiral antenna (diameter $\phi \simeq \lambda$) loaded by a MIS varactor diode (Table I,3). The antennas are

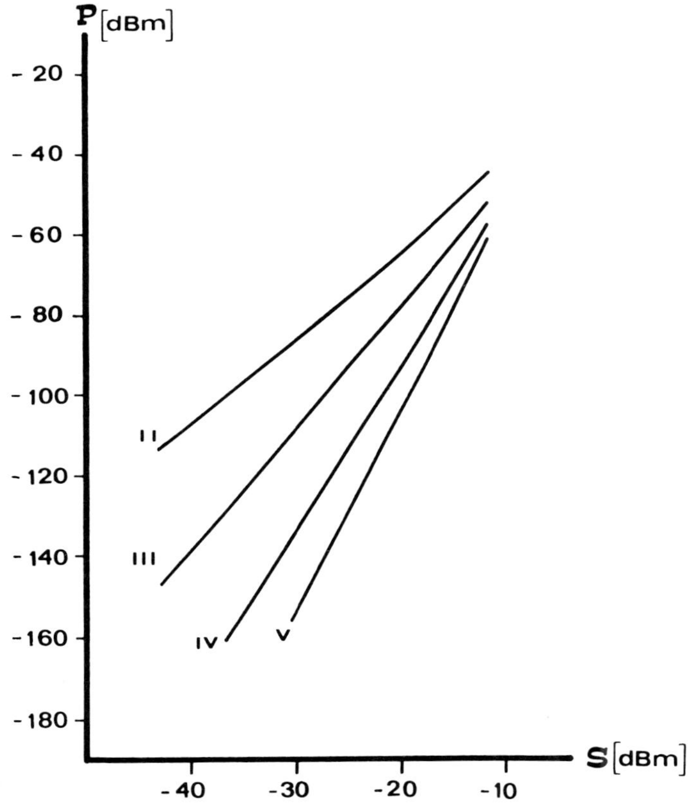

FIGURE 7. Half-wave dipole loaded by backward diode. Total harmonic radiated powers vs. incident power density.

illuminated by a plane monochromatic wave at f = 300 MHz. The total radiated power on the harmonic frequencies from $2f$ to $5f$ is displayed vs. the incident power density.

Figure 9 refers to a small thin loop (Ω = 10, ϕ = 0.1 *m*) loaded by a hyper-abrupt diode (Table I,4). The latter may tune the loop on a frequency whose value depends on the incident field strength as shown in curve (1). Curve (2) displays the total 2nd harmonic radiated power, at an incident power density of -20 *dBm*, vs. the frequency of the incident wave.

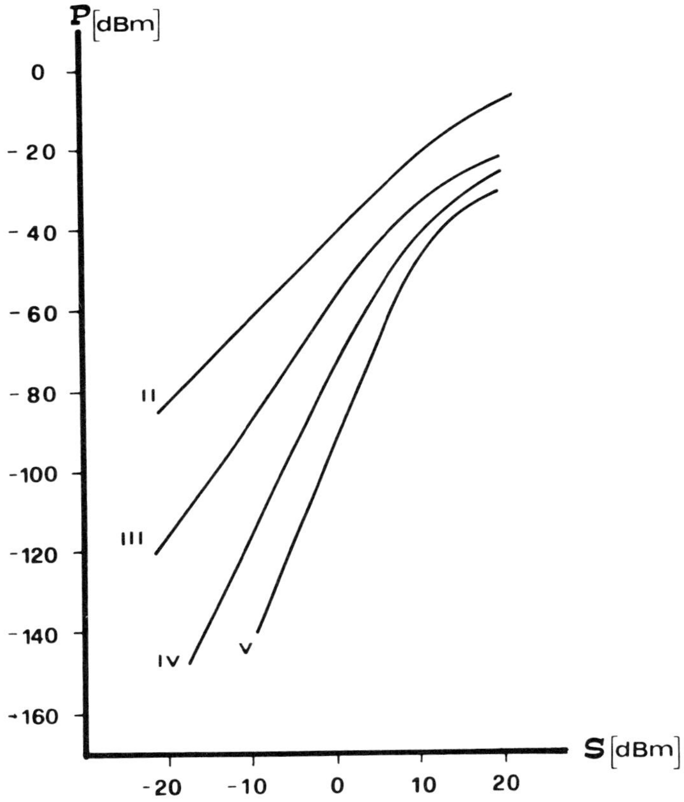

FIGURE 8. Self-complementary plane-sheet spiral antenna loaded by MIS varactor diode. Total harmonic radiated powers vs. incident power density.

VII. CONCLUSIONS

Nonlinearly loaded antennas, e.g., wire, loop or spiral antennas connected to a nonlinear load, e.g., a diode, have been considered. With some limitations, discussed in Section I, an equivalent circuit approach is possible. From a mathematical viewpoint, we are faced with a nonlinear integral equation of

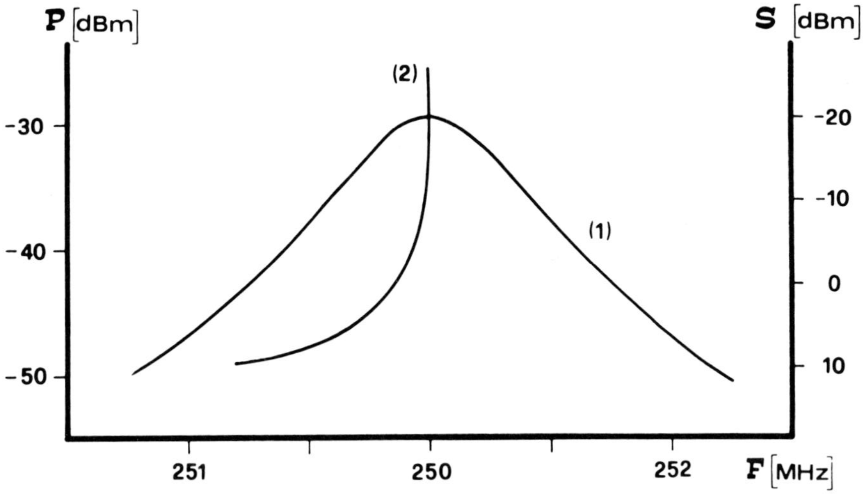

FIGURE 9. Small loop loaded by hyper-abrupt diode. Total 2nd harmonic radiated power and resonant frequency vs. incident power density.

Volterra type. The latter may be solved numerically or analytically, provided an analytical functional relationship between load current and voltage is known. Accordingly, the whole Section II of this chapter has been devoted to nonlinear load modelling. It is shown that nonlinear junction capacitance of semiconductor diodes cannot be neglected, even at frequencies below 1 MHz. It is also suggested that some devices (e.g., backward, MIS and possibly SCLD diodes) may be promising whenever use is made of nonlinearity.

The analytical Volterra series solution is reviewed under Section III, together with some general procedures for obtaining Volterra kernels. Kernels up to fifth order are tabulated in Appendix A. Convergence of Volterra series has been considered in Section IV, wherein a number of interesting results are heuristically obtained. The series is convergent provided a simple inequality holds which, in the case of memoryless circuits, merely states that the product of the generator linear resistance times the maximum differential conductance of the load, in the range of the excitation, is less than unity. An upper bound for truncation error is also given.

Numerical techniques are reviewed in Section V, including the method of moments, the step-by-step integration, and the harmonic balance technique, which has been recently applied for the first time to *NLA* by the present authors. A minimization algorithm, based on Fibonacci-golden section rule, is presented in Appendix B in the frame of the last method.

Finally, some examples of nonlinearly loaded antennas are illustrated in Section VI.

APPENDIX A

Successive Volterra Kernels

The formal complexity of Volterra kernels rapidly increases with their order. Systematic procedures do exist, however, for their computation [68]. Alternatively, they can be measured, on the basis of the generalized impulse response concept [69].

For the reader's convenience, we are collecting under Table II the formal expression of the first five kernels, $\hat{H}_i(\cdot)$, wherein ω_i is the angular frequency, $\hat{Z}(\omega)$ is the (linear) impedance of the equivalent circuit of fig. 2, and $\hat{y}_k(\omega) = \gamma_k + j\omega\xi_k$, the non-linearity being described by:

$$F(v_d(t)) = \sum_{2}^{\infty}{}_k \left(\gamma_k + \xi_k \frac{d}{dt}\right) v_d^k(t) \ . \tag{A.1}$$

The general kernel, $\hat{H}_n$, is the sum of several terms, $\hat{\Xi}_{nk}$, representing the contribution of the k-th degree term in (A.1) as follows:

$$\hat{\Xi}_{nk} = -\frac{1}{n!}\hat{y}_k(\omega_1 + \ldots + \omega_n)\hat{Z}(\omega_1 + \ldots + \omega_n)\hat{\Sigma}_n(\omega_1,\ldots,\omega_n). \tag{A.2}$$

To calculate $\hat{\Sigma}_n$, all k-partitions of n are first determined:

$$m_1 + m_2 + \ldots + m_k = n \ . \tag{A.3}$$

Let m_i be present M_i times in (A.3). Then assuming that all the ω_j are distinct, there are

$$\frac{n!}{\prod_i M_i! \ \prod_i m_i!}$$

distinct products of k factors $\hat{H}_i(\cdot)$:

$$\hat{H}_{m_1}(\cdot)\hat{H}_{m_2}(\cdot) \ \ldots \ \hat{H}_{m_k}(\cdot)k! \ \prod_i M_i! \tag{A.4}$$

differing at least for the exchange of two distinct arguments

TABLE II. Volterra kernels up to fifth order.

$$H_2(\omega_1,\omega_2) = -Z(\omega_1+\omega_2)Z(\omega_1)Z(\omega_2)y_2(\omega_1+\omega_2)$$

$$H_3(\omega_1,\omega_2,\omega_3) = -\left\{ \frac{2!2!}{3!}\left[Z(\omega_1)H_2(\omega_2,\omega_3) + Z(\omega_2)H_2(\omega_1,\omega_3) + \right.\right.$$
$$\left. + Z(\omega_3)H_2(\omega_1,\omega_2)\right] y_2(\omega_1+\omega_2+\omega_3) + \frac{3!}{3!} Z(\omega_1)Z(\omega_2)Z(\omega_3) \cdot$$
$$\left. y(\omega_1+\omega_2+\omega_3) \right\} Z(\omega_1+\omega_2+\omega_3)$$

$$H_4(\omega_1,\omega_2,\omega_3,\omega_4) = -\left\{\left\{ \frac{2!2!2!}{4!} \left[H_2(\omega_1,\omega_2)H_2(\omega_3,\omega_4) + H_2(\omega_1,\omega_3)\cdot\right.\right.\right.$$
$$\left. H_2(\omega_2,\omega_4) + H_2(\omega_1,\omega_4)H_2(\omega_2,\omega_3)\right] + \frac{2!3!}{4!}\left[Z(\omega_1)H_3(\omega_2,\omega_3,\omega_4) + \right.$$
$$\left.\left. Z(\omega_2)H_3(\omega_1,\omega_3,\omega_4) + Z(\omega_3)H_3(\omega_1,\omega_2,\omega_4) + Z(\omega_4)H_3(\omega_1,\omega_2,\omega_3)\right]\right\}\cdot$$
$$y_2(\omega_1+\omega_2+\omega_3+\omega_4) + \frac{3!2!}{4!}\left[H_2(\omega_1,\omega_2)Z(\omega_3)Z(\omega_4) + H_2(\omega_1,\omega_3)\cdot\right.$$
$$Z(\omega_2)Z(\omega_4) + H_2(\omega_1,\omega_4)Z(\omega_2)Z(\omega_3) + H_2(\omega_2,\omega_3)Z(\omega_1)Z(\omega_4) +$$
$$\left. + H_2(\omega_2,\omega_4)Z(\omega_1)Z(\omega_3) + H_2(\omega_3,\omega_4)Z(\omega_1)Z(\omega_2)\right] y_3(\omega_1+\omega_2+\omega_3+\omega_4) +$$
$$\left. + \frac{4!}{4!} Z(\omega_1)Z(\omega_2)Z(\omega_3)Z(\omega_4)\, y_4(\omega_1+\omega_2+\omega_3+\omega_4) \right\} Z(\omega_1+\omega_2+\omega_3+\omega_4)$$

$$H_5(\omega_1,\omega_2,\omega_3,\omega_4,\omega_5) = -\left\{\left\{\frac{2!3!2!}{5!}\left[H_2(\omega_1,\omega_2)H_3(\omega_3,\omega_4,\omega_5) + H_2(\omega_1,\omega_3)\cdot\right.\right.\right.$$
$$H_3(\omega_2,\omega_4,\omega_5) + H_2(\omega_1,\omega_4)H_3(\omega_2,\omega_3,\omega_5) + H_2(\omega_1,\omega_5)H_3(\omega_2,\omega_3,\omega_4) +$$
$$+ H_2(\omega_2,\omega_3)H_3(\omega_1,\omega_4,\omega_5) + H_2(\omega_2,\omega_4)H_3(\omega_1,\omega_3,\omega_5) + H_2(\omega_2,\omega_5)\cdot$$
$$H_3(\omega_1,\omega_3,\omega_4) + H_2(\omega_3,\omega_4)H_3(\omega_1,\omega_2,\omega_5) + H_2(\omega_3,\omega_5)H_3(\omega_1,\omega_2,\omega_4) +$$
$$\left. + H_2(\omega_4,\omega_5)H_3(\omega_1,\omega_2,\omega_3)\right] + \frac{2!4!}{5!}\left[Z(\omega_1)H_4(\omega_2,\omega_3,\omega_4,\omega_5) + Z(\omega_2)\cdot\right.$$
$$H_4(\omega_1,\omega_3,\omega_4,\omega_5) + Z(\omega_3)H_4(\omega_1,\omega_2,\omega_4,\omega_5) + Z(\omega_4)H_4(\omega_1,\omega_2,\omega_3,\omega_5) +$$
$$\left.\left. + Z(\omega_5)H_4(\omega_1,\omega_2,\omega_3,\omega_4)\right]\right\} y_2(\omega_1+\omega_2+\omega_3+\omega_4+\omega_5) + \left\{\frac{3!2!2!}{5!}\right\} Z(\omega_1)\cdot$$
$$\left[H_2(\omega_2,\omega_3)H_2(\omega_4,\omega_5) + H_2(\omega_2,\omega_4)H_2(\omega_3,\omega_5) + H_2(\omega_2,\omega_5)H_2(\omega_3,\omega_4)\right] +$$

TABLE II continued.

$$Z(\omega_2)\Big[H_2(\omega_1,\omega_3)H_2(\omega_4,\omega_5) + H_2(\omega_1,\omega_4)H_2(\omega_3,\omega_5) + H_2(\omega_1,\omega_5)\cdot$$
$$H_2(\omega_3,\omega_4)\Big] + Z(\omega_3)\Big[H_2(\omega_1,\omega_2)H_2(\omega_4,\omega_5) + H_2(\omega_1,\omega_4)H_2(\omega_2,\omega_5) +$$
$$H_2(\omega_1,\omega_5)H_2(\omega_2,\omega_4)\Big] + Z(\omega_4)\Big[H_2(\omega_1,\omega_2)H_2(\omega_3,\omega_5) + H_2(\omega_1,\omega_3)\cdot$$
$$H_2(\omega_2,\omega_5) + H_2(\omega_1,\omega_5)H_2(\omega_2,\omega_4)\Big] + Z(\omega_5)\Big[H_2(\omega_1,\omega_2)\cdot$$
$$H_2(\omega_3,\omega_4) + H_2(\omega_1,\omega_3)H_2(\omega_2,\omega_4) + H_2(\omega_1,\omega_4)H_2(\omega_2,\omega_3)\Big]\Big\} + \frac{3!3!}{5!}\cdot$$
$$\Big[H_3(\omega_1,\omega_2,\omega_3)Z(\omega_4)Z(\omega_5) + H_3(\omega_1,\omega_2,\omega_4)Z(\omega_3)Z(\omega_5) + H_3(\omega_1,\omega_2,\omega_5)\cdot$$
$$Z(\omega_3)Z(\omega_4) + H_3(\omega_1,\omega_3,\omega_4)Z(\omega_2)Z(\omega_5) + H_3(\omega_1,\omega_3,\omega_5)Z(\omega_2)Z(\omega_4) +$$
$$+ H_3(\omega_1,\omega_4,\omega_5)Z(\omega_2)Z(\omega_3) + H_3(\omega_2,\omega_3,\omega_4)Z(\omega_1)Z(\omega_5) + H_3(\omega_2,\omega_3,\omega_5)\cdot$$
$$Z(\omega_1)Z(\omega_4) + H_3(\omega_2,\omega_4,\omega_5)Z(\omega_1)Z(\omega_3) + H_3(\omega_3,\omega_4,\omega_5)Z(\omega_1)\cdot$$
$$Z(\omega_5)\Big]\Big\{ y_3(\omega_1+\omega_2+\omega_3+\omega_4+\omega_5) + \frac{4!2!}{5!}\Big[H_2(\omega_1,\omega_2)Z(\omega_3)Z(\omega_4)Z(\omega_5)\cdot$$
$$+ H_2(\omega_1,\omega_3)Z(\omega_2)Z(\omega_4)Z(\omega_5) + H_2(\omega_1,\omega_4)Z(\omega_2)Z(\omega_3)Z(\omega_5) +$$
$$+ H_2(\omega_1,\omega_5)Z(\omega_2)Z(\omega_3)Z(\omega_4) + H_2(\omega_2,\omega_3)Z(\omega_1)Z(\omega_4)Z(\omega_5) + H_2(\omega_2,\omega_4)\cdot$$
$$Z(\omega_1)Z(\omega_3)Z(\omega_5) + H_2(\omega_2,\omega_5)Z(\omega_1)Z(\omega_3)Z(\omega_4) + H_2(\omega_3,\omega_4)Z(\omega_1)$$
$$Z(\omega_2)Z(\omega_5) + H_2(\omega_3,\omega_5)Z(\omega_1)Z(\omega_2)Z(\omega_4) + H_2(\omega_4,\omega_5)Z(\omega_1)Z(\omega_2)\cdot$$
$$Z(\omega_3)\Big] y_4(\omega_1+\omega_2+\omega_3+\omega_4+\omega_5) + Z(\omega_1)Z(\omega_2)Z(\omega_3)Z(\omega_4)Z(\omega_5)\cdot\frac{5!}{5!}$$
$$y_5(\omega_1+\omega_2+\omega_3+\omega_4+\omega_5)\Big\} Z(\omega_1+\omega_2+\omega_3+\omega_4+\omega_5).$$

between two kernels of different order. Note that the factorial coefficient accounts for permutations between factors and symmetry of the kernels with respect to permutations of their arguments.

Summing up all terms (A.4) obtained from all partitions (A.3), the sum $\hat{\Sigma}_n$ is then obtained.

APPENDIX B

Minimization Procedures

Let $f(x)$ be a unimodal function in $(0,L)$, i.e., a function with a single minimum at $x = x_o \in (0,L)$. Given any two points, say $x = a$ and $x = b$, $a < b$, within $(0,L)$, either $f(a) < f(b) \Longrightarrow x_o \in (0,b)$ or $f(a) \geq f(b) \Longrightarrow x_o \in (a,L)$. The original search interval is thus reduced to $(0,b)$ or (a,L); it is therefore convenient to assume:

$$b = L - a . \tag{B.1}$$

The procedure may be iterated until the search interval becomes smaller than a fixed $\delta > 0$.

Any rule for choosing "a" establishes a correspondence between L, the original search interval length and k, the number of functional evaluations needed to reduce the size of the search interval to δ or less. A rule will be optimal if k is minimum for any given L, i.e., if L is maximum for any given k.

Let us call $L(k)$ the correspondence between k and L relative to the optimal rule. Suppose: $f(a) < f(b) \Longrightarrow x_o \in (a,L)$. The next step will be to evaluate $f(x)$ somewhere in (a,L), and compare with $f(b)$. There are $k-1$ functional evaluations still available in (a,L). So we shall impose:

$$L(k-1) \leq L(k) - a . \tag{B.2}$$

On the other hand if: $f(a) \geq f(b) \Longrightarrow x_o \in (0,b)$, may be

$x_o \in (O,a)$. No evaluation of $f(x)$ is already available in (O,a), and we are left with only $k-2$ functional evaluations there. Accordingly, we shall impose:

$$L(k-2) \leq a\,. \tag{B.3}$$

From (B.2 and 3) we get:

$$L(k) \leq L(k-1) + L(k-2)\,. \tag{B.4}$$

No reduction of the original search interval length is possible, of course, for $k < 2$. Accordingly:

$$L(O) \leq \delta \;;\; L(1) \leq \delta\,. \tag{B.5}$$

In the spirit of the optimum algorith, we let:

$$L(k) = L(k-1) + L(k-2)\,, \tag{B.6}$$

$$L(O) = L(1) = \delta\,, \tag{B.7}$$

whence, in view of (B.1,2 and 3):

$$\left.\begin{aligned} a &= L(k-2) \\ b &= L(k-1) \end{aligned}\right\} \tag{B.8}$$

Relations (B.6 and 7) define a Fibonacci[7] number suite, whereas (B.8) establish the required optimal rule for choosing "a".

Note that, for $k \gtrsim 5$, we have:

$$\frac{L(k-1)}{L(K)} \simeq 0.618 \tag{B.9}$$

(golden section of the unit-segment).

Relations (B.6 and 9) define the so called golden-section rule for choosing "a", yielding:

$$\left.\begin{aligned} a &= 0.382L \\ b &= 0.618L \end{aligned}\right\}\,. \tag{B.10}$$

[7] *Lionardo Fibonacci, Italian mathematician, medical doctor and wizard, was born and died in Pisa in the 13th centure. His numbers are relevant to the aspect of leaves.*

REFERENCES

1. Schuder, J.C. and Stoeckle, H., "The silicon diode as a receiver for electrical stimulation of body organs," *Trans. Am. Soc. Artific. Intern. Organs 10*, 366-370 (1964).
2. Schuder, J.C. and Stoeckle, H., "The silicon diode as an experimental cardiac pacemaker receiver," Circulation *30*, Suppl. III, 167-171 (1964).
3. Lawton, J.G., "Concept for a microwave search instrument for the location of avalanche victims," V. Eigenmann Foundation Symposium, Sülden, (1975).
4. Pinto, I., "Radar armonico a microonde per la ricerca di sepolti da valanga," In print on Alta Frequenze (1980).
5. Higa, W.H., "Spurious signals generated by electron tunnelling in large reflector antennas," *Proc. IEEE 63*, 306-313 (1975).
6. Banta, E.D., "Spurious responses in linear arrays using nonlinear elements," *IEEE Trans. Antennas and Propagat. 12*, 129-136 (1964).
7. Merewether, D.E. and Ezell, T.F., "The interaction of cylindrical posts and radiation-induced electric field pulses in ionized media," *IEEE Trans. Nucl. Sci. 21*, 3-13 (1974).
8. Bucci, O.M. and Franceschetti, G., "Input admittance and transient response of spheroidal antennas in dispersive media," *IEEE Trans. Antenna Propagat. 22*, 526-536 (1974).
9. Bucci, O.M., De Bonitatibus, A. and Savarese, C., "Transient response of coaxially driven spheroidal antennas-theory," Alta Frequenza *47*, 3-20 (1978).
10. Marcuvitz, N., "Microwave Handbook" Ch. 4. McGraw-Hill, New York, (1951).
11. Sarkar, T.K. and Weiner, D.D., "Analysis of mild nonlinearities in electromagnetic systems using the Volterra series approach," URSI Spring Meeting, Seattle, June 18-22 (1979).
12. Sarkar, T.K., Weiner, D.D. and Harrington, R.F., "Analysis of nonlinearly loaded multiport antenna structures over an imperfect ground plane using the Volterra series method," *IEEE Trans. Electromagn. Compatib. 20*, 278-287 (1978).
13. Sarkar, T.K., Weiner, D.D. and Harrington, R.F., "Effect of imperfect ground plane on nonlinear inhomogeneous antenna," Dept. of Electrical and Computer Eng., University of Syracuse, Tech. Rpt. n. 4, November (1976).
14. Sarkar, T.K., Weiner, D.D. and Harrington, R.F., "Analysis of nonlinearly loaded n-port antenna structures," Dept. of Electrical and Computer Eng., University of Syracuse, Tech. Rpt. n.2, April (1976).
15. Sarkar, T.K. and Weinger, D.D., "Scattering analysis of nonlinearly loaded antennas," *IEEE Trans. Antennas Propagat.24*,

125-130 (1976).
16. Finn, R.M., "Scattering from nonlinearly loaded conducting bodies," Office of Naval Research, Contract N00014-72-C0016, Rpt. n. SURC-TR-72-60, March (1972).
17. Kanda, M., "The characteristics of a traveling-wave, linear antenna with a nonlinear load," URSI Spring Meeting, Seattle, June 18-22 (1979).
18. Landt, J.A., "A time-domain computer code for nonlinear circuits and thin-wire antennas," URSI Spring Meeting, Seattle, June 18-22 (1979).
19. Landt, J.A., Miller, E.K. and Deadrick, F.J., "Time domain modelling of nonlinear loads," Lawrence Livermore Laboratory, Rept. UCRL-52172 (1979).
20. Schuman, H., "Time-domain scattering from a nonlinearly loaded wire," *IEEE Trans. Antennas Propagat. 22*, 611-613 (1974).
21. Liu, T.K., Tesche, F.M. and Deadrick, F.J., "Transient excitation of an antenna with a nonlinear load: numerical and experimental results," *IEEE Trans. Antennas Propagat. 25*, 539-542 (1977).
22. Liu, T.K. and Tesche, F.M., "Analysis of antennas and scatterers with nonlinear loads," *IEEE Trans. Antennas Propagat. 24*, 131-139 (1976).
23. Tesche, F.M. and Liu, T.K., "Transient response of antennas with nonlinear loads," *Electronics Lett. 11*, 18-19 (1975).
24. Pinto, I. and Franceschetti, G., "Harmonic balance techniques applied to the study of nonlinearly loaded antennas," in print.
25. Franceschetti, G. and Pinto, I., "Nonlinearly loaded antennas," URSI Spring Meeting, Seattle, June 18-22 (1979).
26. Simmons, J.G., "Generalized formula for the electric tunnel effects between similar electrodes separated by a thin insulating film," *J. Appl. Phys. 34*, 1793-1803 (1963).
27. Simmons, J.G., "Electrical tunnel effect between dissimilar electrodes separated by a thin insulating film," *J. Appl. Phys. 34*, 2581-2589 (1963).
28. Simmons, J.G., "Low-voltage current-voltage relationship of tunnel junctions," *J. Appl. Phys. 34*, 238-239 (1963).
29. Knauss, H.P. and Breslow, R.A., "Current-voltage characteristics of tunnel junctions," *Proc. IRE 50*, 1834-1838 (1964).
30. van der Ziel, A., "Solid State Physical Electronics" Sect. 15-2. Prentice-Hall, (1968).
31. Champlin, K.S., Anderson, D.B. and Gunderson, P.D., "Charge carrier inertia in semiconductors," *Proc. IEEE 52*, 677-685 (1964).
32. Dickens, L.S., "Spreading resistance as a function of frequency," *IEEE Trans. Microwave Theory Techn. 15*, 101-109 (1967).
33. Moll, J.L., "The evolution of the thoery of the current-voltage characteristic of pn junctions," *Proc. IRE 46*, 1076-1091 (1958).

34. Sze, S.M., "Physics of Semiconductor Devices" Sect. 8.5. Wiley Interscience, New York (1969).
35. Miller, S.L., "Avalanche breakdown in Germanium," *Phys. Rev. 99*, 1234-1241 (1955).
36. Sze, S.M., "Physics of Semiconductor Devices" Sect. 3.5. Wiley Interscience, New York, (1969).
37. Spirito, P., "Avalanche multiplication factors in Ge and Si abrupt junctions," *IEEE Trans. Electronic Devices 21*, 226-231 (1974).
38. Norwood, M.M. and Shatz, E., "Voltage variable capacitor tuning - a review," *Proc. IEEE 56*, 788-796 (1968).
39. Lawrence, H. and Warner, R.M., Jr., "Diffused junction depletion layer capacitance," *Bell Sys. Tech. J. 39*, 389-403 (1960).
40. O'Hearn, W.F. and Chang, Y.F., "An analysis of frequency dependence of the capacitance of abrupt pn junction semiconductors," *Solid St. Electr. 13*, 473-483 (1970).
41. Sze, S.M., "Physics of Semiconductor Devices" Sect. 3.4. Wiley Interscience, New York, (1969).
42. Scanlan, J.O., "Analysis and Synthesis of Tunnel Diode Circuits" Ch. 2. Wiley Interscience, New York, (1966).
43. Schurmer, H.V., "Microwave Semiconductor Devices" Ch. 5. Pitman, New York, (1971).
44. Sze, S.M., "Physics of Semiconductor Devices" Sec. 3.7(7). Wiley Interscience, New York, (1969).
45. Lindner, R., "Semiconductor surface varactors," *Bell Sys. Tech. J. 41*, 803-831 (1962).
46. Sze, S.M., "Physics of Semiconductor Devices" Sect. 8.8. Wiley Interscience, New York, (1969).
47. Bussgang, J., Ehrman, L. and Graham, B., "Nonlinear systems with multiple inputs," *Proc. IEEE 66*, 1088-1119 (1974).
48. Wiener, N., "Response of a nonlinear device to noise," MIT Rad. Lab. Rept. 129, V-16, Cambridge, Mass., April (1942).
49. Tricomi, F.G., "Differential Equations" Ch. 1. Wiley Interscience, New York, (1969).
50. "Nonlinear system modelling and analysis, with applications to communications receivers," Rome Air Development Center, Rept. AD-766 278, Ch. 2, June (1973).
51. Landt, J.A., et al., "WT-MBA/LLL1B: a computer program for the time domain response of thin wire structures," Lawrence Livermore Lab. Rept. LLL UCRL-51585 (1974).
52. Sayre, E.P. and Harrington, R.F., "Time-domain radiation and scattering by thin wires," *Appl. Sci. Res. 26*, 413-443 (1972).
53. Sayre, E.P., "Transient response of wire antennas and scatterers," Dept. of Electrical Engineering, Thesis, Syracuse University (1969).
54. Latham, R.W. and Lee, K.S.H., "On the transient response of an infinite cylindrical antenna," *Radio Sci. 5*, 715-723 (1970).

55. Bucci, O.M., "A rigorous calculation of the capacitance of prolate spheroids," Alta Frequenza *13*, 283-290 (1974).
56. Nakhla, M.S., Vlach, J., Gopal, J. and Singhal, K., "Distortion analysis of transistor networks," *IEEE Trans. Circuits and Systems 25*, 99-106 (1978).
57. Nakhla, M.S. and Vlach, J., "A piecewise harmonic balance technique for determination of the periodic response of nonlinear systems," *IEEE Trans. Circuits and Systems 23*, 85-91 (1976).
58. Lindenlaub, J.C., "An approach for finding the sinusoidal steady-state response of nonlinear systems," Proc. 7th Ann. Allerton Conf. on Circuits and Systems, Chicago (1969).
59. Nakhla, M.S., "Steady state analysis of nonlinear periodic systems," Ph.D. Dissertation, Univ. Waterloo, Ontario, Canada (1975).
60. Baily, E., "Steady-state harmonic analysis of nonlinear networks," Ph.D. Thesis, Stanford University (1968).
61. Beneŝ, V.E. and Sandberg, I.W., "Applications of a theorem of Dubrovskii to the periodic responses of nonlinear systems," *Bell Sys. Tech. J. 43*, 2855-2874 (1974).
62. Fletcher, R. and Powell, M.J., "A rapidly convergent descent method of minimization," *Comp. J. 6*, 163-183 (1963).
63. Powell, M.J., "An efficient method for finding the minimum of a function of several variables without calculating derivatives," *Comp. J. 7*, 155-162 (1964).
64. Johnson, S.M., "Best exploration for maximum is Fibonaccian," Rand Corp. Rept. P856 (1956).
65. Brooks, S.H., "A discussion of random methods for seeking maxima," *Operational Res. 6*, 244-251 (1958).
66. Khas'minskii, R.Z., "Application of random noise in optimization," Problemi Pederachy Informatii *1*, 113-115 (1965).
67. Vaysbord, E.M., "Covergence of random search," *Eng. Cybernetics 3*, 44-48 (1968).
68. Reddy, P.K. and Reddy, D.C., "Volterra kernels in nonlinear systems," *Electronics Lett. 18*, 426-427 (1973).
69. Arach, J., "Measurement of Wiener kernels by means of random signals," *IEEE Trans. Automatic Control*, 123-125 (1970).

NONLINEAR OSCILLATIONS (LIMIT CYCLES) IN PHYSICAL AND BIOLOGICAL SYSTEMS

F. Kaiser

Institut für Theoretische Physik
Universität Stuttgart

I. INTRODUCTION

I.1. Nonlinear Systems

Nonlinear systems can exhibit a behaviour which is completely absent in the regime of linear dynamics. Both the development and maintenance of such a behaviour (e.g., temporal, spatial and functional order, which lead to self-organizing phenomena in physics, chemistry and biology) seem to be provided by a general mechanism: nonlinear dissipation. To describe the nonlinear regimes of well-organized structures, the concept of *dissipative structures* has been introduced by Prigogine [1,2]. Furthermore, numerous nonlinear systems from quite different fields of research show striking similarities in their behaviour when they are in their ordered state or when they pass from the disordered to the ordered one. This strongly indicates that the inherent dynamics of these systems is determined by the same basic principles of cooperation. A systematic development of the existing analogies has led to Haken's concept of *synergetics* [3,4]. An excellent review on cooperative behaviour in nonlinear systems may be found in a series of books and conference reports [vid. ref. 5-10].

ISBN 0-12-709660-4

Dissipative structures can be understood as phenomena arising beyond the domain of stability of steady states on the thermodynamic branch. Dissipation as an organizing factor can only occur under certain conditions. Necessary prerequisites are: open systems, nonlinear dynamics and stabilization of the ordered structure far away from thermal equilibrium by a suitable energy or matter input.

Let us consider a system, which is initially in thermodynamic equilibrium. Usually, an ordered state is reached when the temperature of the system is lowered. Examples are the transitions from a para- to a ferromagnetic state or from normal to a superconducting situation. These examples belong to the domain of equilibrium phase transitions. An energy input into these systems would lead to a temperature increase without a cooperative behaviour and without a transition to a new state. However, pronounced cooperative phenomena may occur in physical and nonphysical systems if the energy input does not lead to a thermalization of the additional energy within the system. If we start from a stable equilibrium state to which an external stimulus (forces or fluxes) is supplied, the system will respond by leaving its equilibrium state. Two different responses have to be distinguished.

a) The external "force" is sufficiently small. The new nonequilibrium state which the system reaches is already determined by the equilibrium properties of the unperturbed system. This is the region where linear response theories, fluctuation-dissipation theorems and Onsager's linear irreversible thermodynamic can be applied [2,11,12].

b) The external "force" is sufficiently large. After running through one or a whole sequence of nonequilibrium states in the linear regime (i.e., thermodynamic branch of steady states) the system's response may become nonlinear. The linear theories lose their validity, the system gets unstable. These instabilities in the nonlinear domain far from thermal equilibrium manifest themselves in a splitting (i.e., bifurcation) of the thermal branch into stable and unstable branches of non-equilibrium steady states. Generalized thermodynamic concepts for open systems [2,11,12] and theories of nonlinear differential equations must be applied [13-16] for a description of this completely new behaviour.

A fundamental thermodynamic prerequisite for the creation and stabilization of dissipative structures far from equilibrium is a negative entropy flow which overcompensates the effect of entropy production. The occurrence of nonequilibrium instabilities in nonlinear and open systems may be viewed as nonequilibrium phase transitions since there exists a number of profound analogies between these transitions and those in equilibrium systems.

Well known examples from nonlinear physics are: lasers, tunnel diodes, Gunn oscillators, Josephson junctions, superradiance, superfluorescence and optical bistability, instabilities in fluid dynamics, etc. [4,17]. Furthermore, instabilities seem to play a dominant role in chemical and biochemical kinetics, in biological rhythm, in neuronal activity, in population dynamics -- [6,11]. Some examples will be discussed in section III.

I.2. Nonlinear Phenomena

The most suitable description of nonlinear phenomena in systems far from thermal equilibrium is via nonlinear differential equations. Steady state solutions of this set of kinetic equations represent the possible types of dissipative structures. Transitions from one type of solutions to another one can be caused by internal or external fluctuations. An additional external drive can lead to further bifurcations and the stabilization of new ordered structures. A detailed investigation of the nonlinear differential equations is an extremely complex task. Some methods have reached a suitable level of applicability: stability analytical methods [13-17], bifurcation theory

[15,18,19], self-organization in nonequilibrium systems [11], catastrophe theory [20] and the concept of synergetics [6]. However, there exists a great overlap between these different concepts and methods.

Nonlinear oscillations and hysteresis behaviour seem to be the most important phenomena in nonlinear systems. Examples for both types of dissipative structures can be found in all disciplines where nonlinear dynamics play a significant role. A hysteresis type of solution represents a multiple steady state, including a bistable behaviour; a nonlinear oscillation corresponds to an oscillating steady state (vid. Fig. 1). Single and multiple steady state solutions as well as nonlinear oscillations belong to the class of temporal structures. Other types of nonlinear phenomena are spatial structures and several types of spatio-temporal structures, e.g., pulses, solitons and spiral waves. However, it should be emphasized that this chapter is in no way intended to be a comprehensive summary of nonlinear phenomena. A detailed literature may be found in the above cited books. We shall restrict ourselves to a discussion of temporal structures of oscillating type, i.e., order in time and their possible bifurcations in quite a number of different disciplines.

I.3. Nonlinear Oscillations

I.3.1. Linear Systems with a Periodic Force. Let us start with a linear system, which is externally driven by a periodic force. A well known example is the externally driven damped harmonic oscillator, which reads in its normalized form:

$$\ddot{x} + \eta\dot{x} + x = F(t), \tag{1}$$

($\cdot = \frac{d}{dt}$, η = damping constant, $F(t)$ = external stimulus).

The resulting motion is a superposition of the transient and the steady state solution. For $F(t) = F_0 \cos\lambda t$ one gets

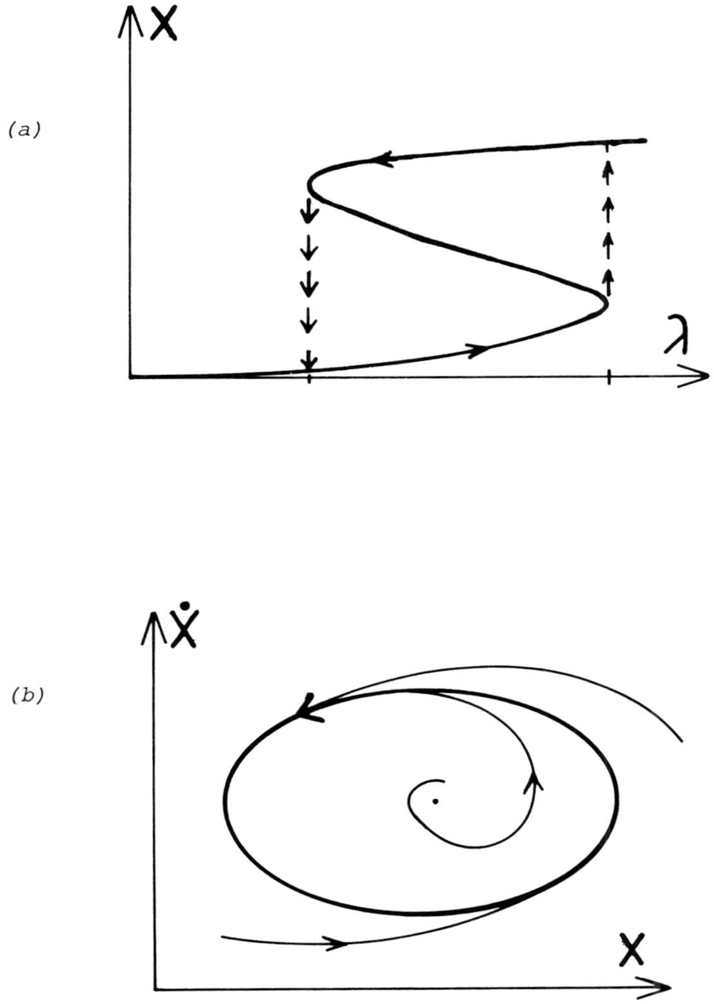

FIGURE 1. Hysteresis (a) and limit cycle type (b) of steady state solutions. x and ẋ are the relevant variables, λ is a bifurcation parameter.

$$x = e^{-\frac{1}{2}\eta t}\left(ae^{i\Lambda t} + be^{-i\Lambda t} \right)$$

$$+ F_0((1-\lambda^2)^2 + \lambda^2\eta^2)^{-\frac{1}{2}} \cos(\lambda t - \psi), \quad (2)$$

$$\left(\Lambda^2 = 1 - \frac{1}{4}\eta^2 \ , \ \tan\psi = \frac{\eta\lambda}{1-\lambda^2}\right) .$$

The transient solution corresponds to the free oscillation, which is damped out in the case of an asymptotic stable system (i.e., $\eta > 0$), $\eta = 0$ corresponds to undamped oscillations in conservative systems. The steady state solution represents the forced oscillation. It vanishes for $F_0 = 0$, which means that without external drive no oscillating solution is possible. This behaviour is valid for all linear oscillators, only oscillations with the external frequency λ can survive.

I.3.2. Nonlinear Systems with a Periodic Force.

a. *Non self-oscillating system*

If a non self-oscillating system is exposed to a periodic force with frequency λ, forced transient and steady state oscillations can occur, including harmonic, higher harmonic and subharmonic oscillations. Furthermore, almost periodic and nonperiodic oscillations are possible. Details depend on the strength of the external force, the initial conditions and the inherent nonlinearity of the system.

As an example we take the linearly damped, anharmonic oscillator:

$$\ddot{x} + \eta\dot{x} + x - \alpha x^3 = F(t) \quad (3)$$

Multiplying equation (3) by x and integrating over the unknown period T of the nonlinear oscillations one gets

$$\eta < \dot{x}^2 >_T = < F(t)\dot{x} >_T \quad (4)$$

$< F(t)\dot{x} >_T$ is a measure for the energy input per period T into the nonlinear oscillation. If its value per period is zero, no oscillation is possible, indicating that equation (3) represents nonlinear oscillations which are damped out for vanishing external force.

b. *Self-oscillating system*

The most interesting case is the external perturbation of a self-oscillating system. A system of this type is defined as a system which produces a periodic process at the expense of non-periodic source of energy. A great variety of oscillating solutions results. Besides transients and beat oscillations, harmonic, superharmonic and subharmonic entrainment are possible, furthermore almost periodic solutions and chaos. The latter is defined as a bounded non-periodic and time-dependent solution. Examples of these different types of oscillating solutions are presented in section V.

As a purely mathematical example we take the harmonic or anharmonic nonlinearly damped oscillator

$$\ddot{x} + f(x) + \mu(x^2 - 1)\dot{x} = F(t) \tag{5}$$

Applying the same procedure as in example (3), we get

$$<\mu(x^2 - 1)\dot{x}^2>_T = <F(t)\dot{x}>_T \tag{6}$$

Both $<F(t)\dot{x}>_T = 0$ and $F(t) = 0$ do not necessarily require

$$<\dot{x}^2>_T = 0 \tag{7}$$

but

$$<x^2\,\dot{x}^2>_T = <\dot{x}^2>_T \tag{8}$$

which leads for a nearly harmonic oscillation (i.e., $x = a\cos\omega t$) to $a_0 = 2$, the well-known steady state limit cycle amplitude of the unperturbed Van der Pol oscillator. Since for $x > a_0$ and $x < a_0$ equation (5) exhibits per period a positively and a negatively damped behaviour, respectively, the existence of a limit cycle with amplitude a_0 is conclusive.

The distinction between self-socillatory and non-self-oscillatory systems is of great importance with respect to the possible behaviour of an externally driven system. Only a system of self-sustained oscillations can exhibit a combination of both free and forced oscillations, which can lead to entrainment, synchronization to the external frequency, quenching of the inherent oscillation and phase-locking. A detailed mathematical description is given in the next section, examples from physical, chemical and biological systems will be presented in section III.

II. LIMIT CYCLE OSCILLATIONS

II.1. Self-sustained Oscillations of Limit Cycle Type

The announcement of singular points and limit cycles by H. Poincaré about one hundred years ago opened a new field in the analysis of nonlinear dynamic systems [21]. A limit cycle is a self-sustained oscillation, i.e., a periodic solution of a self-oscillatory system (vid. section I.3.2b). From a stability point of view a limit cycle trajectory is a closed path periodic solution which is not a member of a continuous family of closed trajectories but a structurally stable solution of a system of ordinary differential equations [13-16].

A limit cycle is determined by internal parameters of the nonlinear kinetic equations. It possesses both, a fixed frequency and a fixed amplitude and is independent of initial conditions. However, it is often a very crude task from a mathematical point of view to find out whether a nonlinear system possesses a stable limit cycle or not. Some mathematical criteria may provide necessary and sufficient conditions for the existence and uniqueness of limit cycles, e.g., the Bendixson criterion and the Liénard theorem [13]. The domain of existence of a limit cycle is given by the Poincaré-Bendixson criterion [15], the frequency near the bifurcation by the Hopf-bifuraction theorem [19] and the direction of bifurcation by the focal value theory [18b]. Apart from some approximative functions the analytic structure of the limit cycle is hard to find. Furthermore, a system exhibiting oscillations of limit cycle type must be structurally stable and at least one unstable singular point of node- or focus-type must lie inside the closed trajectory, onto which the trajectories unwind from both sides (see figure 1). Since a conservative system is not structurally stable, it cannot exhibit this type of oscillations. A rather general nonlinear oscillation is given by

$$\ddot{x} + \eta f(x)\dot{x} + g(x) = 0 \tag{9}$$

The nonlinear functions $f(x)$ and $g(x)$ are assumed to have the following structure:

$$f(x) = a + b\,x + c\,x^2 + \ldots$$

$$g(x) = A\,x + B\,x^2 + C\,x^3 + \ldots$$

Equation (9) represents a prototype of a nonlinearly damped, anharmonic oscillator, including the simplified examples (1), (3) and (5) with $F(t) = 0$. The steady state solutions are

$$\ddot{x}_{ss} = 0 = \dot{x}_{ss}\,, \qquad x_{ss} = const.,$$

where x_{ss} is a solution of $g(x_{ss}) = 0$. Calculating the first non-vanishing focal value [vid. ref. 18b], one finds

$$\alpha_3 = |const| \cdot \eta(Bb - Ac) \tag{10}$$

with $\eta > 0$, only for $Bb - Ac < 0$ a stable limit cycle can exist. $b = c = 0$ corresponds to a linear damping term which excludes the existence of a limit cycle. This manifests again that self-sustained oscillations cannot exist in linearly damped anharmonic systems. $\eta = 0$ (i.e., conservative system) leads to $\alpha_3 = 0$ which rules out the existence of limit cycles. A further inspection of equation (10) shows that at least the combinations $(Bx^2 + bx\dot{x})$ or

$(Ax + \gamma x^2 \dot{x})$ are required for the existence of a unique and stable limit cycle of equation (9). Similar investigations can be performed with other types of self-oscillatory systems.

II.2. Perturbation of Limit Cycle Oscillations

A limit cycle can be perturbed by noise (i.e., random fluxes or forces) or by a systematic stimulus (i.e., externally driven). Furthermore, parameter changes or perturbation through coupling to other oscillations may lead to an altered behaviour. Qualitatively, the following scheme can be given:

$$Ł\ C(A_s,\ \omega_s) \xrightarrow{\lambda, F, \phi} \begin{cases} L\ C(A, \omega) \\ \text{other types of steady states and instability} \end{cases}$$

A limit cycle with steady state amplitude A_s and frequency ω_s is perturbed by a stimulus of frequency λ and amplitude F together with a phase shift ϕ. The nonlinear response is a new limit cycle oscillation with amplitude A and frequency ω, phase-shifted, including harmonic ($\omega = \lambda$), superharmonic ($\omega = n\lambda$, $n = 2, 3, 4, \ldots$) and subharmonic ($\omega = \frac{1}{n}\lambda$, $n = 2, 3, 4, \ldots$) entrainment. Other types are almost periodic or nonperiodic solutions. In addition, a complete collapse of the oscillation may result, which leads to the onset of propagating pulses, travelling waves etc. Both, the region and limits of entrainment also depend on the strength of the nonlinearity which is present in the kinetic equations under consideration.

An exact theoretical analysis for nonlinear oscillators is hard to realize for both the unperturbed and the perturbed case. Approximation methods must be applied, which usually start with a combination of both free and forcing oscillations, including appropriate higher and subharmonics [15,22]. If these methods fail, only numerical integration procedures are applicable.

So far, we have only considered a single limit cycle oscillation. If several oscillations are present, coupling between different oscillations must be considered. Various coupling mechanisms have to be taken into account, e.g., direct coupling of the dynamic variables of individual oscillators, indirect coupling (i.e., coupling to the output of the other oscillators) or diffusive coupling.

Unsolved is the question, whether experimentally observed almost periodic oscillations and chaotic behaviour are caused by internal or external random fluctuations or by the intrinsic nonlinear nature of a purely deterministic dynamic system combined with a limit cycle coupling.

III. EXAMPLES FOR LIMIT CYCLE OSCILLATIONS

There exists an immense number of both theoretical models and experimental results which exhibit oscillatory behaviour. In this chapter we try to present a small selection of them, which is to be hoped to give a suitable review on these subjects.

III.1. Van der Pol Oscillator (VdP)

Van der Pol's equation [23]

$$\ddot{x} + \mu(x^2 - 1)\dot{x} + x = 0 \qquad (11)$$

can perhaps be regarded as the fundamental example of a nonlinear ordinary differential equation, which possesses self-sustained oscillations as periodic solutions. It can be used as a prototype of limit cycle oscillations and as an illustration for studying entrainment.

Equation (11) has been considered in connection with the subject of triode oscillations [24] and of electrical circuits with a neon-tube [25]. For $\mu \ll 1$, equation (11) represents a rather sinusoidal oscillation and for $\mu \gg 1$ an oscillation of relaxation type exists. Both types of oscillations have been analyzed extensively in the technical literature. Furthermore, by applying the Liénard theorem [13] it can be shown, that a unique stable limit cycle exists for all values of $\mu > 0$. The forced VdP oscillator (i.e., $F(t)$ on the right hand side of equation (11)) exhibits near resonance the typical harmonic entrainment phenomenon [26,27]: only forced oscillations are present with an amplitude larger than that of the free oscillation. Far from resonance, small amplitude oscillations which are a combination of both, free and forced oscillations, exist. Subharmonic entrainment was found experimentally in the neon-tube system [28]. The importance of the VdP oscillator to model biological oscillations will be discussed in section III.3.

III.2. Some experimental Results

In the meantime there exists an increasing number of fascinating experiments which exhibits nonlinear oscillations. Well-known examples from physics are ultrashort pulses of a strongly pumped laser [29], self-pulsing in bistable absorption systems [30] and electromagnetic oscillations in nonlinear media [31]. Further examples are oscillations in tunnel diodes, hydrodynamic oscillations, the parametric and the Gunn oscillator [vid. ref. 4,5].

Oscillations in chemical and biochemical systems are the subject of an intensive investigation. For recent reviews on chemical and biochemical instabilities leading to oscillations we refer to references [8], [11] and [32-35], where the relevant literature is presented. The Bray, the Belousov-Zhabotinski and the Briggs-Rauscher reactions exhibit various oscillations on a single chemical reaction chain level [36]. On a subcellular level, oscillations of NADH in the glycolytic cycle of metabolism and periodic synthesis of cyclic-AMP have been observed [37]. Oscillations have also been measured in artificial membranes [38]. An example is the Teorell-oscillator [39], which exhibits sustained electrical oscillations.

Oscillatory phenomena at a macroscopic dimension are frequently observable in the living world. We think, first of all, of rhythmic behaviour on an inter- and supercellular level, e.g., spontaneous and induced periodicities of bioelectric nature, such as the propagation of excitations along nerve membranes [40], the slow oscillations in the activity of the neurons in the cortex (i.e., electroencephalographic activity, EEG), especially α-rhythm of the human brain waves [41], the electrical activities recorded in the gastro-intestinal tract of humans and animals [42] etc. Under certain conditions the cells of the heart exhibit a spontaneous periodicity, an example to mention is the simultaneous firing of cardiac pacemaker cells [43].

Most developed organisms possess a multitude of endogenous rhythms which persist as free running oscillations even in the absence of external stimuli. Circadian rhythms are the most important examples, they may be considered as those biological rhythms which under constant conditions continue to exhibit oscillations with a period of about one day. Systems which belong to this class of oscillators are called circadian clocks or *biological clocks*. Circadian rhythms are subject to entrainment by environmental periodicities, e.g., light, temperature, noise, etc. Furthermore, very weak electromagnetic fields in the extreme low frequency region (10 Hz) can influence the period of the circadian rhythm of humans, where the latter are kept under constant conditions of light and temperature.

The period of the free running rhythm depends on genetic factors and prevailing environmental conditions. Pacemakers seem to be the underlying mechanism on a cellular and molecular level. Presumably the observed biological cycles may consist of a great number of coupled pacemaker loops. A detailed discussion of biological clocks, their physical and biochemical nature and their possibility to be entrained by external means can be found in a series of excellent books, e.g., [44-47]. Biological oscillators span a wide range of frequencies and wave-like phenomena.

Certain neural phenomena have frequencies in fractions of seconds up to milliseconds, biochemical oscillators are mainly in the frequency region of some minutes. Some other biological oscillators may be measured in months and years, e.g., population cycles in ecological systems [49]. In between there are the circadian rhythms.

However, it should be emphasized that most of the oscillatory phenomena in physical, chemical and biological systems are intimately linked with spatial order and spatial structures, leading to spatio-temporal activities or to spatial patterns.

III.3. Theoretical Models Exhibiting Oscillating Phenomena

Models for oscillatory behaviour and entrainment have been the subject of many investigations in the field of mechanical as well as of electrical engineering [vis. ref. 15]. We want to restrict ourselves to a few typical examples from physics, chemical and biological kinetics and biological rhythmic behaviour.

III.3.1. Models for Self-sustained Oscillations. Necessary requirements for a model system to exhibit a limit cycle oscillation have been discussed in the foregoing sections. Very briefly repeated, the existence of self-sustained oscillations of dynamical systems is only possible in open systems with dissipation, nonlinear interactions and structurally stable steady states. The latter ones must be stabilized far from thermal equilibrium, which means that some modes of behaviour are partially thermally decoupled from the rest of the system which may be viewed as a heatbath. The "far from equilibrium conditions" correspond to irreversibility and hence to unidirectionality of the inherent process.

Well-known examples from physics are the Laser rate equation [4], the Maxwell-Bloch equations to describe phenomena in nonlinear optics [17], the Lorentz equations for a description of turbulence [50], the retarded Josephson equation to describe the full dynamics of a Josephson junction [51], nonlinear electric field equations for Gunn-instabilities [52], to mention only a few of them.

Periodic chemical reaction models have been discussed for more than thirty years. Bonhoeffer was the first who stressed the importance of relaxation oscillations for chemical reactions [53]. Chemical oscillations should be created and maintained by two intrinsic trigger processes. Higgins investigated feedback and autocatalytic reaction mechanism, including the necessary mechanism for a chemical reaction to exhibit oscillations [32]. Temperature oscillations in chemical reactors [54] and electro-chemical oscillatory behaviour [39] represent further examples. A mathematical model for an abstract chemical reaction was developed by Prigogine and his group [1]. The so-called Brusselator seems to be the best studied reaction model with respect to temporal and statial order [11]. However, a realistic chemical system with Brusselator dynamics is still lacking. Necessary and sufficient conditions for a chemical reaction model system with only two dynamic variables are given in [55]. The oscillatory Belousov-Zhabotinski reaction found its mathematical description in the Oregonator model, which is based on the Field-Körös-Noyes (FKN) mechanism [56]. This model led to a full agreement between the main features of the *B-Z* reaction predicted by the FKN mechanism and experimental observations.

A number of hypothetic enzyme reaction models, including positive and negative feedback reaction mechanism and end product inhibition has been established by several authors. For a recent review see references [11,37]. Phenomenological models for glycolytic oscillations have been proposed to describe these metabolic periodicities, and reaction schemes for regulatory processes to model the periodic synthesis of cyclic-AMP have been suggested (for details vid. ref. 37).

Mathematical models for the control of mitosis, based on a

relaxation type of oscillations of a modified Brusselator system [57] and for catabolic repression [58] have been investigated in some detail. The Jacob-Monod model for induction and repression of cellular growth [59] and Goodwin's model for an enzyme synthesis control system including the functioning and differentiation of higher organisms [60] should be mentioned since these models have played some pioneering work in the development of oscillatory enzyme reactions and cellular control.

Prebiotic evolution has been described in a series of papers in Eigen's theory of selforganization of matter [61]. Ecological cycles have been modeled by several types of nonlinear differential and integro-differential equations, including delay and diffusion [49,62]. A well-known example is the Lotka-Volterra prey-predator system [63]. However, this system is a conservative one, exhibiting an infinite and continuous family of solutions. It is rather interesting to note, that the introduction of time delay leads to a stable and unique limit cycle solution, replacing the pathological neutrally stable cycles by a regular oscillation.

A description of biological clocks is practically not existing. Both, the relevant variables and the inherent nonlinear dynamics are not known. A suitable model should be based on a group of mutually coupled oscillations, each of which must be a stable limit cycle. Its frequency could be very much larger compared with that of the resulting clock, since it may be based on fast physiological processes. Fortunately, at least qualitative similarities exist between the behaviour of coupled oscillators and circadian rhythm.

III.3.2. Multiple, Coupled and Entrained Limit Cycles. Multiple limit cycle oscillations exhibit additional phenomena, the most interesting of which are threshold and excitability: a stable steady state or a stable limit cycle is separated from a second stable limit cycle by an unstable one, which divides the

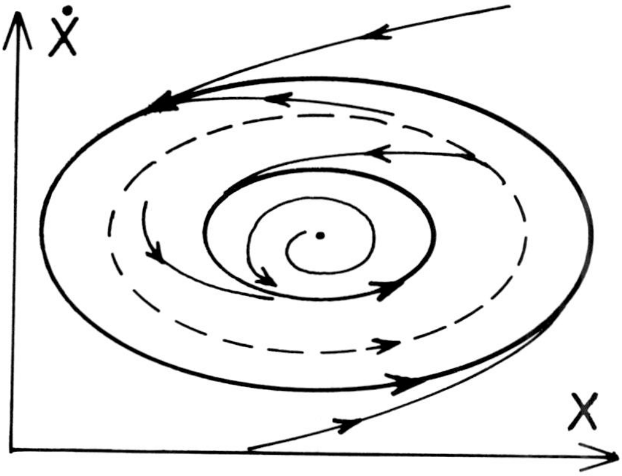

FIGURE 2. Steady state solutions of a multiple limit cycle system. The solid closed curves represent stable oscillations, the dashed one an unstable limit cycle. The arrows on the trajectories indicate the direction, the system runs with increasing time.

space of the variables into a small amplitude (or zero amplitude) oscillation and into a large one. A finite excitation (i.e., threshold) is required to drive the system from the stable small oscillation to the large amplitude oscillatory behaviour (i.e., excitability). An example of multiple limit cycles is given in Figure 2.

A well-known example is the generalized van der Pol oscillator [24], which describes the dynamics of a valve generator with the resonant circuit in the grid or in the anode circuit [15]. The afore mentioned model of catabolic repression [58] of metabolism and Selkov's model of a special enzyme-substrate reaction including two enzymes display multiple limit cycles.

Coupled or externally driven limit cycle oscillators have been widely used for a modelling of chemical, biochemical and biological functions. The coupling of two Brusselators in series without backcoupling exhibits a whole series of oscillating phenomena, i.e., synchronization, multiple periodic oscillations, subharmonic entrainment and almost periodic behaviour [64]. Diffusive coupling of two Brusselators served as a model for mitosis control [57]. For a harmonically driven Brusselator soft and hard mode instabilities and the typical behaviour of an entrained nonlinear oscillator have been found [65]. Two linearly coupled van der Pol oscillators have been the basis for a comparison of coupled biological oscillators for a description of population phenomena with a system of synchronized mathematical oscillators [66]. A single externally driven VdP oscillator has been investigated as an entrainment model for systems displaying adaptive behaviour [67]. RLC coupled symmetrical [68] and asymmetrical [69] VdP oscillators have been a successful model to describe the spontaneous electrical "slow-wave" rhythm of the gastro-intestinal tracts.

In all these models both, the VdP oscillator and the Brusselator scheme are functioning as purely mathematical models, which exhibit a single stable limit cycle. The inherent self-sustained oscillation serves as the starting point for coupling and entrainment, which both are necessary for a modelling of more complicated oscillatory phenomena. However, it should be stressed that a physical or chemical basis for the nonlinear mechanism leading to self-sustained oscillations is highly desirable, when one wants to describe the experimental findings of oscillating nature.

An example of this type of model formation is the modified and simplified Oregonator with a periodic external drive, called Kyoter [70]. Since it is based on the FKN mechanism [vid section III.3.1], it may be considered as a modelling beyond a mathematical level. Furthermore, it leads to additional results, which may be looked for in further experiments.

III.3.3 Periodic Activity of Neurons and Nervous Systems.
a. Neuronal Spikes

About thirty years ago, Hodgkin and Huxley constructed a phenomenological model for the generation and propagation of impulses in neuronal activity [40]. Oscillatory phenomena of the resulting Hodgkin-Huxley equation (HH) under space and current clamped conditions have been reported [71]. The mathematical investigation of this and related models has been mainly concentrated on establishing the existence and stability of pulse-like solutions of the dynamic equations with a diffusion term.

A rather simplified set of model equations has been developed [72,73]. The resulting nonlinear kinetic equations (i.e., FitzHugh-Nagumo equations (FHN)) represent a two-variable system which has the structure of the Bonhoeffer-van der Pol system (BVP). The latter is an extended version of the VdP oscillator including both, a decay and a source term, the so-called passivated iron-wire model for pulse propagation [74].

Much work has been devoted to the FHN model. Compared with the HH equations, this model is less closely related to experimental data, but it exhibits all typical effects and is easier to be investigated from a mathematical point of view. Again most studies have been concerned with space dependent problems, i.e., propagating pulses, trains of impulses, solitary waves, etc, since these types of solutions are required for a description of neuronal activities. Bifurcation phenomena, leading to periodic oscillations, have been found for the space-independent equations [75a,b]. All these phenomenological model equations are based on a simulation of the axon as a cylindrical electrical cable with a conducting core.

A coupling of BVP or FHN oscillators might be interesting with respect to a simulation of more complex oscillation patterns. However, such models are an insufficient description, since they are based on a purely mathematical level. A model of this type of treatment has been studied. It consists of two nearly identical neurons, which are assumed to be electrotonically coupled [76]. For two BVP neurons, which are coupled by diffusion and which are externally stimulated, only two kinds of stable periodic solutions exist: an in-phase and a 180-degree out-of-phase solution of the coupled system.

b. Rhythmic Activities in the Brain

Theoretical models for large-scale oscillatory phenomena in the nervous system have been worked out. Wilson and Cowan presented a model for a description of the experimentally found oscillatory response of the mammalian brain to an external electric stimulation [77,78]. The occurrence of self-sustained oscillations in certain parts of the brain at characteristic frequencies, which are not related to external stimuli, is a rather interesting fact. The origin of these periodic electric oscillations, exhibited as α, β, δ and θ rhythms in the EEG with frequencies mainly from 5 to 20 Hz, have been the subject of some theoretical work on brain waves. Wiener submitted the hypothesis that the brain can be regarded as a set of nonlinear and coupled oscillators, each of which can be entrained by the others [79]. Dewan has stressed the analogies between EEG activity and the phenomena related with limit cycles [80]. He proposed a model for brain waves in terms of a generalized VdP oscillator. In addition, a description of the peak-dip shape of an EEG power spectrum by a system of entrainable nonlinear oscillators with an external harmonic drive can reveal some analogies with experiments [81]. Again, all these models are not related to any known physiological data.

More realistic models should be based on the rather interesting new findings of brain research. The interaction of neurons in the central nervous system through graded electrotonic potentials rather than through spike transmission alone [82], synchronization in EEG pattern [83] and entrainment of oscillatory neural activity [84] are recent experimental results. Models for interconnected neurons, which include the collective behaviour of neural nets, may be viewed as alternative and new approaches [85-87]. Arguments that slow EEG activity might participate in an order maintaining function may serve as a first developmental step for a description of the fundamental role, which brain waves seem to play in the central nervous system [88].

The cerebral cortex of all vertebrates produces the well-known electrical rhythmic activity, externally recorded in the EEG. It has long been considered to be a noisy or fluctuating background without a direct physiological role in brain function, though detailed computer analyses of the EEG have indicated that correlated and synchronized states between EEG and behavioural states might exist [89].

The question of whether the EEG may have a physiological and a function maintaining role in brain tissue is closely related to the problem of possible interactions between intrinsic or environmental electric fields and cerebral tissues. Effects of weak electromagnetic fields on the behaviour of both, humans and animals have been observed [90]. As a result, the function of the mammalian central nervous system can be modified by electric gradients in cerebral tissue that are substantially smaller than those occurring in postsynaptic excitation. Adey has proposed that a hierarchy of excitatory processes may exist, in which synaptic mechanism present but one level [91]. Both, the neuronal spikes created in the axon hillock of the soma and slow neuronal wave activity generated in the dendrites seem to be incorporated in the joint generation of the EEG. This speculative cerebral cooperativity serves as a working model to describe effects of electromagnetic fields on brain functions [92].

Brain interactions with weak fields by cooperative mechanism and consequently the effectiveness of extremely weak tissue electric gradients (possibly below those of thermal gradients) have been the starting point for some rather speculative physical models and descriptions. We want to restrict to a few of them, namely, to those, which include nonlinear oscillations. Kaczmarek

[93] stressed the importance of limit cycles in modeling the calcium binding situation in biological membranes. He presented a cooperative calcium ion binding model, where very weak electromagnetic oscillating fields could influence the rate constants for the initial Ca^{2+} binding. Though exhibiting an extreme sensitivity to the frequency of the external stimulus, the model does not completely explain the experimental findings, i.e., frequency and intensity windows of both, the increased and decreased Ca^{2+} efflux.

A multiple limit cycle model for a description of these experimental findings has been suggested by the author [93]. The model (a generalized VdP oscillator) exhibits two stable oscillations and an unstable one [vid. Fig. 2]. Depending on both, the frequency and the intensity of the external drive, a series of instabilities may occur, the most important of which is the suppression of the small amplitude oscillation for a sufficiently large perturbation. For higher frequencies and appropriate intensities, the large amplitude oscillation also gets suppressed. Both, the frequency and intensity windows can be explained with these transitions. However, since a physical basis for this mathematical model is still lacking (but under investigation), details of the calculations will be omitted. Fröhlich [94] has suggested that the high sensitivity of the brain and of other biological systems to weak electromagnetic signals seems to imply that these systems possess a storage mechanism based on nonlinear oscillations of limit cycle type. This concept will be presented in some detail in the next section. The existence of low-level electromagnetic field effects in biological systems is still a matter of an extensive investigation. The experiments are mainly concerned with nonionizing radiation in the ELF, the radio-frequency and the micro-wave region [91-95]. The interpretation of the observed effects is highly controversal. Both, thermal and nonthermal effects may contribute to the resulting behaviour.

The complete problem of weak electromagnetic field interactions with biological systems must be evaluated in the light of modern physics. The role, which the concepts of transport instabilities and dissipative structures can presumably play will be discussed in the subsequent section. Special emphasis will be put on non-linear oscillations of limit cycle type and on their modelling on a physical basis.

IV. COHERENT OSCILLATIONS IN BIOLOGICAL SYSTEMS

IV.1. Long Range Coherence

Biological systems exhibit stability in a way, in which some modes of behaviour remain very far from thermal equilibrium. To give an interpretation of this obvious fact, one should primarily look for basic physical ideas, which may serve as a starting point to establish physical theories for biological systems. The functional complexity of biological materials requires the application of macroscopic theories. These theories should rather describe the collective properties of the considered system than its behaviour on a molecular level. Obviously, the great success of molecular biology arises from the discovery of the detailed structure of biomolecules such as DNA and enzymes. However, since no systematic spatial order was found, new concepts such as order of motion seem to be appropriate. Only the combination of both, the local chemical kinetics and some collective phenomena are considered as a suitable description on a physical basis.

From the point of view of theoretical physics it must be realized that in contrast to most materials usually treated in physics, biological materials possess extraordinary properties as a consequence of a long evolutionary process. Furthermore, it should be emphasized that biological materials can only exhibit their typical properties, when they are "fed," i.e., when energy passes through them, leading to a stabilization of the "active" state.

In physics, order of motion as well as other types of order are connected with the existence of macroscopic wave functions, which implies the existence of phase correlations over macroscopic distances and of coherent states. In 1967, Fröhlich suggested that long range coherence might well describe order in biological functions, too [97]. The possible relevance of long range phase correlations in biological systems is rather tempting and will be discussed subsequently.

IV.2. Coherent Oscillations

Striking material properties common to most biological systems are their extraordinary dielectric properties. Most interesting are the electric fields in cell membranes, which have average values of the order of 10^5 V/cm. These fields play a basic role in nerve conduction, but their function in the membranes of other cells is nearly unknown.

By these internal fields the whole membrane is strongly polarized. In addition, an oscillation of parts of the membrane is connected with corresponding electric vibrations. The longest of the longitudinal electric vibrations can yield an electric dipole vibration of the whole system. Fröhlich has made two suggestions, which can be taken as the physical basis for a theoretical investigation [96,97]:

Postulate 1: long wavelength electric vibrations are very strongly and coherently excited in biological systems, when the latter are active, i.e., when metabolic energy is available.

Postulate 2: biological systems have metastable states with a very high electric polarization.

Although these suggestions are rather speculative, they have received some theoretical support, mainly by Fröhlich's "vibrational model" and by his "high-polarization model." These phenomenological models have got some additional justification by a microscopic treatment of these models [vid. ref. 98 for details]. Both, the transition to a highly-polarized state and to a "Bose-condensation-like excitation" of at least one mode can only occur, when the external energy supply exceeds critical threshold values. In addition, strong nonlinear interactions are absolute necessities. All these types of excitation can have far reaching consequencies for the behaviour of biological systems. If, for example, the coherently excited vibration represents an oscillating giant dipole, then long range and frequency selective forces may be activated [99]. Furthermore, Fröhlich's proposal has received increasing support through some rather exciting experimental results. Some evidence for the existence of coherent oscillations in the proposed frequency region (10^{10} - 10^{12} Hz) has been found in bacteria and in yeast, including frequency effects at small intensities, sharp resonances and threshold behaviour [100-104]. Highly excited sharp Anti-Stokes lines in Raman spectra are an indication of a strong excitation of some modes [103,105]. "In vivo" experiments with micro-dielectrophoresis point to the existence of oscillatory cellular dipole fields [106].

It should be emphasized again that Fröhlich's suggestions are heavily based on general results of the physics of dielectric material. Yet, the consequences of these physical laws for biological systems are by no means unique and a matter of further laborious investigations.

IV.3. Limit Cycle Model for Brain Waves

Recently, Fröhlich has extended his concept to give a possible explanation of the extraordinary high sensitivity of certain biological systems to very weak electric and magnetic fields. Details of these experiments may be found in reference [91].

From the above mentioned considerations and model calculations, he has derived a further postulate,

Postulate 3: the selective long range interaction in conjunction with the existence of highly polar metastable states and of coherent electric vibrations may be decisive for the establishment of the well-known low frequency oscillations in the brain (i.e., EEG activity).

This postulate contains a specific statement of a more general concept. Quite generally, we apply three further hypotheses, which may be stated as follows:

- observable biological oscillations must be stable limit cycles, coupled sets of limit cycles or entrained ones,
- stable limit cycles model coherent oscillations of active systems,
- stable limit cycles are a suitable basis for a description of the effects of external fields on biological systems.

The limit cycle concept reveals a possible explanation of the experimentally found specific sensitivities of biological systems to a weak external stimulus by the following mechanism: a limit cycle represents a storage of energy, e.g., metabolic energy in a biological situation. This inherently stored energy makes possible the creation of a response signal, though only a very small amount of external energy is available. The only function of the externally supplied energy is to start a response signal, if a certain threshold is reached. This trigger action is highly frequency and intensity dependent. Details of the resulting transport instability depend on the specific system.

Through the mechanism of an externally driven internal oscillation, the existence of nonthermal effects becomes plausible. The relevance of this concept will be demonstrated by Fröhlich's brain wave model.

Details of the model are given in reference [94]. Assuming a selective long range interaction for biomolecules (vid. Post. 3), an additional supply of substrates can lead to a collective enzyme-substrate reaction in the Greater Membrane of the brain. The possibly resulting periodic chemical processes (i.e., periodic activation and deactivation of polar enzymes) is connected with an electric oscillation. Parts of the system may then exhibit low frequency coherent oscillations over macroscopic areas. These oscillations might well be related to the rhythmic activity of parts of the brain (i.e., brain waves).

The brain wave model is described by a set of nonlinear differential equations, which in a transformed version reads [94]:

$$d_t\nu = \gamma\sigma + (c^2 e^{-r^2\nu^2} - d^2)\nu + \alpha A\sigma\nu + F(t) \tag{12}$$

$$d_t\sigma = -\beta\nu - \alpha A\sigma\nu \tag{13}$$

with

$$\nu = N - \gamma/\alpha A \; ; \; \sigma = S - \beta/\alpha A \; ; \; \alpha,\beta,\gamma > 0.$$

N, A are the concentrations of excited and unexcited enzyme molecules, respectively, S is the number of substrate molecules per unit volume. $\gamma\sigma$ results from the assumed long range interaction, $\beta\nu$ and $\alpha A\sigma\nu$ originate from the nonlinear enzyme substrate reactions. Aside from these "chemical" terms we have an additional "dielectric" term, which consists of two parts: a term describing the system's tendency to become "ferroelectric," i.e., $c^2 e^{-r^2\nu^2}\nu$ and one for frictional losses (electrical resistances), i.e., $-d^2\nu$. $F(t)$ results from the interaction of the system with its surrounding and needs not necessarily be an external electromagnetic field. It may well represent external or internal chemical influences of a systematic or a random nature.

Equations (12,13) have been discussed in detail in a series of papers [107-109], where special emphasis has been on all possible types of steady state solutions. In addition, the relevant bifurcation schemes, the existence of a limit cycle, its collapse and the onset of different kinds of travelling waves have been investigated.

The most relevant situation is the existence of a self-sustained oscillation for $F(t) = 0$, with amplitude ν_o and frequency ω_o. A necessary prerequisite is $c^2 > d^2$. The resulting oscillation is a low frequency one, the high frequency behaviour is hidden in the parameters α, β and γ.

We only want to present those results, which we have received by an application of a numerical integration procedure for a harmonic external stimulus, i.e., $F(t) = F_o \cos\lambda t$ (underlying details are given in references 107-109).

a. *Time-independent stimulus*

Both $\lambda = 0$ and $F(t) = \langle F(t)\rangle_t$ belong to this class of perturbations ($\langle\ldots\rangle_t$ denotes a suitable averaging procedure).

Starting with $F_o = 0$, an intensity increase of the external stimulus leads to a decrease of both, the frequency and amplitude of the internal oscillation. For $c^2 < d^2 + \beta$, one gets a second order limit cycle bifurcation, i.e., at F_{c2} the amplitude is zero though the frequency is still finite (hard mode transition). A nonoscillating steady state solution of focus type is present for $F_o > F_{c2}$, which for $F_o \geq F_{c3}$ bifurcates into a stable node (soft mode transition). However, for $c^2 > d^2 + \beta$ (i.e., strong "ferroelectric" tendency compared with the loss process), a first order limit cycle bifurcation occurs. This means that a zero frequency, finite amplitude oscillation gets unstable, yielding a stable node for $F_o > F_c$. For $F_{c1} < F_o < F_c$, relaxation oscillations are present, bifurcating at F_{c1} from a rather sinusoidal type of oscillations.

Both, the amplitude ν_λ and frequency ω_λ of the resulting oscillations as a function of F_o are drawn in Figure 3. Figure 4 shows the related oscillations and phase plane diagrams for $c^2 \gtrless d^2 + \beta$.

As a general result, the limit cycle is suppressed, if the strength of the external time-independent field is driven beyond a certain critical value. Furthermore, the resulting nonoscillating steady state is unstable with respect to a space dependent perturbation. Thus a spatio-temporal structure gets built up.

(a)

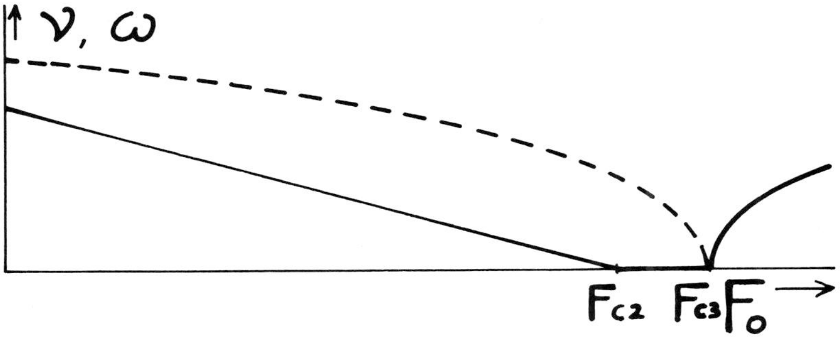

(b)

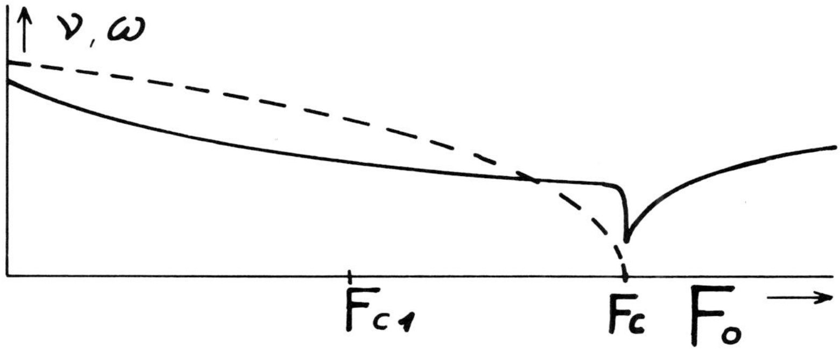

FIGURE 3. Amplitude ν(——) and frequency ω(---) of the limit cycle as a function of the external stimulus F_o for a time independent perturbation. (a) $c^2 < d^2 + \beta$: hardmode transition, $F_o = F_{c2}$. (b) $c^2 > d^2 + \beta$: softmode transition with nonzero amplitude ν (for details see text).

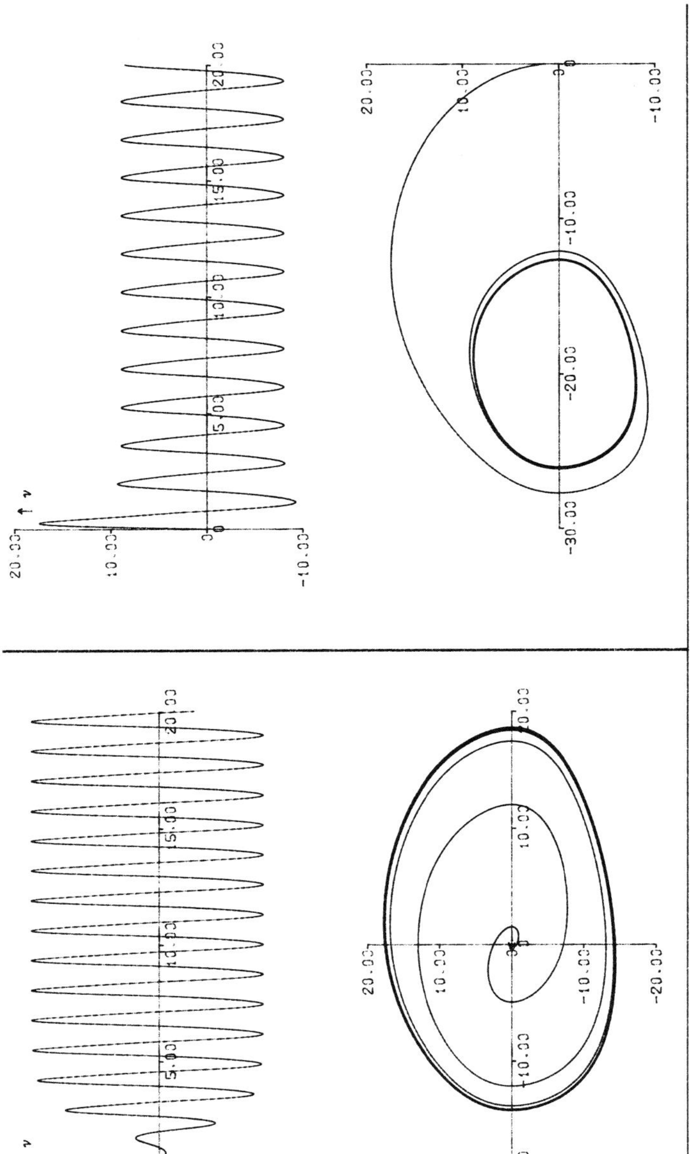

Figure 4a

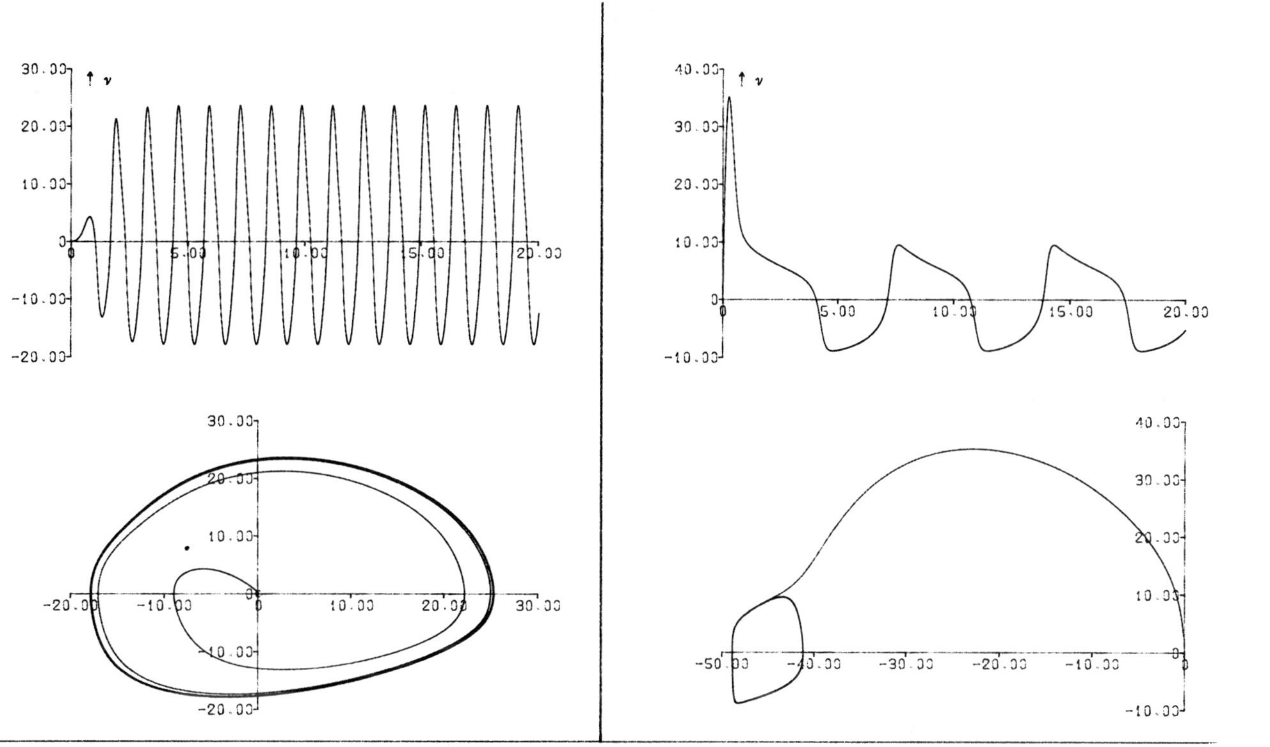

Figure 4b. Solutions of equations 12 and 13 (computer integration procedure) for a time independent stimulus, i.e., $\lambda = 0$. The oscillation patterns show ν as a function of time t, the corresponding phase plane diagrams present ν as a function of σ. (a) $c^2 < d^2 + \beta$; left hand side: $F_o = 0$; right hand side: $0 < F_o < F_{c^2}$. (b) $c^2 > d^2 + \beta$; left hand side: $F_o = 0$; right hand side; $F_{cl} < F_o < F_c$.

b. Time-dependent stimulus

Let us start with the free oscillation of limit cycle type, the frequency and amplitude of which are ω_o and ν_o, respectively. Depending on both, the frequency and amplitude of the external drive (i.e., λ and F_o), the typical features of a nonlinear resonance behaviour for self-sustained oscillations are exhibited. The results may be summarized as follows:

1. For $\omega_o - \delta < \lambda < \omega_o + \varepsilon$ (i.e., near resonance), complete entainment is achieved. The forced oscillation is a proper limit cycle with frequency λ and amplitude ν_λ (with $\nu_\lambda > \nu_o$). Both δ and ε depend on the strength of the perturbation, F_o. For increasing F_o, both values increase, which means that the region of entrainment increases and thus the region, where the external oscillation dominates.
2. For $\lambda \approx \omega_o + \varepsilon$ an oscillation with the external frequency results, the amplitude of which is far below ν_o. The resulting forced oscillation is partially suppressed. This quenching effect at the limit of entrainment is usually exhibited by limit cycles.
3. Outside the entrainment region ($\lambda < \omega_o - \delta$ and $\lambda > \omega_o + \varepsilon$), almost periodic oscillations occur. This type of oscillations possesses a whole set of frequencies and amplitudes, the averages of which are approximately ω_o and ν_o. The free oscillation dominates on the average. Furthermore, typical beat oscillations appear near the boundary of entrainment. Both, the resulting frequency and amplitude as a function of the external frequency λ are drawn in Figure 5 for a fixed value of F_o.
4. For λ fixed, one gets quite a series of transitions, when F_o increases. The following scheme presents the qualitative behaviour:

$$\begin{matrix} \omega_o & & \{\omega \gtrless \omega_o\} & & \lambda & & \lambda \\ \nu_o & \longrightarrow & \{\nu \gtrless \nu_o\} & \longrightarrow & \{\nu \gtrless \nu_o\}^{\pm} & \longrightarrow & \nu_\lambda \end{matrix} \qquad \text{(I)}$$

(a)

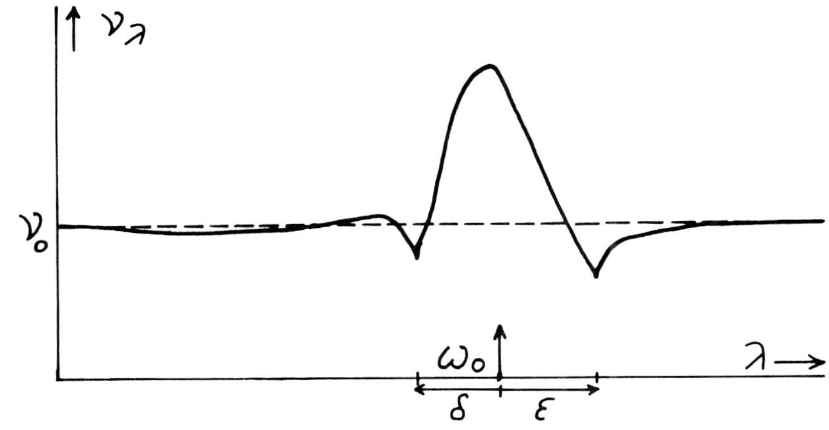

(b)

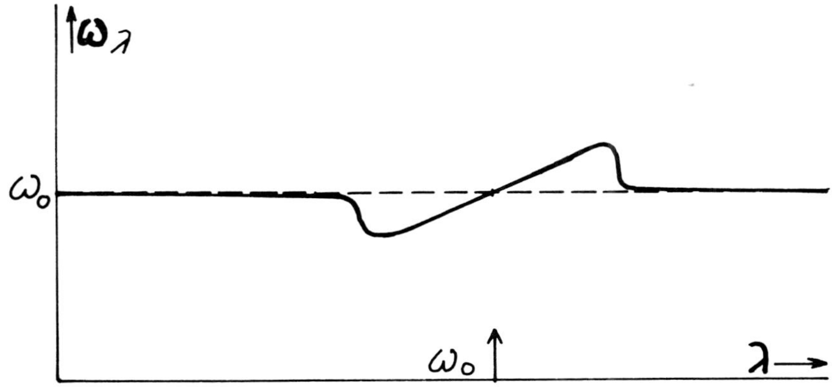

FIGURE 5. Resulting steady state amplitude ν_λ (a) and frequency ω_λ (b) of the perturbed limit cycle as a function of the external stimulus, $F = F_o \cos\lambda t$, for F_o fixed (i.e., $F_o = F_f$). ω_o and ν_o are the frequency and amplitude, respectively, of the unperturbed limit cycle. Both a typical nonlinear resonance (a) and frequency entrainment region (b) are exhibited.

The free oscillation (ω_o, ν_o) is perturbed, yielding almost periodic oscillations $(\{\omega\},\{\nu\})$. For a further increasing F_o, additional oscillations are built up $(\lambda > \omega_o, \{/\}^+)$ or some oscillations die out $(\lambda < \omega_o, \{/\}^-)$. This result demonstrates the transition from the region of nearly free oscillations to that of entrained ones. Subsequently, all amplitudes are stabilized to ν_λ; as a result, only the forced oscillation (λ, ν_λ) is present (complete entrainment). If F_o is increased a little bit more, the oscillations get unstable, in addition no stable nonoscillating solution can exist.

5. For increasing amplitude F_o the maximum of the resonance curve in the entrainment region is more and more shifted to lower frequencies. This behaviour explains, why the lowest intensity, which leads to a collapse of the oscillation, is found well below ω_o, the frequency of the free oscillation.

In Figure 6 several resonance curves are drawn.

Some typical externally perturbed oscillations are shown in Figures 7 and 8. Figure 7a exhibits almost periodic oscillations and the result, when some oscillations have died out, whereas in Figure 8a complete entrainment is already achieved for small intensities. Special emphasis should be given to the existing large amplitude transients, which finally are responsible for the collapse. In Figure 9 the steady state limit cycle amplitude ν_λ is presented as a function of the external field F_o. Starting with $F_o = 0$, the amplitude increases continuously from its unperturbed value ν_o. However, at the transition point from the almost periodic oscillations to the entrained ones, a drastic amplitude jump to smaller values occurs. Beyond this critical region, the amplitude again increases. Finally, the oscillation collapses in the same way as near resonance.

It should be emphasized that both, an adiabatic switching on of the external perturbation and the inclusion of a phase shift between internal and external oscillation will only lead to very

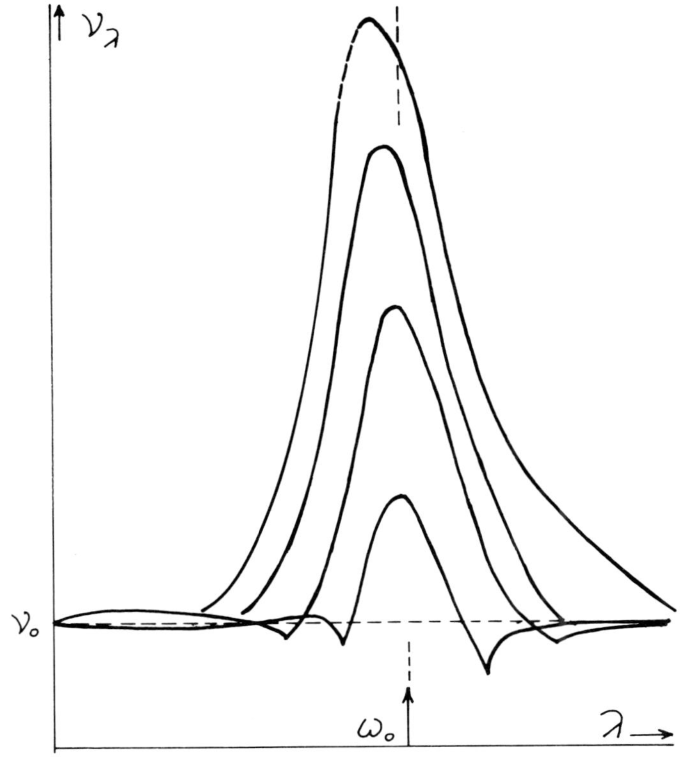

FIGURE 6. Resulting steady state amplitude ν_λ as a function of the external stimulus $F = F_o \cos\lambda t$ for four different values of F_o ($F_o = n F_f$; n = 1, 2, 3, 4). The dashed part of the upper curve corresponds to the onset of instabilities, i.e., limit cycle collapse (vid. Figure 5 for details).

slight changes in the resulting behaviour; consequently the details of these calculations are omitted, the statements 1-5 of this section remain valid.

In conclusion, these results show that a wide range of nonlinear oscillating behaviour can result, among which the sharp nonlinear resonances, the existence of an anti-resonance like behaviour (partial quenching) and the occurrence of beatmodes should be noticed.

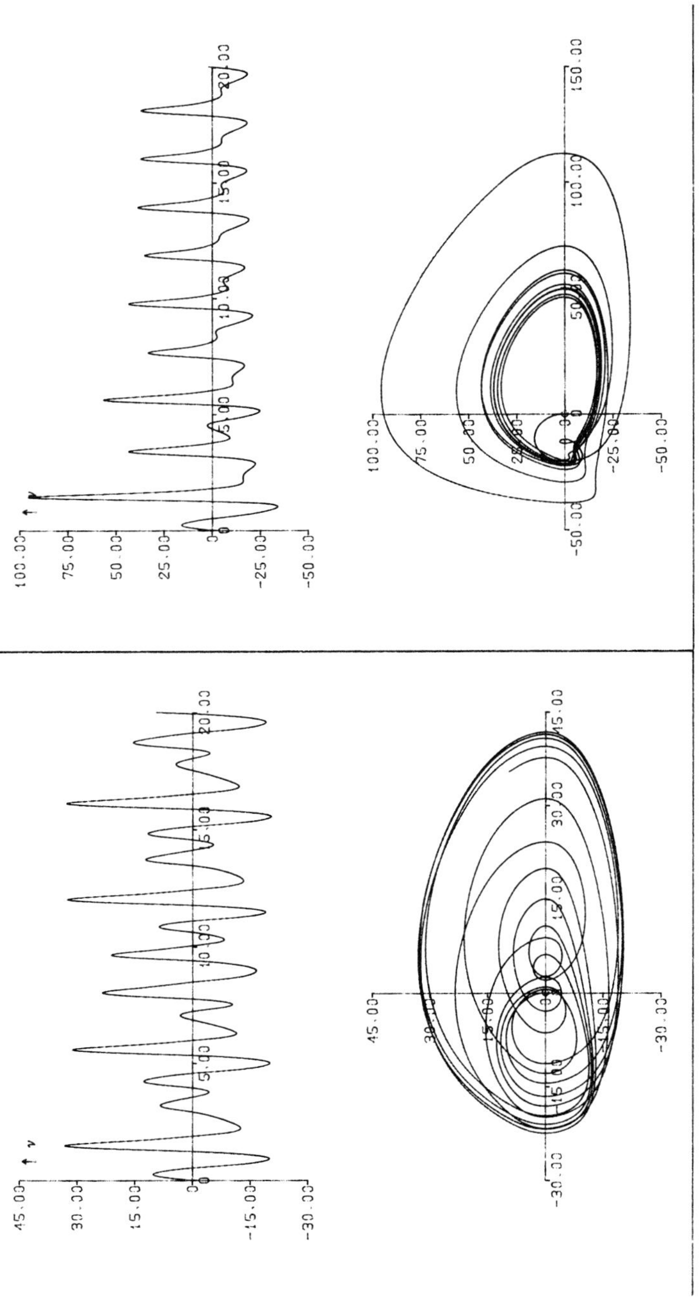

v
45.00
30.00
15.00
0
-15.00
-30.00
5.00
10.00
15.00
20.00
100.00
75.00
50.00
25.00
-25.00
-50.00
150.00

Figure 7a.

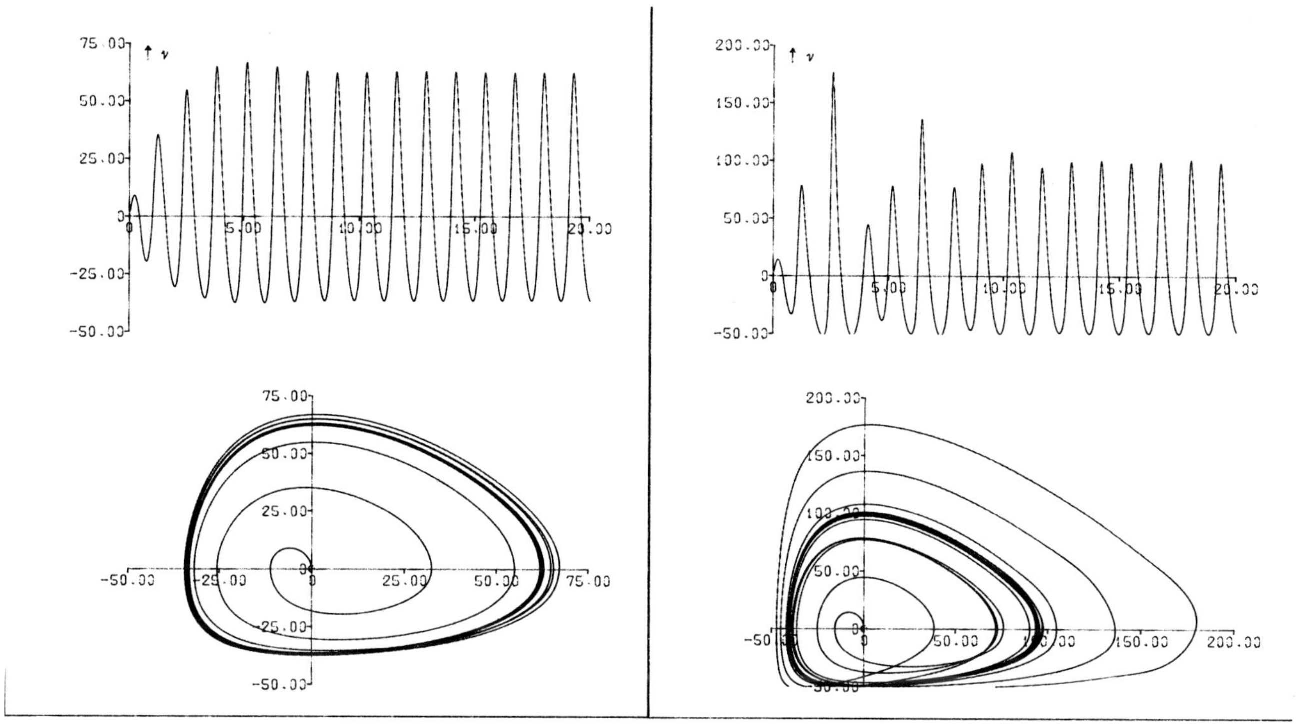

Figure 7b.

Figure 7. Computer integrated equations 12 and 13 for a time dependent stimulus $F=F_o \cos \lambda t$. (a) $\lambda = 2/3\ \omega_o$; left hand side: $F_o = F_s$; right hand side: $F_o = 2F_s$. (b) $\lambda = \omega_o$; left hand side: $F_o = F_s$; right hand side: $F_o = 2F_s$. (ω_o = frequency of unperturbed limit cycle, $F_s = 1/2\ F_c^2$).

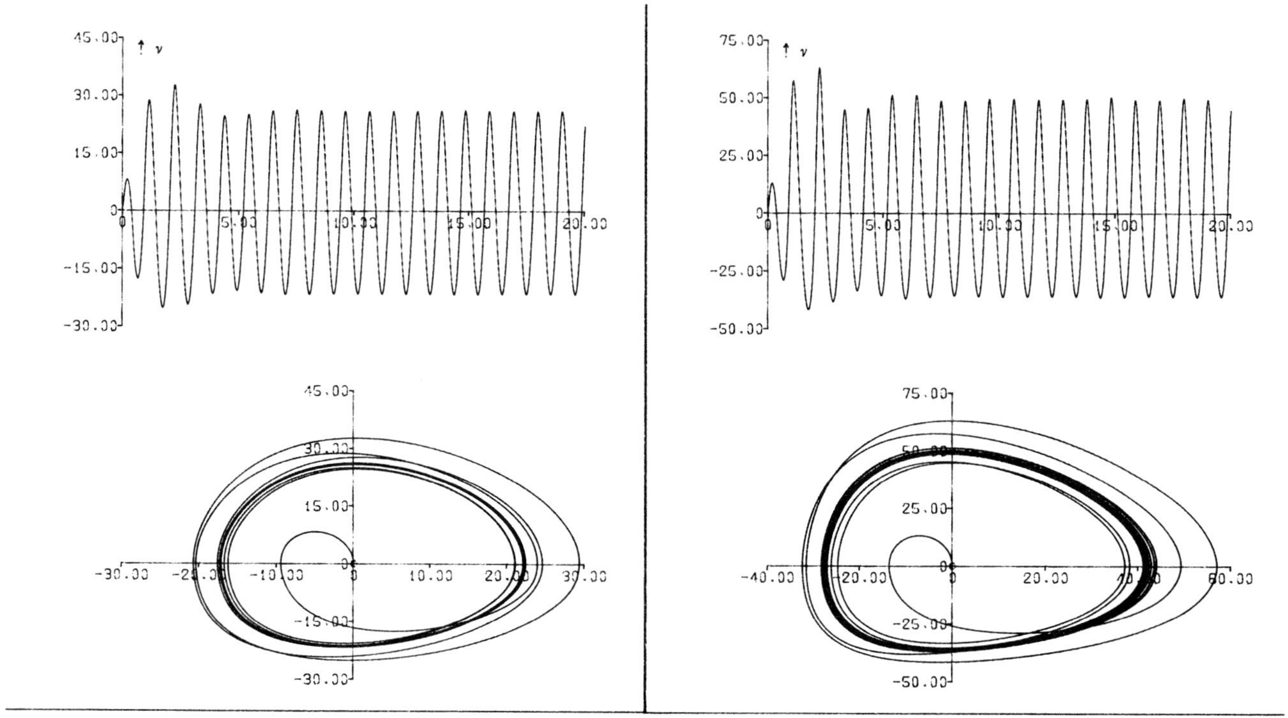

45.00
30.00
15.00
0
-15.00
-30.00
↑ ν
5.00
10.00
15.00
20.00
75.00
50.00
25.00
-25.00
-50.00
-10.00
-20.00
10.00
-40.00
20.00
60.00

Figure 8a.

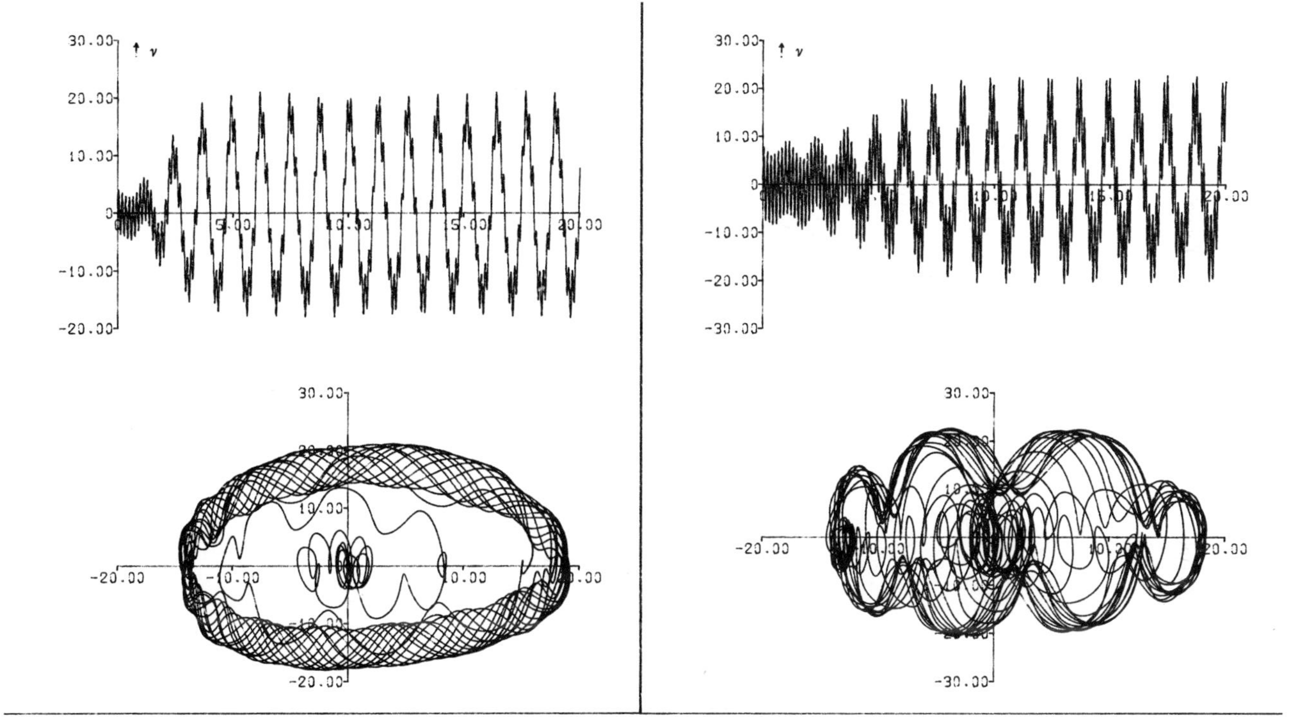

Figure 8b.

Figure 8. Figure 7 continued. (a) $\lambda = 1.2\omega_0$; *left hand side:* $F_0 = F_s$; *right hand side :* $F_0 = 2F_s$. *(b)* $\lambda = 8\omega_0$; *left hand side:* $F_0 = 3F_s$; *right hand side:* $F_0 = 6F_s$.

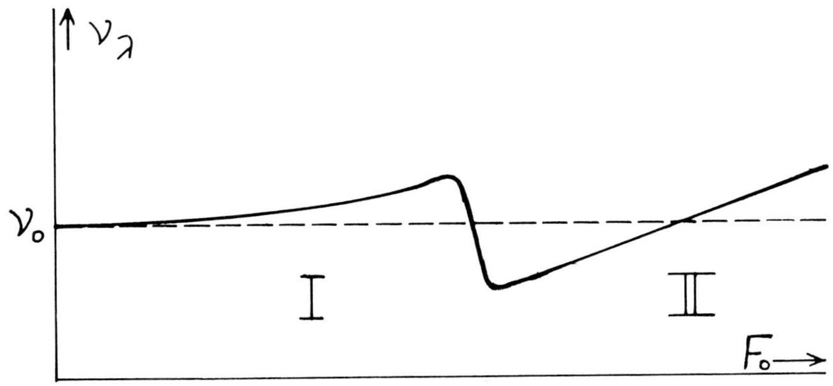

FIGURE 9. Resulting limit cycle amplitude ν_λ as a function of the externally applied intensity for λ fixed, i.e., $\lambda >> \omega_0$. Region I corresponds to almost periodic oscillations, region II to entrainment. The amplitude jump occurs at the limit of entrainment, beyond this limit only forced oscillations do exist.

Though the model is highly speculative and a matter of controversal discussions, it might at least serve as a guide line, how both, a biological functional order and its interaction with perturbations (e.g., electromagnetic fields) can be described on a suitable physical basis.

A detailed discussion of biological effects in the microwave region and of related questions is given in a very recent paper of Fröhlich [110].

V. CONCLUDING REMARKS

We would like to comment particularly on four remarkable aspects of nonlinear oscillations of self-sustained type.

First, these oscillations can only occur in systems, where a competition between energy input and energy dissipation and nonlinear interactions are feasible, even when the system is in his stable steady state. This type of dynamics is intimately linked with nonequilibrium constraints and irreversibility. The intrinsic order of these systems is achieved by cooperative phenomena, involving a great number of atoms or molecules. The cooperativity amounts to the establishment of a precise coordination of the dynamics, including interactions of the subsystems over a characteristic time-span scale of macroscopic dimension. Second, a description of nonlinear phenomena must be heavily based on the physical properties of the subsystems. It should include the relevant physical laws governing the interactions between the units as well as with its surrounding. Furthermore, it should be emphasized that the solutions of the kinetic equations depend also on macroscopic features such as the system size and geometry and on boundary conditions. A mathematical description without a suitable accomodation to the physical situation may only be accepted at a very preliminary stage of investigation.

Third, oscillations of limit cycle type reveal a possible description of trigger actions, which seem to be an important step in the creation of instabilities in biological and in other systems. Thus, in the case of a time-periodic behaviour, both, the internal dynamic state and its nonlinear response to a stimulus receive a suitable description. As a consequence of this concept, the extreme sensitivity of certain active biological systems to electromagnetic fields gets a rather plausible explanation, including both, the very spectacular frequency and the intensity dependency found in some experiments.

Fourth, nonequilibrium phenomena such as limit cycles are the result of instabilities and subsequent bifurcations. In order to present a complete treatment of the dynamics and the resulting dissipative structures, one must incorporate the effect of fluctuations. Consequently, a deterministic description must be supplemented by a stochastic one, including probability distributions and stability questions. Furthermore, a statistical treatment of open systems is absolutely necessary for multistable systems and for transitions to multiple states, whereas in the case of an oscillatory state, a deterministic level seems less questionable. Organized states of oscillating nature, whose occurrence critically depends on the properties of fluctuations, require an improved treatment. However, questions concerning the role of noise and fluctuations and the onset of stochastic phenomena, were not the subject of this chapter.

REFERENCES

1. Prigogine, I., *in* "Theoretical Physics and Biology" (M. Marois, ed.), pp. 23-31. North-Holland, Amsterdam, (1969).
2. Glansdorff, P. and Prigogine, I., "Thermodynamic Theory of Structure, Stability, and Fluctuations" J. Wiley, New York, (1971).
3. "Synergetics" (H. Haken, ed.) B.G. Teubner, Stuttgart, (1973).
4. Haken, H., "Cooperative phenomena in systems far from thermal equilibrium and in non-physical systems," *Rev. Mod. Phys. 47*, 67-121 (1975).
5. "Cooperative Effects" (H. Haken, ed.) North-Holland, Amsterdam, (1974).
6. Haken, H., "Synergetics, an Introduction" Springer-Verlag, Berlin, (1978).
7. "Synergetics, a Workshop" (H. Haken, ed.) Springer-Verlag, Berlin, (1978).
8. "Synergetics, far from Equilibrium" (A. Pacault and C. Vidal, eds.) Springer-Verlag, Berlin, (1979).
9. "Pattern Formation by Dynamic Systems and Pattern Recognition" (H. Haken, ed.) Springer-Verlag, Berlin, (1979).
10. "Cooperative Phenomena" (H. Haken and M. Wagner, eds.) Springer-Verlag, Berlin, (1973).
11. Nicolis, G. and Prigogine, I., "Self-organization in Nonequilibrium Systems" J. Wiley, New York, (1977).
12. Nicolis, G., "Irreversible thermodynamics," *Rep. Prog. Phys. 42*, 225-268 (1979).
13. LaSalle, J. and Lefshetz, S., "Stability by Ljapunov's Direct Method" Academic Press, New York, (1961).
14. Hayashi, C., "Nonlinear Oscillations in Physical Systems" McGraw-Hill, New York, (1964).
15. Andronov, A.A., Vitt, A.A. and Khaikin, S.E., "Theory of Oscillators" Pergamon Press, Oxford, (1970).
16. Hirsch, M.W. and Smale, S., "Differential Equations, Dynamical Systems, and Linear Algebra" Academic Press, New York, (1974).

17. "Coherence in Spectroscopy and Modern Physics" (F.T. Arecchi, R. Bonifacio and M.O. Scully, eds.) Plenum Press, New York, (1978).
18. Andronov, A.A., Leontovich, L.A., Gordon, I.I. and Maier, A.G.,
a. "Qualitative Theory of Second-order Dynamic Systems"
b. "Theory of Bifurcations of Dynamic Systems on a Plane" J. Wiley, New York, (1973).
19. Marsden, J.E. and McCracken, M., "The Hopf-bifurcation and its Application" Springer-Verlag, Berlin, (1976).
20. Thom, R., "Stabilité Structurelle et Morphogénèse" Benjamin, New York, (1972).
21. Poincaré, H., "Les Méthodes Nouvelles de la Mécanique Célèste" Gauthiers-Villars, Paris, (1892).
22. Struble, R.A., "Nonlinear Differential Equations" McGraw-Hill, New York, (1962).
23. van der Pol, B., "On relaxation oscillations," *Phil. Mag. Ser. 7, 2*, 978-992 (1926).
24. Appleton, E.V. and van der Pol, B., "On a type of oscillation-hysteresis in a simple triode generator," *Phil. Mag. Ser. 6, 43*, 177-193 (1922).
25. van der Pol, B. and van der Mark, J., "The heartbeat considered as a ralaxation oscillation and an electrical model of the heart," *Phil. Mag. Ser. 7, 2*, 763-775 (1926).
26. van der Pol, B., "Forced oscillations in a circuit with nonlinear resistance," *Phil. Mag. Ser. 7, 13*, 65-79 (1927).
27. Holmes, P.J. and Rand, D.A., "Bifurcations of the forced van der Pol oscillator," *Quart. Appl. Math. 1*, 495-509 (1978).
28. van der Pol, B. and van der Mark, J., "Frequency demultiplication," Nature *3019/120*, 363-364 (1927).
29. Haken, H. and Ohno, H., "Onset of ultrashort laser pulses: first or second order phase transition?" *Opt. Comm. 26*, 117-118 (1978).
30. Bonifacio, R., Gronchi, M. and Lugiato, L.A., "Self-pulsing in bistable absorption," *Opt. Comm.* (preprint).
31. Demchenko, V.V. and El-Siragy, N.M., "On stability of electromagnetic oscillations in media with cubic nonlinearity," Physica *67*, 333-346 (1973).
32. Higgins, J., "The theory of oscillating reactions," *Ind. Ing. Chem. 59*, 18-62 (1967).
33. Nicolis, G. and Portnow, J., "Chemical oscillations," *Chem. Rev. 73*, 365-384 (1973).
34. Hess, B. and Boiteux, A., "Oscillatory phenomena in biochemistry," *Ann. Rev. Biochem. 40*, 237-258 (1971).
35. "Biological and Biochemical Oscillators" (B. Chance, E.K. Pye, B. Hess and A. Ghost, eds.) Academic Press, New York, (1973).
36. Noyes, R.M. and Field, R.J., "Oscillatory chemical reactions," *Ann. Rev. Phys. Chem. 25*, 95-119 (1974).
37. Hess, B., Goldbeter, A. and Lefever, R., "Temporal, spatial and functional order in regulated biochemical and cellular

systems," *Adv. Chem. Phys. 38*, 363-412 (1978).
38. Franck, U.F., "Kinetic feedback processes in physico-chemical oscillatory systems," *Faraday Symp. 9* (1974).
39. Shashoua, V.E., "Electrical oscillatory phenomena in protein membranes," *Faraday Symp. 9* (1974).
40. Hodgkin, A.L. and Huxley, A.F., "Quantitative description of membrane current and its application to conduction and excitation in nerve," *J. Physiol. 117*, 500-544 (1952).
41. Bullock, T.H., "Introduction to Nervous System" W.H. Freeman, San Francisco, (1977).
42. Duthie, H.L., "Electrical activity of gastro-intestinal smooth muscle," *GUT 15*, 669-681 (1972).
43. Moe, G.K., Rheinboldt, W.C. and Abildoskov, J.A., "A computer model for atrial fibrillation," *Am. Heart J. 67*, 200-220 (1964).
44. "Circadian Clocks" (J. Aschoff, ed.) North-Holland, Amsterdam, (1965).
45. Brown, F.A., Hastings, J.W. and Palmer, J.D., "The Biological Clock" Academic Press, New York, (1971).
46. Bünning, E., "The Physiological Clock" Springer-Verlag, Berlin, (1973).
47. Pavlidis, T., "Biological Oscillators: Their Mathematical Analysis" Academic Press, New York, (1973).
48. Hastings, J.W. and Schweiger, H.-G., "The Molecular Basis of Circadian Rhythms" Dahlem-Konferenzen, Berlin, (1976).
49. May, R.M., "Stability and Complexity in Model Ecosystems" PUP, Princeton, New Jersey, (1973).
50. Lorenz, E.N., "Deterministic nonperiodic flow," *J. Atmos. Sci. 20*, 130-141 (1963).
51. Schlup, W.A., "Critical limit cycles of the retarded Josephson equation," *J. Phys. A., 11*, 1871-1878 (1978).
52. Nakamura, K., "Nonlinear flucutations associated with instabilities in dissipative systems," *Progr. Theor. Phys. 57*, 1874-1885 (1977).
53. Bonhoeffer, K.F., Über periodische chemische Reaktionen," *Z. Elektrochemie 51*, 24-29 (1947).
54. Aris, R., "Elementary Chemical Reactor Analysis" Prentice-Hall, New Jersey, (1969).
55. Tyson, J.J. and Light, J.C., "Properties of two-component bimolecular and trimolecular chemical reaction systems," *J. Chem. Phys. 59*, 4164-4173 (1973).
56. Tyson, J.J., "The Belousov-Zhabotinski Reaction" Springer-Verlag, Berlin, (1976).
57. Tyson, J.J. and Kaufmann, S., "Control of mitosis by a continuous biochemical oscillation," *J. Math. Biol. 1*, 289-310 (1975).
58. Babloyanz, A. and Sanglier, M., *FEBS Lett. 23*, 364-371 (1972).
59. Monod, J. and Jacob, F., "General conclusions: teleonomic mechanism in cellular metabolism, growth and differentia-

tion," *Quant. Biol. 26*, 389-401 (1961).
60. Goodwin, B.C., "Oscillatory behaviour in enzymatic control processes," *Adv. in Enzyme Regulation 3*, 425-438 (1965).
61. Eigen, M. and Schuster, P., "The Hypercycle" Springer, Berlin, (1979).
62. Cohen, C.S. and Rosenblat, S., "Multi-species interaction with hereditary effects and spatial diffusion," *J. Math. Biol. 7*, 231-241 (1979).
63. Volterra, V., "Leçon sur la Théorie Mathématique de la Lutte pour la Vie" Gauthiers-Villars, Paris, (1931).
64. Tyson, J.J., "Some further studies of nonlinear oscillations in chemical systems," *J. Chem. Phys. 58*, 3919-3930 (1973).
65. Tomita, K. and Kai, T., "Entrainment of a limit cycle by a periodic external excitation," *Progr. Theor. Phys. 57*, 1159-1177 (1977).
66. Grasman, J. and Jansen, M.J., "Mutually synchronized relaxation oscillators as prototypes of oscillatory systems in biology," *J. Math. Biol. 7*, 171-197 (1979).
67. Nicolis, J.S., Galanos, G. and Protonotarios, E.N., "A frequency entrainment model with relevance to systems displaying adaptive behaviour," *Int. J. Control 18*, 1009-1027 (1973).
68. Linkens, D.A., "Stability of entrainment conditions for RLC coupled van der Pol-oscillators used as a model for intestinal electrical rhythm," *Bull. Math. Biol. 39*, 359-362 (1977).
69. Linkens, D.A., "The method of harmonic balance applied to coupled asymmetrical van der Pol-oscillators for intestinal modelling," *Bull. Math. Biol. 41*, 573-589 (1979).
70. Tomita, H., "Chaos and its description," *in* "Pattern Formation by Dynamic Systems and Pattern Recognition" (H. Haken, ed.), pp. 90-97. Springer-Verlag, Berlin, (1979).
71a. Troy, W.C., "Oscillation phenomena in the Hodgkin-Huxley equations," *Proc. Roy. Soc. Edinb. 74A*, 23-32 (1975).
b. Troy, W.C., "The bifurcation of periodic solutions in the Hodgkin-Huxley equations," *Q. Appl. Phys. 36*, 73-82 (1978).
72. FitzHugh, R., "Impulses and physiological states in theoretical models of nerve membrane," *Biophys. J. 1*, 445-466 (1961).
73. Nagumo, J., Arimoto, S. and Yoshizawa, S., "An active pulse transmission line simulating nerve action," *Proc. IRE 50*, 2061-2070 (1962).
74. Bonhoeffer, K.F., "Activation of passive iron as a model of excitation in nerve," *J. Gen. Physiol. 32*, 69-91 (1948).
75a. Troy, W.C., "Bifurcation phenomena in FitzHugh's nerve conducting equation," *J. Math. Anal. 54*, 678-691 (1976).
b. Hadeler, K.P., an der Heiden, U. and Schuhmacher, K., "Generation of the nervous impulse and periodic oscillations," *Biol. Cyb. 23*, 211-218 (1976).
76. Kawata, M., Sokabe, M. and Suzuki, R., "Synergism and

antagonism of neurons caused by an electrical synapse, *Biol. Cyb. 34*, 81-89 (1979).
77. Wilson, H.R., "Mathematical models of neural tissue," *in* "Cooperative Effects" (H. Haken, ed.), pp. 247-262. North-Holland, Amsterdam, (1974).
78. Cowan, J.D., "Neurosynergetics," *in* "Synergetics, a Workshop" (H. Haken, ed.), pp. 228-240. Springer-Verlag, Berlin, (1978).
79. Wiener, N., "Cybernetics" MIT Press, Cambridge, Mass., (1948).
80. Dewan, E.M., "Nonlinear oscillations and electroencephalography," *J. Theor. Biol. 7*, 141-159 (1964).
81. Kreifeldt, J., "Ensemble entrainment of self-sustaining oscillators: a possible application to neural signals," *Math. Biosci. 8*, 425-436 (1970).
82. Schmitt, F.O., Dev, P., and Smith, B.H., "Electrotonic processing of information by brain cells," Sciences *193*, 114-130 (1976).
83. Elul, P., "Relation of neural waves to EEG," *Neurosci. Res. Prog. Bull. 12*, 97-101 (1974).
84. Wall, C., Kozak, W.M. and Sanderson, A., "Entrainment of oscillatory neural activity in the cat's lateral geniculate nucleus," *Biol. Cyb. 33*, 63-75 (1979).
85. Harth, E.M., Csermaly, T.J., Beek, B. and Lindsay, R.D., "Brain functions and neural dynamics," *J. Theor. Biol. 26*, 93-120 (1970).
86. Taylor, J.G., "Spontaneous behaviour in neural networks," *J. Theor. Biol. 36*, 513-528 (1972).
87. Freeman, W.J., "Mass Action in the Nervous System" Academic Press, New York, (1975).
88. Glasman, R.B. and Malamut, B.L., "Does the brain actively maintain itself?" *Bio Systems 9*, 257-268 (1977).
89. Adey, W.R., "The influence of impressed electrical fields at EEG frequencies on brain and behaviour," *in* "Behaviour and Brain Electrical Activity" Plenum Press, New York, (1974).
90. Adey, W.R., "Experiment and Theory in Long Range Interactions of Electromagnetic Fields at Brain Cell Surfaces" Plenum Press, New York, (in press).
91. "Brain Interactions with Weak Electric and Magnetic Fields" (S.M. Bawin and W.R. Adey, eds.) *MIT Neurosci. Res. Progr. Bull. 15*, 1-107 (1977).
92. Kaczmarek, L.K., "Cation binding models for the interaction of membranes with EM fields," *MIT Neurosci. Res. Progr. Bull. 15*, 54-60 (1977).
93. Kaiser, F., "Coherent oscillations in biological systems, workshop on the mechanism of microwave biological effects," University of Maryland, May 14-16 (1979), (to be published).
94. Fröhlich, H., "Possibilities of long and short range electric interactions of biological systems," *MIT Neurosci. Res. Progr. Bull. 15*, 67-72 (1977).

95. Baransky, S. and Czerski, P., "Biological Effects of Microwaves" Dowden, Hutchinson and Ross, Stoudsbourg (1976).
96. Fröhlich, H., *in* "Theoretical Physics and Biology" (M. Marois, ed.), pp. 13-22. North-Holland, Amsterdam, (1969).
97. Fröhlich, H., "Long range coherence and energy storage in biological systems," *Int. J. Quant. Chem. 2*, 641-649 (1968).
98. Kaiser, F., "Boltzmann equation approach to Fröhlich's vibrational model of Bose condensation-like excitations of coherent modes in biological systems," *Z. Naturforsch. 34a*, 134-146 (1979).
99. H. Fröhlich, "Selective long range dispersion forces between large systems," *Phys. Lett. 39A*, 153-154 (1972).
100. Devyatkov, N.D., "Influences of mm-band electromagnetic radiation on biological objects," *Usp. Fiz. Nauk 110*, 452-469 (1973).
101. Berteaud, J., Dardalhon, M., Rebeyrotte, N. and Averbeck, D., "Action d'un rayonnement électromagnetique à longueur d'onde millimétrique sur la croissance bactérienne," *C.R. Acad. Sci. 281D*, 843-846 (1975).
102. Webb, S.J. and Stoneham, M.E., "Resonances between 10^{11} and 10^{12} Hz in active bacterial cells as seen by Laser Raman spectroscopy," *Phys. Lett. 60A*, 267-268 (1977).
103. Webb, S.J., Stoneham, M.E. and Fröhlich, H., "Evidence for nonthermal excitation of energy levels in active biological systems," *Phys. Lett. 63A*, 407-408 (1977).
104. Grundler, W. and Keilmann, F., "Nonthermal effects of millimeter microwaves on yeast growth," *Z. Naturforsch 33c*, 15-22, (1978).
105. Drissler, F. and MacFarlane, R.M., "Enhanced anti-Stokes Raman scattering from living cells of Chlorella Pyrenoidosa, *Phys. Lett. 69A*, 65-67 (1978).
106. Pohl, H., "Micro-dielectrophoresis of dividing cells," (preprint), (1979).
107. Kaiser, F., "Limit cycle model for brain waves," *Biol. Cyb. 27*, 155-163 (1977).
108. Kaiser, F. "Coherent oscillations in biological systems I: bifurcation phenomena and phase transitions in an enzyme-substrate reaction with ferroelectric behaviour," *Z. Naturforsch 33a*, 294-304 (1978).
109. Kaiser, F., "Coherent oscillations in biological systems II: limit cycle collaps and the onset of travelling waves in Fröhlich's brain wave model," *Z. Naturforsch 33a*, 418-431 (1978).
110. Fröhlich, H., "The biological effects of microwaves and related questions," *in* "Advances in Electronics and in Electron Physics" Academic Press (in print).

NONLINEAR INTERACTIONS OF ELECTROMAGNETIC WAVES WITH BIOLOGICAL MATERIALS

Frank S. Barnes
Chia-lun J. Hu

Department of Electrical Engineering
University of Colorado
Boulder, Colorado

INTRODUCTION

There has been an interest in the effect of electromagnetic waves on biological materials and processes for a long time. Perhaps the earliest experiments were by Volta in the stimulation of a frog's nerve with an electric current. However, in spite of this long history, it is remarkable how limited our knowledge is of the electrical fields on biological processes and materials. One of the recent stimuli for study has been the concern over microwave safety and the effect on the population as a whole of a wide variety of radio and microwave sources such as FM transmitters, CB's, and industrial RE furnaces [1]. The most readily observed effect of the application of microwaves and radio waves to biological materials is heating. With appropriate measurements of the conductivity, dielectric constant, heat capacity, thermal conductivity, blood flow, and their distribution in space, it is relatively straight forward but numerically complicated to calculate the temperature distribution in a biological body as a function of the incident power density, time of exposure and other environmental factors. A great deal of effort has been

ISBN 0-12-709660-4

expended in generating reasonable models of the boundary value problems associated and the distribution of energy deposited in models of this type [2-4]. However, the principal purpose of this chapter will be to review other phenomena which may occur as a result of the application of RF or microwave energy to biological materials. The phenomena which we wish to examine are of four types:

(1) The possibility of rectification in membranes and its effect on the chemical concentration balances that are maintained across the biologically active surface;

(2) The effect of electric fields on the orientation of long chain molecules and the distribution of ions around them with resulting anisotropy and nonlinearity in the electrical conducting processes;

(3) The effect of short pulses of high intensity;

(4) A brief examination of some possible nonlinear interactions which are associated with the quantum phenomena in this range of frequencies.

I. THE BIOLOGICAL MEMBRANE AS A RADIO OR MICROWAVE RECTIFIER

A distinguishing feature of biological membranes is their ability to maintain concentration gradients of various charged particles across the membrane. The present physiological models allow us to visualize this as a phospholipid bilayer into which large protein molecules are inserted or attached to the surface. Current hypotheses indicate that these molecules are involved in forming channels which selectively allow the transport of a wide variety of substances (see Figure 1) [5]. It appears that the electric fields and the charged layer associated with the boundaries of the membrane play an important role in controlling the transport of substances through the membrane and in some cases

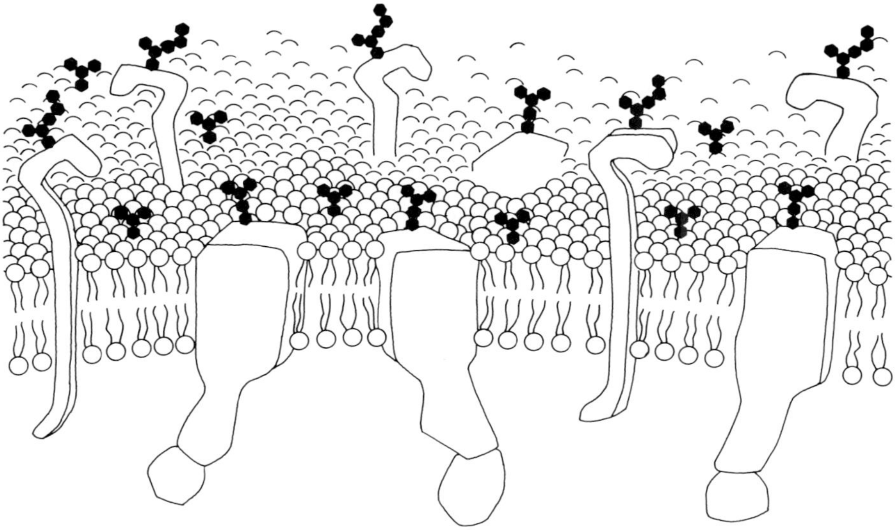

FIGURE 1. Model of the plasma membrane includes proteins and carbohydrates as well as lipids. Integral proteins are embedded in the lipid bilayer; peripheral proteins are merely associated with the membrane surface. The carbohydrate consists of monosaccharides, or simple sugars, strung together in chains that are attached to proteins (forming glycoproteins) or to lipids (forming glycolipids). The asymmetry of the membrane is manifested in several ways. Carbohydrates are always on the exterior surface and peripheral proteins are almost always on the cytoplasmic, or inner, surface. The two lipid monolayers include different proportions of the various kinds of lipid molecule. Most important, each species of integral protein has a definite orientation, which is the same for every molecule of that species (redrawn from "The Assembly of Cell Membranes," by H.F. Lodish and J.E. Rothman, Scientific American, vol. 240, no. 1, January 1979, pp. 48-63).

there are threshold voltages across the membrane which will open and close ion channels for Na and K. Electrically, it appears that both long chain molecules with dipole moments and ions such as potassium, calcium, sodium, etc., play an important role in the function of the membrane. At low frequencies, the I-V characteristic for cell membranes has been measured for a variety of cells and a typical curve is shown in Figure 2 [6]. An I-V curve of this type is predicted for a system which maintains a potential barrier and system of charged particles where the current carriers have a Boltzmann distribution of energy [7].

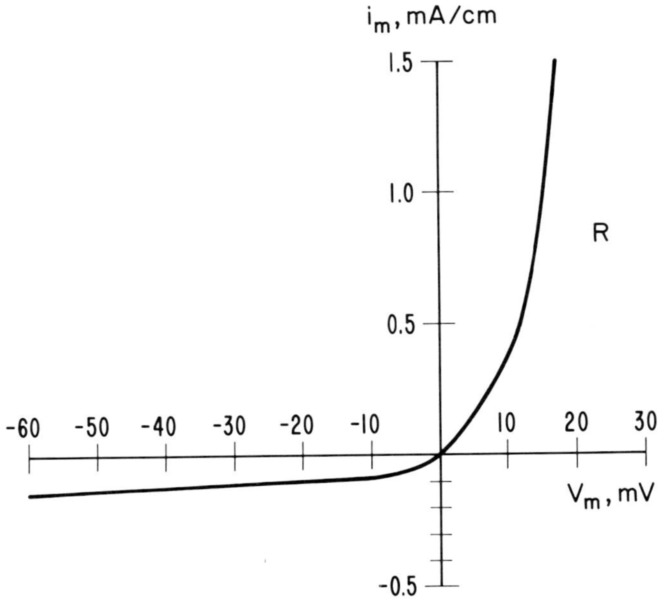

FIGURE 2. The steady-state current-potential characteristic for a squid axon membrane as calculated from longitudinal data for the axon (ref: Cole, [6]).

That is, the concentration of charged particles on the two sides of the barrier is related by

$$C_2 = C_1\left(e^{\frac{q\phi}{\eta KT}}\right)$$

where

q is the charge on the particle,

ϕ is the potential height of the barrier,

K is Boltzmann's constant,

T is the absolute temperature, and

η is a factor which takes into account the geometry and any generation and recombination of charged particles that may occur in the bilayer region.

The current-voltage curve in Figure 2 may be approximated by

$$I = I_o\left(e^{\frac{q\phi}{\eta KT}} - 1\right) .$$

Thus, at low frequencies, these characteristic curves are very similar to those for a p-n junction diode and the application of an ac signal would lead to rectification. A typical cell membrane is biased with the potential of V_o = 50 millivolts. The high frequency characteristics of these membranes have yet to be measured in a definitive way. However, they are expected to be dependent upon the transit time of the charged particles through the membranes and the nature of the gating mechanism for the ion channels. Since the membranes are from 100 - 200 Å thick, ion transport at thermal velocities would lead to a cutoff frequency in the vicinity of 10 GHz; however, if the usual mobilities for ions in a fluid apply, the cutoff frequency will be much lower. Because of the large capacitance associated with the membrane of approximately 1 μfarad/cm^2, it has usually been assumed that no applied field at radio or microwave frequencies appeared across the membrane to disturb the potential V_o. However, if we treat the high water content material on either side of the membrane

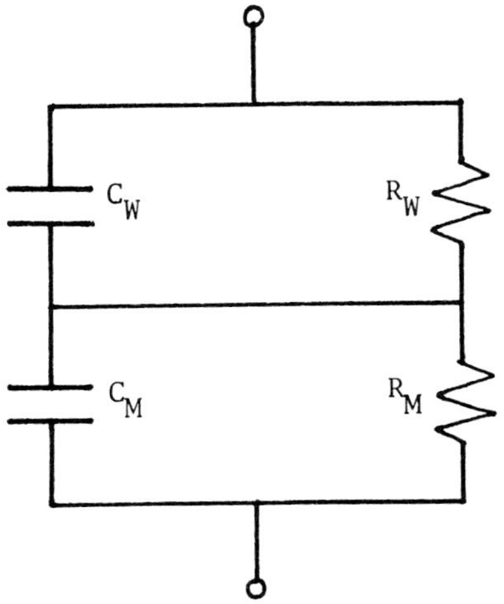

FIGURE 3. Equivalent circuit for the electronic characteristics of the membrane immersed in high water content media where $C_M = \varepsilon_M(A)/L$; $R_M = \rho L/A$; $C_W = \varepsilon_W(A)/d$; and $R_W = \rho_W d/A$ where L is the thickness of the membrane, A is the area, d is the thickness of the high water content material, ρ_W is the resistivity of the high water content material, and ρ_M is the resistivity of the membrane.

and the membrane as a resistor and a capacitor in parallel, as shown in Figure 3, then the following situation occurs as a function of frequency. At low frequencies, the field or the voltage across the membrane and the high water content liquids adjacent to it divide as the ratio of the resistivities of the material with the bulk of the voltage appearing across the membrane. At intermediate frequencies, the membrane resistance is shorted out by the associated capacity. At high frequency, the voltage divides as if there were two capacitors in series and the ratio of the electric fields for a plane wave model is given by

$$E_m = \frac{\varepsilon_w}{\varepsilon_m} E_w$$

where

ε_m is the dielectric constant of the membrane material,

ε_w is the dielectric constant for the high water content tissue,

E_m and E_w are the corresponding electric field strengths. In the 3 GHz region, the dielectric constant for water is approximately 50 and that for hydrocarbons such as oil or fat is approximately 8 or 9 [8]. Thus, the electric field strength in the membranes is approximately six times that in the adjacent high water content material. If we use the high frequency model to compute the applied voltage $V = V_0 + V_1 \cos \omega t$ and then apply it to the low frequency characteristic for lack of a better model or any direct measurements, it leads to DC current and corresponding shifts in the ion concentration balance.

There is some experimental evidence that the foregoing theory may apply, with appropriate corrections for the geometry of the cell membranes [9]. Seaman and Wachtel have demonstrated a shift in the firing rate for pacemaker cells in Aplysia upon irradiation in the 1.5 GHz region (see Figure 4) [10]. It is to be noted that the change in the firing rate in these experiments was opposite to the direction of that obtained upon warming the cells. More recent experiments in Boulder at 500 MHz indicate that the change in firing rate is dependent upon the orientation of the field with respect to the cell, and that we can repeatedly stop the firing of these pacemaker cells. We can also obtain the same changes in firing rate by injecting a DC current in approximately one part in a thousand of the calculated Rf current density for the cell proportion as a whole [11]. The critical experiments necessary to trace out the nature of the nonlinearities of these frequencies, the effects of geometry, and the details of the chemical transport processes as effected by the application of a microwave or radio wave signal have yet to be done. However, the results to date are sufficiently exciting to make this an area for serious study. It is to be noted that the chemical concentration shift or imbalance created by the application of a field which is rectified in this manner goes as the integral of the current or the exposure. Stated another way, the concentration shift is proportional to the DC component of the current and the length of the exposure.

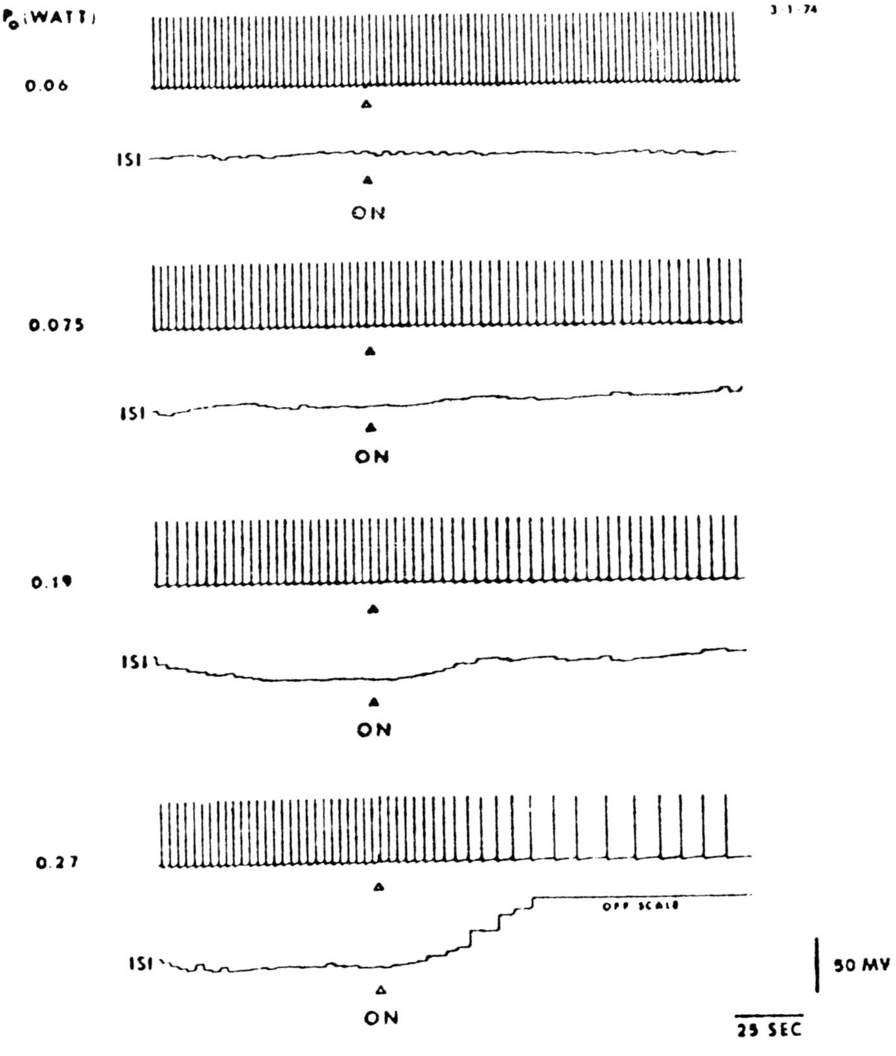

FIGURE 4. An example of the firing pattern changes seen in a pacemaker neuron in response to increasing microwave levels. The Interspike Interval (ISI) recording was achieved using a "periodometer" circuit. This particular case shows the determination of lowest effective absorbed power (LEAP) as $P_{in} = 0.075$ watts. For this experiment, SAR = 94.3 P_{in} watts/kg. Thus, this LEAP is equal to an SAR of 7.1 w/kg.

II. THE ALIGNMENT OF LONG CHAIN MOLECULES

The ability of electric fields to orient biological structures along the field lines has been known for some time. Professor Schwan and his colleagues described the alignment of red blood cells into long chains (pearl chains) in the late 1950's, and developed a theory to describe them (see Figure 5) [12]. A simplified approach to this problem is to consider the problem of a slab with dielectric constant ε_1 and a conductivity σ_1 inserted into the medium with a dielectric constant ε_2 and conductivity σ_2 [13]. Upon the application of an electric field to this geometry, torques will be applied to the dielectric slab in order for it to take on the lowest energy configuration, which will either be at right angles or parallel to the electric field, depending upon the ratio of ε_1 to ε_2. The torque on the slab is given by

$$\tau = -\frac{E_0^2 V}{2}\left\{\frac{\varepsilon_1^2 - \varepsilon_2^2}{\varepsilon_2}\right\} \sin 2\theta_0$$

where

E_0 is the electric field in the bulk material,
ε_1 is the dielectric constant in the bulk material,
ε_2 is the dielectric constant of the slab,
θ is the angle between the field and the long axis of the slab,
V is the volume of the slab,

and if we assume Brownian motion to be the disordering force, then the threshold for the observation of the alignment of the dielectric slab is given by

$$E_{th} = \sqrt{\frac{2kT\varepsilon}{(\varepsilon_0^2 - \varepsilon^2)V}} \; .$$

In addition to larger particles such as platelets and blood cells, this phenomenon applies to long-chain molecules. With the

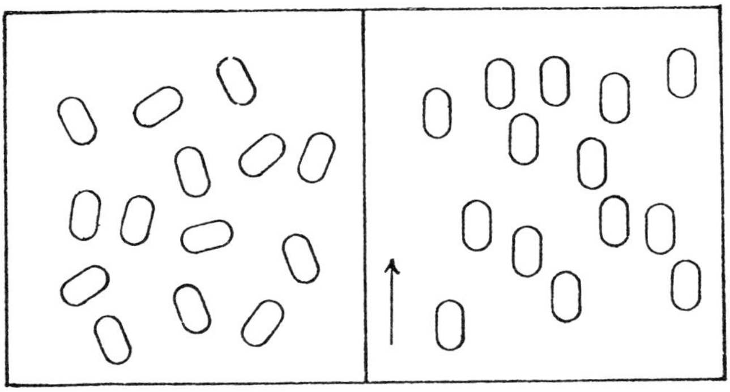

(a)

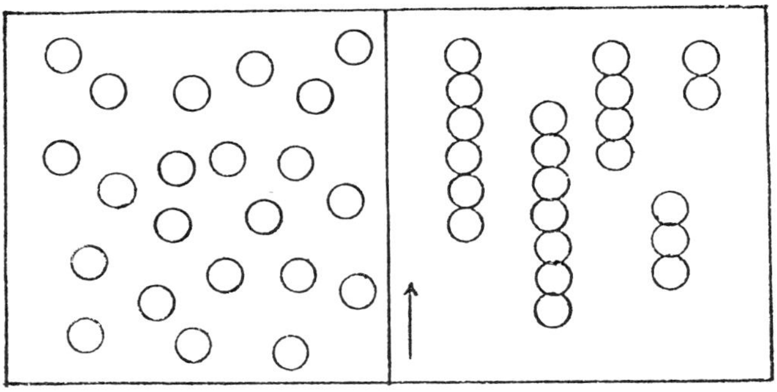

(b)

FIGURE 5. Preferential orientation due to an applied electric field. (a) Kerr effect. (b) Pearl chain effect.

apparatus in Figure 6, we have observed shifts in the birefringence of blood plasma upon the application of electric fields in the range from 30 - 200 volts/cm [14]. There is another aspect of this orientation phenomenon: the conductivity of a liquid containing long chain polymers becomes anisotropic if large fields are applied. Additionally, there is a time dependence of the conductivity of these fluids associated with the time

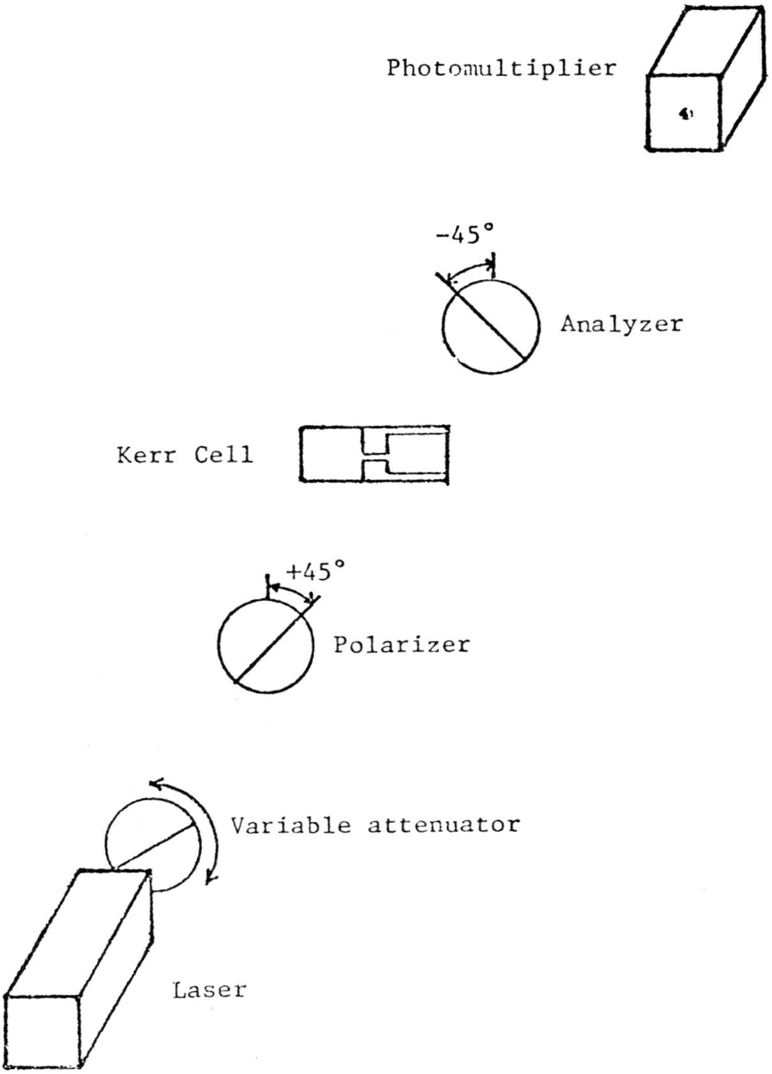

FIGURE 6. Detection system.

that it takes to orient the long chain molecules and the change in the distribution of ions around large molecular ions. With no applied field long chain molecules are expected to be oriented at random. However, upon the application of a significant field, these particles are either rotated or distorted with the reduction of conductivity in one direction and an increase in the other. The time dependence of the conductivity may be the superposition of additional changes including thermal changes in viscosity of the fluid or the mobility of the ions, and changes in the ion distribution.

III. LOW FREQUENCY NONLINEARITIES IN THE CONDUCTIVITY OF ELECTROLYTES

There is a substantial amount of literature on nonlinear effects and dispersion in electrolytic solutions containing long chain molecules. The first group of studies reported in the literature is mostly devoted to the Wien effect, which is to be explained in some detail in the following.

Wien first observed in 1927 to 1931 [15-18] that some electrolytes under high pulsed fields will show an incremental conductivity which is approximately proportional to the strength of the applied field. Quantitatively, his observations can be divided into two groups which have been called the first Wien effect and the second Wien effect. The first Wien effect is generally observed in solutions of strong electrolytes. It is seen that when a high pulsed field (for example, 200 KV/cm and 1 to 10μs duration) is applied to some strong electrolyte, the conductivity of the solution will increase 1-10% and the increment is approximately linear up to a certain limit. On the other hand, around the same time, Gyemant [19], Wien [20], and Schiele [21] also found that *some* electrolytic solutions with much *weaker* ionic strengths showed much *stronger* relative increments of conductivity under low fields, say 1 KV/cm. Since these two groups of data could not be explained numerically by a single mechanism, two different theories were derived. The first effect in strong electrolytes was explained by Joos and Blumentritt [22-24],

Falkenhagen [25,26], and Wilson [27] (and summarized by Eckstrom and Schmelzer [28]) as due to partial breakdown of the ionic-cloud surrounding the heavy ions when high voltage is applied. That is, heavy ions (e.g., polyions) are usually surrounded by lighter ions (or counter ions) of opposite charge. But due to thermal agitation, these counter ions will not be closely attached to the heavy ions. Instead, they will form a cloud around each heavy ion with decreasing density away from the center of the heavy ion. Now, when a *low field* is applied to this ion-cloud system, the heavy ion will be drifted in one direction and its ionic cloud will be slightly deformed and act as a dragging force (this dragging force is also called the electric force of relaxation in the literature) opposite to the movement of the heavy ion. On the other hand, if a strong field is applied, the heavy ion will be dragged towards the edge of the ionic cloud. Therefore, its mobility is increased because the redistribution of the ionic cloud will decrease the dragging force *nonlinearly*. Consequently, the conductivity increases. This is the first Wien effect. On the other hand, when the solution is weakly ionized, the ionic cloud is dilute. Therefore, the dragging effect of the cloud is not significant under a strong field. But the heavy ions will be dissociated (or ionized further) under the applied field, and the charge on each heavy ion will be largely increased. Consequently, its electric mobility will increase and the number of current carriers will also increase, which thus accounts for the large relative increase in the conductivity under moderate field strengths. This is called the second Wien effect (Onsager [29], Mead and Fuoss [30,31], Onsager and Liu [32]). The experimental methods used to investigate Wien effects are mostly pulse excitation methods.

Other studies on Wien effects include polyelectrolyte studies by Bailey and Patterson in 1952 [33] and Wissbrun-Patterson in 1958 [34,35], relaxation and Wien effects on K-polyphosphate and

Na-DNA by Eigen and Schwarz in 1954 [36,37], negative Wien effects on uranyl nitrates and perchlorate in 1965 [38-40] and nerve impulse propagation mechanisms as explained by Wien effects in 1970 - 1976 [41-45]. Most recent studies on Wien effects are mostly theoretical. These include the work by Chen and McIlnory in 1978 [46-47].

Finally, a paper published in 1975 [48] by Grunhagen is worth noting. This paper reports a temporary change of UV absorption strength when a high voltage pulse is applied to some amphiphilic electrolytic solutions. This effect is explained by the second Wien effect or the dissociation of ions under high fields.

The second group of papers we are interested in concerns the impedance dispersion effects of electrolytes under high fields. In 1928 - 29, Debye, Falkenhagen and Williams [49-51] published three theoretical papers explaining the low-field dispersion effects in electrolytes in general. This theory was based on the Debye-Huckel-Onsager statistical theory of electrolytes. A decade later, Falkenhagen and Fleischer [52] generalized this dispersion theory by introducing the following fact. When no field is applied, the ionic cloud surrounding a heavy ion is at thermal statistical equilibrium. If the central heavy ion in this equilibrium system is suddenly removed away from the ionic cloud, then the cloud will relax to uniform distribution in space in time τ. This is called the relaxation time of the ionic cloud. Now if the frequency of an a.c. applied field is greater than $1/\tau$ then the field-perturbed cloud distribution will not have enough time to reach thermal equilibrium with the heavy ion, and the applied field will accumulatively make the cloud more and more relaxed and more and more uniform in space. Therefore, the cloud dragging effect (see the explanation on the first Wien effect) will be reduced significantly. Consequently, the mobility of the ion-cloud system will be increased significantly or the conductivity will have an appreciable jump across this frequency. In 1936, Arnold and Williams [53,54] used a method similar to Wien's pulse comparison method to measure the changes of conductivity and dielectric constant when frequency is swept. They found that the conductivities for KCL, $MgSo_2$, etc., under high fields, are monotonically increasing when frequency is increased from 1.8 mc to 10 mc. The rising conductivities always approach some saturated levels. In 1955, Eigen and DeMaeyer [55] proposed a zero-beat electronic detection method for measuring the dispersion effect *at high voltages*, and Persoons [56] in 1974 realized this scheme by measuring the dispersion effects in tetrabutylammonium picrate in diphenylether. Eigen and Schwartz also published several papers in the period between 1955 and 1959 [57-59] on the dispersion effect of the *nonlinear impedance* of polymer or biological substances such as K-polyphosphate and Na-DNA. They felt that this dispersion effect is due to the internal changes of these giant molecules under *strong* high frequency fields (0.1 mc and up).

All the above studies are for the impedance dispersion effects of electrolytes under high fields (1 KV/cm to 200 KV/cm). The dispersion effects of electrolytes and bio-mediums under *low fields* are numerous. Schwan in 1959 published a paper summarizing some of these studies [60]. Recent dispersion studies on electrolyte solutions include Delbo's [61] work in microwave regions and Minakata's 2-phase theory on dispersion at high frequencies [62]. It is seen that either in the high field or in the low field experiments, the impedance dispersion effect is mostly due to relaxation in the medium or in the ionic clouds. That is, σ or ε is always a monotonic function of frequency. It is also to be noted that many dispersion effects are due to electrode polarization and overvoltages [63,64]. But the dispersion caused by these effects is significant only when the applied voltage is comparable to the overvoltages, which in many cases are below 1 volt. Therefore, these polarization and overvoltage effects may not be significant under high voltage experiments.

IV. SOME EXPERIMENTAL DATA ON THE EFFECTS OF LARGE PULSED MICROWAVE FIELDS ON CELLS

A small amount of experimental work has been done on the effect of short pulses on microwaves on cells [65,66,67]. Fout, et al, have shown that the cell damage to zebra fish embryos by microsecond pulses at 2.7 GHz at field strengths from 5 KV/cm to 10 KV/cm and repetition rates of 330 pulses per second leads to damage which is morphologically different than that which is seen by cooking the cells in hot water. (See Figures 7 and 8.) These electron micrographs indicate that the cell membranes and the mitochondria membranes are broken or destroyed by the large

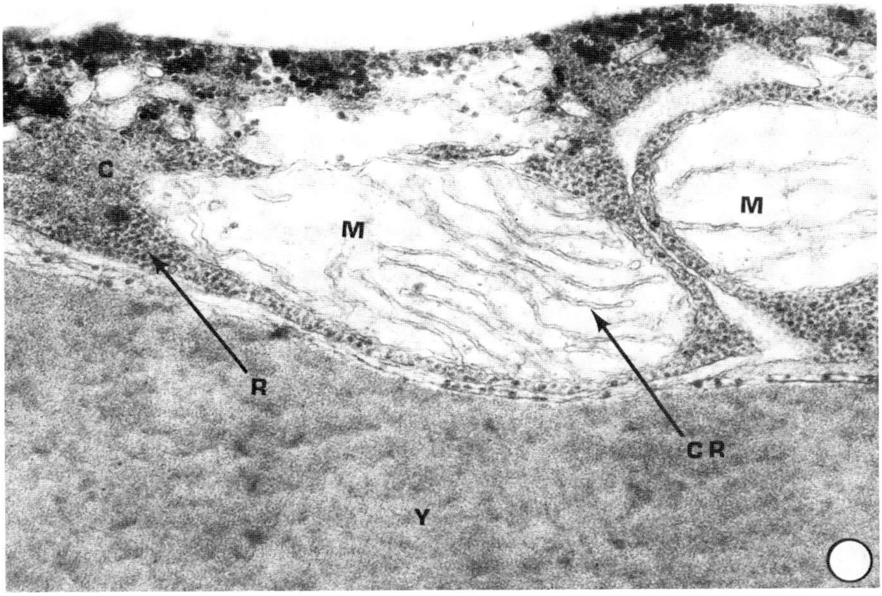

FIGURE 7. Electron micrograph of a normal, unexposed embryo showing part of a cell with intact mitochondria (M). The lower grey area is the yolk (Y). C, cytoplasm; CR, cristae; R, ribosomes. X 42,500.

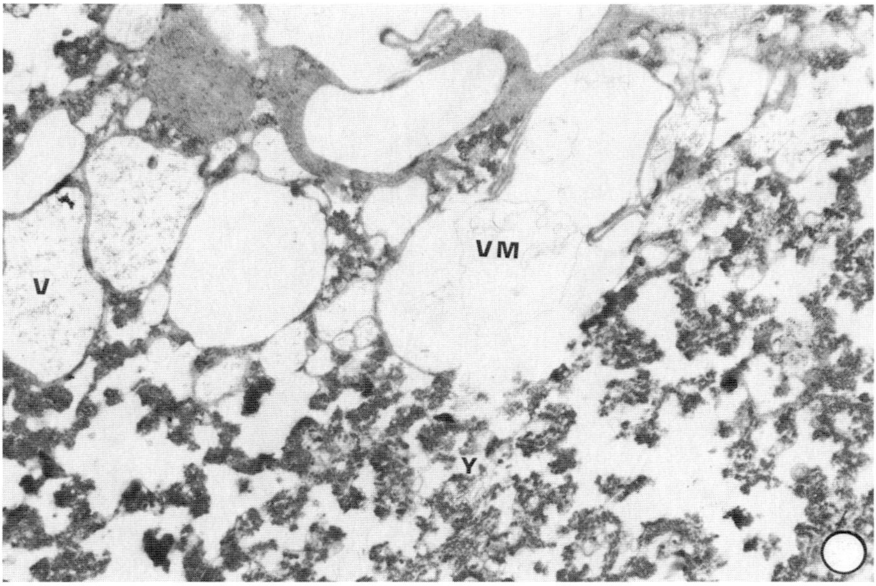

FIGURE 8. Electron micrograph of an embryo exposed to 10 kV/cm microwaves for 14 seconds. The vesicles (V) represent disintegrating mitochondria (VM) and cell membranes. The yolk (Y) has also fragmented and does not have the same homogeneous structure as in the unexposed embryo shown in Figure 7. X 21,250.

microwave fields in a way which is not observed when cells are raised to the same or higher temperatures by heating in hot water for the same length of time than that which is calculated for the microwave exposures. Similar work [67, Figures 9 and 10] on transformed brain cells in a different exposure system shows similar membrane damage for exposures to fields greater than 1.7 KV/cm and calculated temperature rise of less than 4°C.

The calculated temperature rise for the cells as a whole is a function of the geometry of the exposure system, the peak power input, the pulse repetition rate, and the average cell conductivity and dielectric constant. The estimated peak temperatures for the whole cells in the foregoing experiments are below those

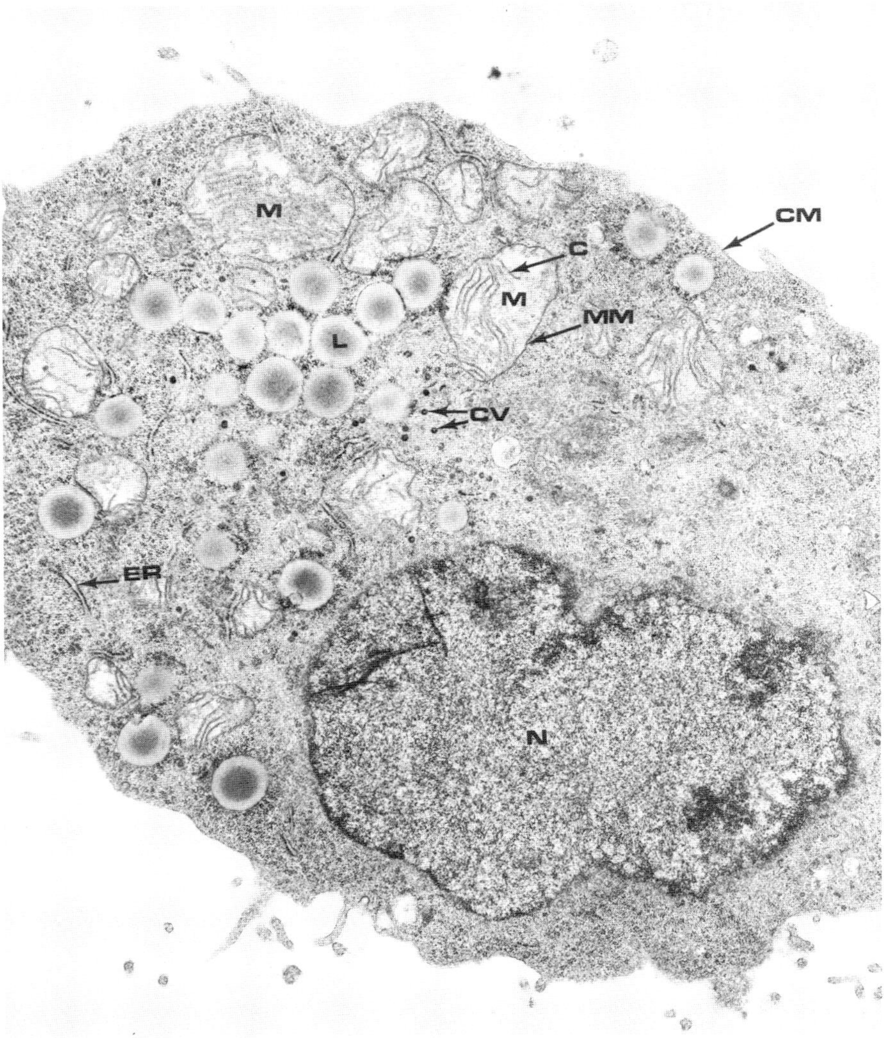

FIGURE 9. Electron micrograph of an unexposed, control cell for comparison with the exposed cells. Mitochondria in these cells normally have few cristae (C). Note the intact mitochondrial membrane (MM). CM, cell membrane; ER, endoplasmic reticulum; L, lipid; N, nucleus; V, A-type virus particles. X 13,298.

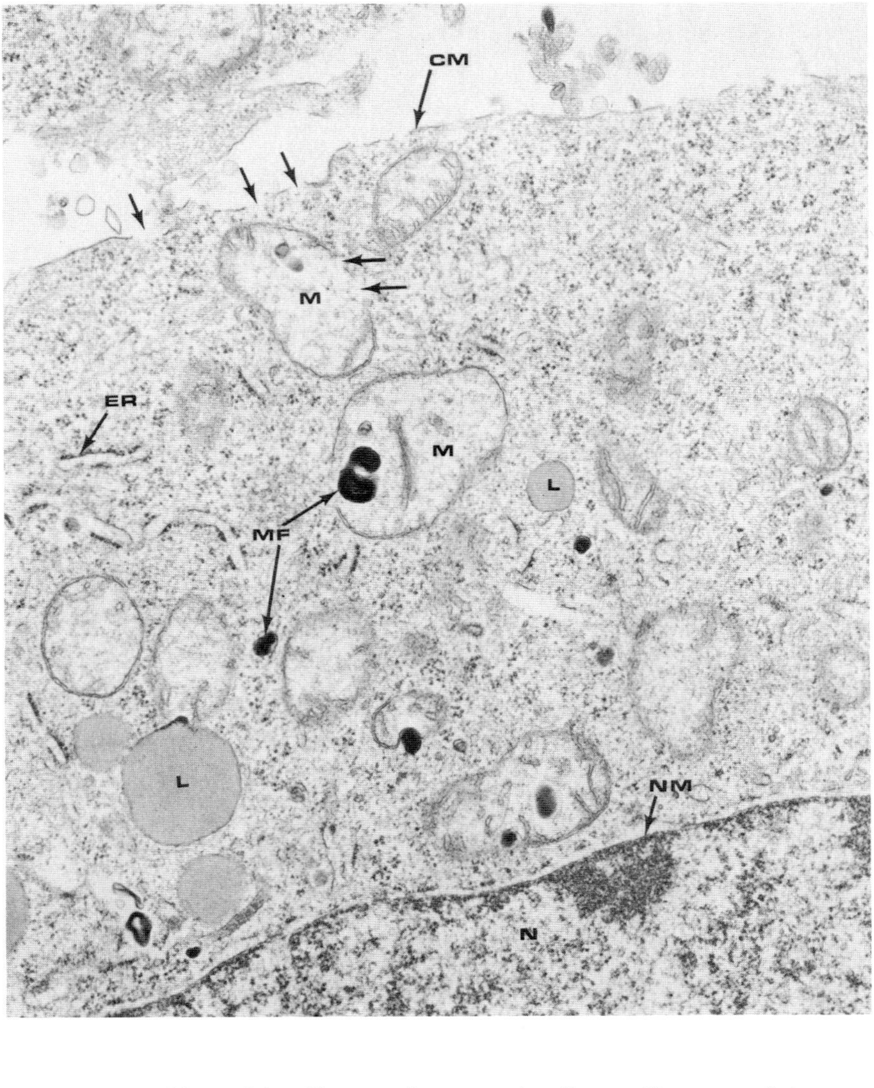

FIGURE 10. This figure shows part of a cell exposed to microwaves at a level of 1.7 kV/cm² for 30 seconds. Note the numerous breaks in the cell and mitochondrial membranes (arrows). Mitochondria (M) and the cytoplasm show numerous myelinated figures (MF) which are an indication of membrane breakdown. Note loss of cristae. CM, cell membrane; ER, endoplasmic reticulum; L, lipid, N, nucleus; NM, nuclear membrane is intact. X 32,618.

required for killing the same cells in hot water with the same exposure times as shown in Figures 11, 12 and 13 for the zebra fish embryos. Additionally, the estimated threshold temperature for death is lowered when the cell temperature before exposure is lowered to 0°C.

The time constant for a possible differential temperature rise for the membrane above the surrounding high water content material can be estimated from

$$\tau = \frac{x^2}{K'}$$

where

x is the thickness of the membrane,

K' is the thermal conductivity.

For $x = 200$ Å and $K = 1.4 \times 10^{-3}$ cm^2/sec this yields $\tau = 3 \times 10^{-10}$ sec which is so short as to make a differential temperature rise between the membranes and their surroundings under exposure to microsecond pulses very small, independently of reasonable differences in the conductivities and dielectric constants. A reasonable explanation of the damaging mechanism for these experiments is yet to be developed.

V. QUANTUM LIMITS

There has been considerable discussion of the possible importance of quantum phenomena as a mode for the interaction of electromagnetic waves in radio and microwave regions with biological materials. This interaction could take place in several forms. The first would be a narrow band resonance which might lead to change in the population distribution of a set of molecular or ion quantum states. It is then assumed that the excited molecule or ion would interact with its environment in a different way than the unexcited molecule by taking on a different configura-

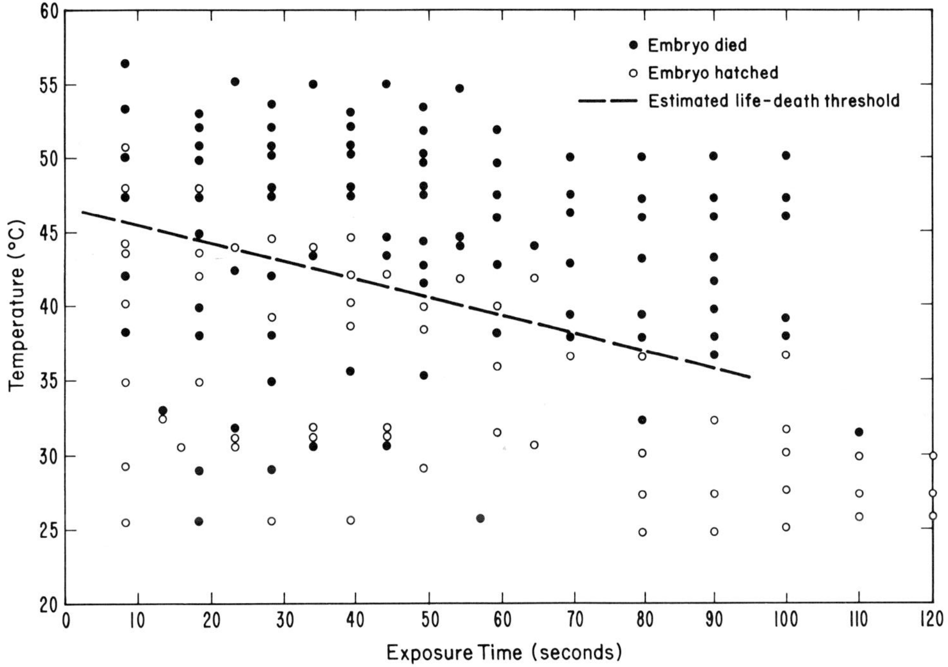

FIGURE 11. Experimental data on threshold for killing zebra fish embryos in hot water as a function of time.

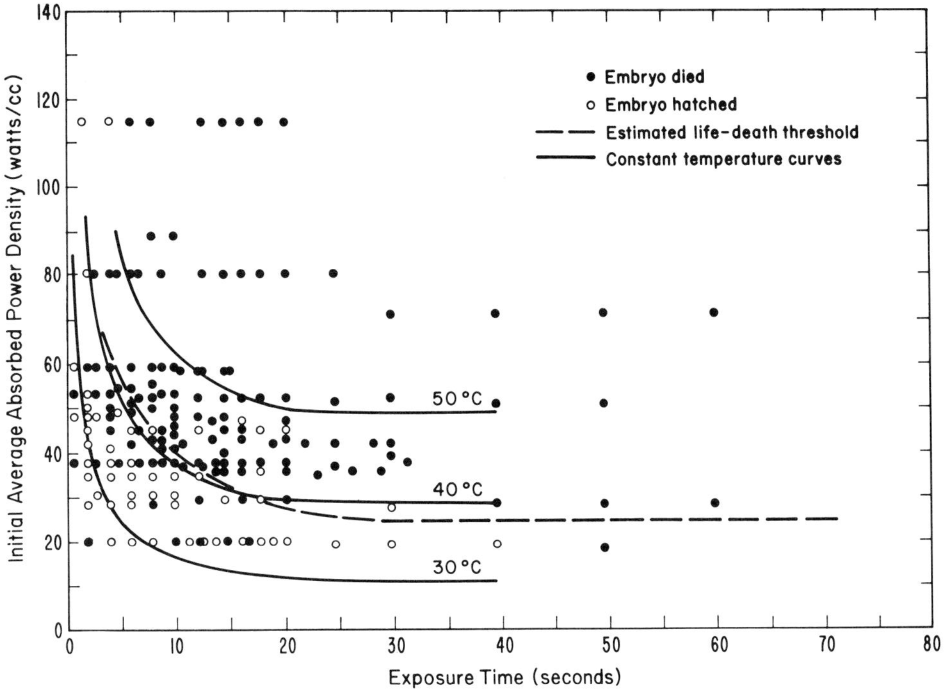

FIGURE 12. Average power densities and calculated temperature rises for the exposure of zebra fish embryos to microsecond pulses at 330 cycles/second with zebra fish embryos as a function of time. The initial temperature of the system is approximately 22°C.

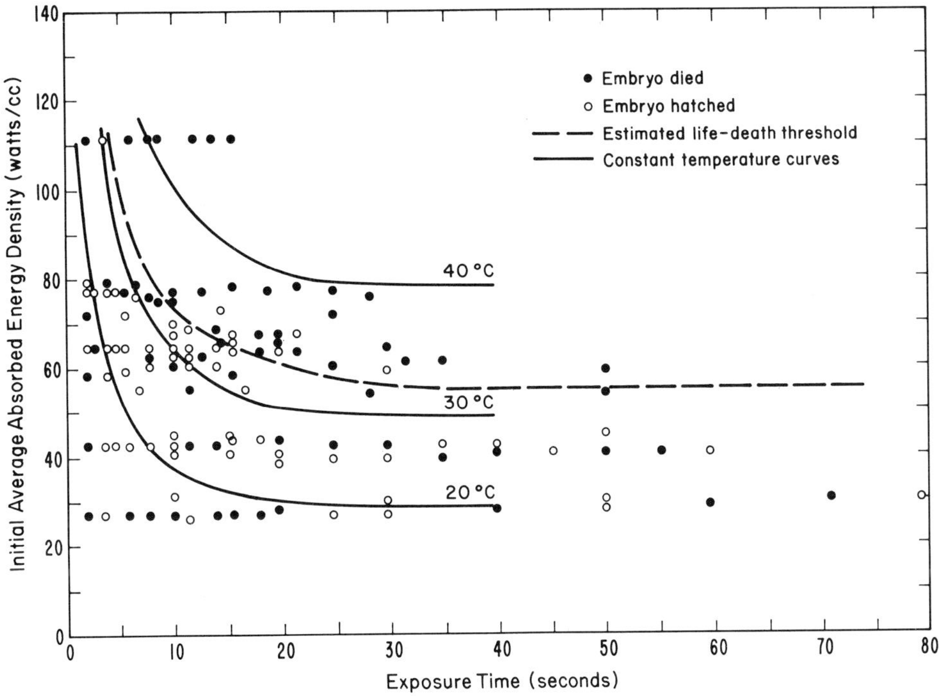

FIGURE 13. Average power densities and calculated temperature rises for the exposure of zebra fish embryos to microsecond pulses at 330 cycles/second with zebra fish embryos as a function of time. The initial temperature of the system is approximately 0°C.

tion or undergoing a chemical change. A second possibility might be the transmission of energy from one part of the *biological* system to another by means of an excited molecule which moves through the biological system. A third is the usual broadband dielectric loss or conductive mechanism where the relaxation of the molecules or ions from the excited states takes place so rapidly that thermal equilibrium makes the local temperature an appropriate way to describe the system.

Let us examine the case of a narrow band resonance where a single pair of energy levels is involved in taking energy out of the electromagnetic wave and coupling it into a molecule of the biological system. For frequencies less than 10 GHz or x-band the energies per photon will be less than 6.6×10^{-24} joules or 4.14×10^{-5} ev. This corresponds to an equivalent temperature $T_{eq} = hf/K = .48°K$, which is a very small amount of energy. Thus, any microwave or radio frequency signal is going to correspond to a very large number of photons per second.

If it is an excited molecule that is to be the basis for the importance of microwave irradiation then the population difference between these two kinds of molecules will be important. At thermal equilibrium the population ratio between states N_2 and N_1 is given by

$$\frac{N_1}{N_2} = exp\ hf/KT$$

where

h is Plank's constant

f is the frequency

K is Boltzmann's constant, and

T is the absolute temperature.

At 310°K and 10 GHz, this corresponds to a population difference of .15% or 1.5 parts in a thousand. Thus, if there is to be a significant population difference the biological molecules are unlikely to be in thermal equilibrium. Living biological systems are not in thermal equilibrium with their surroundings and there are several configurations of energy states where the small microwave quantums of energy might be important, such as in Figure 14B and 14C. In this case, 14B, the basic energy required to create an excited state is on the order of an electron volt and would be obtained from a chemical or an optical process. The microwave photon would provide the additional energy or satisfy an additional selection rule to get to states N_3 or N_4 from N_2 which would in turn allow a biologically important change to take place. Another possible configuration of states is shown in Figure 14C. In this case, the energy levels N_1, N_2, ..., N_m must be equally spaced so that as the microwave signal is applied it saturates the transitions between the levels and tries to equalize the populations between N_1 and N_m. If there now is a metastable state N_0 which can be filled from N_m, this can provide a way to store energy, which could in turn cause a change in the biological system. The intensity or power density required to saturate a transition of this kind for a two-level system is given by

$$I_s(f) = \frac{8\pi n^2 hf\Delta f}{\lambda^2}$$

where

n is the index of refraction
h is Plank's constant
f is the frequency
λ is the wave length, and
Δf is the line width.

For a homogeneously broadened line the population difference $\Delta N = N_1 - N_2$ or the absorption coefficient $\gamma(f)$ saturates as

$$N_2 - N_1 = \frac{\Delta N_0}{1 + I/I_s(f)}$$

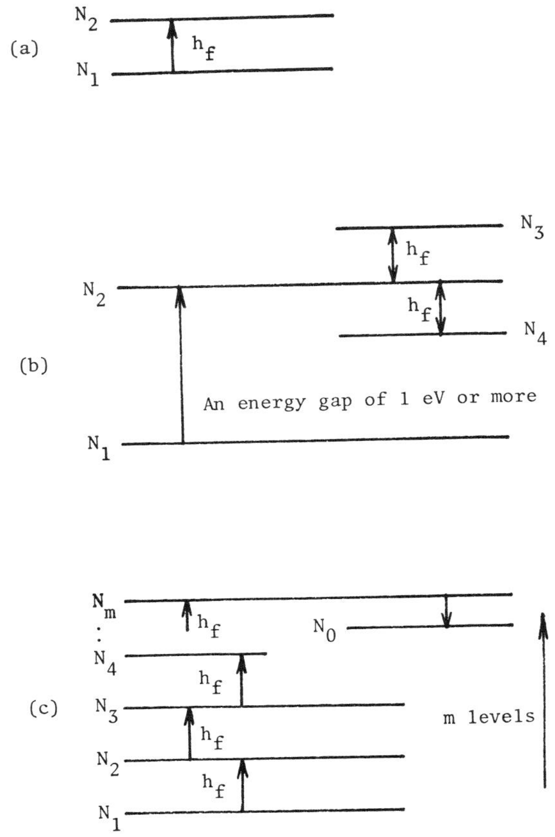

FIGURE 14. Energy level diagrams for possible quantum systems in biological materials.

or

$$\gamma(f) = \frac{\gamma_0(f)}{1 + I/I_s(f)}$$

where ΔN_0 and $\gamma_0(f)$ are the low intensity values.

$$\gamma_0 = \frac{\Delta N_0 \gamma^2}{8\pi n^2 t_s} \quad g(f)$$

where t_s is the spontaneous emission lifetime and $g(f)$ is the line shape factor. If we assume a dielectric constant of 50 for H_2O and $f = 10$ GHz, then $I_s(f) = 9.25 \times 10^{-22}$ watts/cm^2. This means that for a line width of 10 MHz we reduce the population difference by a factor of two with an input power of approximately 10^{-14} watts/cm^2. For inhomogeneously broadened lines

$$\gamma(f) = \frac{\gamma_0}{\sqrt{1 + I/I'_s}}$$

where

$$I'_s = \frac{4\pi^2 n^2 hf\Delta f}{\lambda^2}$$

and the numbers are nearly the same. This is a very low power density. A simple estimate indicates that it will take on the order of m times this signal strength to saturate this kind of a pumping chain to yield a population $N_m = N_0/m$. However, the extrapolation from $m = 3$ to 10^5 is so large that it requires both a more thorough theoretical study and direct experimental evidence before it should be taken seriously as a way of achieving biologically important molecules or ions in excited metastable states. However, this same low saturation power means it will be difficult to detect changes in the absorbed power by looking at a signal which is transmitted through the biological material. The nonlinear absorption characteristic makes the transition nearly transparent at power densities significantly larger than $I_S(f)$ even though the γ_0 is large. It is interesting to note that the ratio of I_S to the noise power due to black body radiation is approximately given by $h\nu/KT = 1.54 \times 10^{-3}$ so that a system at thermal equilibrium is well into the saturated region. The noise power will drive transitions between the states at a rate which is determined by the matrix element for the transition and inversely proportional to the spontaneous emission lifetime. Thus, for a system utilizing a single quantum transition to remain with a population imbalance far from thermodynamic equilibrium a very small matrix element is required. In contrast, the relatively large signals from most microwave and radio sources lead naturally to multiple quantum transitions.

There is a case where the foregoing description of the population distribution may not be applicable. This is where the ground state is degenerate and may contain two or more phases which are separated by a large energy barrier. A classic example of this kind of situation are the domains in a ferromagnetic material. For example, in a doughnut such as those used for computer core memories, the domains may be polarized either clockwise or counter-clockwise in states which are energetically the same. However, switching from one state to the other requires an energy many times KT and thus each polarization is very stable. We may find a similar situation in membranes where a large protein molecule may be inserted through the membrane in either direction but the probability of switching directions is very small. In these cases, the orientation of the proteins will depend on the past history, and in particular, on how the membrane was formed. Natural membranes are asymmetrical but artificial bilipid membranes are usually symmetrical.

Among the quantum states which might exist are a set of vibrational modes in long chain molecules which have been discussed by Frönlich [68-70]. If one takes the length of a molecule such as

DNA and divides by the velocity of sound, one predicts a series of vibrational modes which would lie in the millimeter wave region [71]. In addition to predicting these vibrational models, Fröhlich and others have proposed that there may exist some coherent modes of vibration for these molecules where the population of the excited states undergoes a Bose-Einstein condensation such that a large fraction of the population lies in a single excited state. In this case, the population distribution would be described by an expression of the form [68]:

$$\bar{n}_i = \{exp[B(\hbar\,\omega_i - \mu) - 1]\}^{-1}$$

where μ is a chemical potential, $\bar{n}_i$ is the average occupation number for the state, ω_i is the angular frequency, $B = 1/KT$ where K is Boltzmann's constant and T is the temperature. As $\hbar\omega \to \mu$ the population density $\bar{n}_i$ condenses in this state very rapidly and in a very nonlinear way. μ is proportional to the power density and thus there is a critical power density where this condensation is predicted to take place. This power density leading to the chemical potential is dependent on the number of vibrational states, the difference between the microwave pumping frequency and the transition frequency, the molecular line width, and the material properties. The explanation of μ is too extensive to go into in this brief review. However, under one set of assumptions the predicted critical power density to undergo condensation is about .4 mw/cm^2 at 240 GHz [68]. The experiments necessary to confirm this kind of nonlinear population distribution have not yet been done and will be difficult to do in a direct way because of the very high absorption coefficient of H_2O throughout the millimeter wave region and the need to have both the right frequency and power level for the exposures. A possible method of observing these modes may be to use Raman spectroscopy to translate the observation of the population distribution of these modes with and without microwave excitation into the optical region of the spectra.

In summary, we can say that because of the very small size of a quantum at these frequencies, to obtain a significant difference in population between the excited and unexcited states the system must be far from thermodynamic equilibrium and relatively isolated from its environment by a small transition matrix element. The low saturation power density limits the size of the signal that can be used to look for these states in absorption to a very low value. It is probable that high order nonlinearity and multiple quantum transitions will be required to excite molecules or ions into long-life metastable states with sufficient energy to have an effect on the biochemical reactions.

VI. CONCLUSIONS

We have reviewed several nonlinear interactions between electromagnetic waves and biological materials. Whereas there is a vast array of literature on phenomenological experiments, there have been relatively few studies on the mechanism of interactions. The thermal chemistry of many of the reactions resulting from changes in temperature have been covered in a variety of reviews (Wood, Jolly, Barnes, Johnson and Eyring). In this review we have tried to look at other mechanisms which may influence biological processes. At the level of this review, biological processes are limited to chemical reactions, ion and molecular transport, molecular deformations or orientation, and molecular excitation. We have shown that radio frequency electromagnetic fields are predicted to generate small DC voltage shifts across the cell membranes and a corresponding shift in the ion currents and concentrations. At field strengths of a few volts per centimeter, corresponding to plane waves fields of 10 mw/cm^2, it is not obvious that these shifts in ion current are important. Considerable experimental work needs to be done to determine the coefficients for this rectification, the cutoff frequency, the effects of geometry on the net current flows, and variation among cells of different types. Only after this has been done will we be able to determine the biological importance of these small ion currents.

Secondly, we have shown that moderate *rf* fields can apply rotational torques to molecules with either dipole moments or induced dipole moments. If these molecules are not rigid this force may distort their shape. Changes in shape could effect the opening and closing of ion channels, the matching keys in genetic process, or antibody-antigen reactions to name a few unproven possibilities. However, to the authors' knowledge, none of these effects have yet been demonstrated. Coupled with changes in orientation or changes in shape, there may be changes in the conductivity of the solutions containing these large molecules. These changes have been measured at high fields but it is not clear that they have any biological significance at low fields. The exception to this may be in the vicinity of the membranes where high fields are known to exist.

The analysis of the population distributions for quantum systems shows that single quantum transitions are unlikely to be important because black body radiation is well above the saturation power density. Thus, equilibrium population differences are very small. Multiple quantum transitions under special conditions could lead to significant changes in the population of excited molecular states. To lead to significant changes, the power levels should be well above the black body radiation density and the process is likely to be very nonlinear. So far, to the authors' knowledge, experimental evidence of narrow band frequency sensitivity transition of the type which would be described by simple quantum systems have not been measured reproducibly by a sufficient number of investigators to be generally accepted by the research community. However, there are some experiments which indicate the need for more experimental work [72,73,74]. The problem of switching molecules from one degenerate state to another in a biological environment needs further investigation to see if the quantum approach is a useful way of describing systems of this type.

ACKNOWLEDGMENT

The authors wish to acknowledge the very stimulating conversation with Dr. Herbert Kroemer who suggested the need to consider degenerate ground states and the analogy to ferromagnetism.

REFERENCES

1. *Proceedings 1977 URSI Bio-Effects Meeting,* Airlie House, Virginia, October 30 - November 4 (1977).
2. Chou, Chung-Kwang and Guy, A.W., "Microwave and RF dosimetry," *The Physical Basis of Electromagnetic Interactions with Biological Systems,* Proceedings of a Workshop, University of Maryland, College Park, Maryland, June 15-17 (1977).
3. Guy, A.W., "5.0 objects in the fields," *Rep. to National Council on Radiation Protection and Measurement,* July (1975).
4. Durney, C.H., et al, *Radiofrequency Radiation Dosimetry Handbook,* 2nd Ed., interim report prepared for USAF School of Aerospace Medicine, Brooks Air Force Base, Texas.
5. Lodish, H.F., and Rothman, J.E., "The assembly of cell membranes," Scientific American, *240,* 1, 51, January (1979).
6. Cole, *Membranes, Ions and Impulses, A Chapter of Classical Biophysics,* University of California Press, p. 155 (1972).
7. Barnes, F.S., and Hu, C.J., "Model for some nonthermal effects of radio and microwave fields on biological membranes," *IEEE Trans. on Microwave Theory and Techniques, MTT-25,* 9, 742, September (1977).
8. Report of National Council on Radiation Protection SC #39, to be published.
9. Wachtel, H., Seaman, R., and Joines, W., "Effects of low-intensity microwaves on isolated neurons," Annals of the New York Academy of Sciences, *247,* 46-62, February (1975).
10. Seaman, R.L., and Wachtel, H., "Slow and Rapid Responses to CW and Pulsed Microwave Radiation by Individual *Aplysia* Pacemakers," Journal of Microwave Power, *13,* 1 (1978).
11. Wachtel, H., private communication.

12. Saito, M. and Schwan, H., "The time constants of pearl-chain formation," *in* "Biological Effects of Microwave Radiation," *1*, Plenum Press, New York (1960).
13. Hu, Chia-lun J. and Barnes, F.S., "A simplified theory of pearl chain effects," *Rad. and Environm. Biophys. 12* (1975).
14. Caldwell, Richard R., "A new technique for investigating birefringence in systems of macromolecules," Master's Thesis, unpublished, University of Colorado (1977).
15. Malsch, J. and Wien, M., "A null method of measuring resistance to transient currents," *Ann. d. Physik, 83*, 305-326, June 16 (1927).
16. Wien, M., "Potential effect of the conductivity of electrolytes in lower fields," *Ann. d. Physik, 85.7*, 795-811, May 2 (1928).
17. Wien, M., "Potential effect of electrolytic conductivity in very strong fields," *Ann. d. Physik, 1.3*, 400-416, February 9 (1929).
18. Wien, M., "Effect of voltage on the conductivity of strong and weak acids," *Phys. Zeits. 32*, 545-547, July 15 (1931).
19. Gyemant, A., "Strong electrolytes," *Phys. Zeits. 29,* 289-293, May 15 (1928).
20. See ref. 18.
21. Schiele, J., "Influence of voltage on the conductivity of strong and weak acids," *Ann. d. Physik, 13*, 811-830, June 2 (1932).
22. Joos, G., "Strong electrolytes and the dependence of conductivity on voltage," *Phys. Zeits. 29,* 570, August 15 (1928).
23. Blumentritt, M., "Behavior of dilute electrolytes with high field strengths," *Ann. d. Physik, 85.7*, 812-830, May 2 (1928).
24. Blumentritt, M., "A more accurate calculation of Wien's tension effect in electrolytes," *Ann. d. Physik, 1.2*, 195-215, January 19 (1929).
25. Falkenhagen, H., "Theory of the complete curve for the Wien effect," *Phys. Zeits. 30,* 163-165, March 15 (1929).
26. Falkenhagen, H., "Dependence of conductivity of strong electrolytes on the voltage," *Phys. Zeits. 32,* 353-365, May 1 (1931).
27. Wilson, W.S., Ph.D. Dissertation, unpublished, Yale University (1936).
28. Eckstrom, H.C. and Schmelzer, C., Chemical Reviews, *24*, 367, (1936).
29. Onsager, L., "Deviations from Ohm's law in weak electrolytes," *J. Chem. Phys. 2*, 599-615, September (1934).
30. Mead, D.J. and Fuoss, R.M., "Dependence of conductance on field strength. Part I. Tetrabutylammonium in dimethyl ether at 50°," *Am. Chem. Soc. J. 61*, 2047-2053, August (1939).
31. Mead, D.J. and Fuoss, R.M., "Dependence of conductance on field strength. Part II. Tetrabutylammonium in diphenyl ether at 50°," *Am. Chem, Soc. J. 62*, 1720-1723, July (1940).

32. Onsager, L. and Liu, C.T., "Theory of the Wien effect in weak electrolytes," *Z. Physik. Chem. 228*(516), 428-432 (1965).
33. Bailey, F.E., Patterson, A., Jr. and Fuoss, R.M., "Wien effects in polyelectrolytes," *J. Am. Chem. Soc. 74*, 1845 (1952).
34. Wissbrun, K. and Patterson, A., Jr., "A study of the high field conductance of three polyelectrolytes," *J. Polymer Sci. 33*, 235 (1958).
35. Wissbrun, K. and Patterson, A., Jr., "A theory for the high field conductance of polyelectrolytes," *J. Polymer Sci. 33*, 249 (1958).
36. Eigen, M. and Schwarz, G., *Z. Phys. Chem. (N.F.), 4*, 380 (1955).
37. Eigen, M. and Schwarz, G., *in* "Electrolytes" (B. Pesce, ed.), figure 6, pp. 309-335, Pergamon Press (1965).
38. Spinnler, J.F. and Patterson, A., Jr., "The Wien effect in uranyl ion solutions I. Uranyl nitrate and perchlorate from 5 to 65°. Negative Wien effects," *J. Phys. Chem. 69-2*, 500-508 (1965).
39. Spinnler, J.R. and Patterson, A., Jr., "The Wien effect in uranyl ion solutions II. Uranyl fluoride from 5 to 65°," *J. Phys. Chem. 69-2*, 508-513 (1965).
40. Spinnler, J.R. and Patterson, A., Jr., "The Wien effect in uranyl ion solutions III. Uranyl sulfate from 5 to 65°," *J. Phys. Chem. 69-2*, 513-517 (1965).
41. McIlroy, D.K., "A mathematical model of the nerve impulse at the molecular level," *Math. Biosci. 7*, 313 (1970).
42. McIlroy, D.K., "Analysis of the enzyme model of the nerve," *Math. Biosci. 8*, 109 (1970).
43. McIlroy, D.K., "The Σ transform of enzyme activity: Application to the enzyme model of the nerve," *Math. Biosci. 8*, 417 (1970).
44. McIlroy, D.K., "Deductions from the enzyme model of the nerve," *Math. Biosci. 9*, 135 (1970).
45. McIlroy, D.K. and Mason, D.P., "Wien dissociation in very low intensity electric fields," *J. Chem. Soc., London, Faraday Trans. #2, 72*(3), 591 (1976).
46. Chan, M.-S., "Wien effect in mixing strong electrolytes," *J. Chem. Phys. 68*, 5442-5447 (1978).
47. McIlroy, D.K., *Proc. Roy. Soc. London, Sec. A, 359*, 303-317 (1978).
48. Grunhagen, H., *Chem. and Phys. of Lipids, 14*, 201 (1975).
49. Debye and Falkenhagen, H., "Dispersion von leitfähigkeit und dielektrizitätskon stante bei starken elektrolyten," *Physik. Z. 29*, 121, 401 (1928).
50. Falkenhagen, H., "Zur theorie der gesamtkurve des Wien-effektes," *Physik. Z. 30*, 169 (1929).
51. Falkenhagen, H. and Williams, J.W., "The frequency dependence of the electrical conductance of solutions of strong

electrolytes," *J. Phys. Chem. 33*, 1121 (1929).
52. Falkenhagen, H. and Fleischer, H., *J. Phys. Chem. 39*, 305 (1938).
53. Arnold, O.M. and Williams, J.W., "The resonance and capacity behavior of strong electrolytes in dilute aqueous solution I. Method for the simultaneous observation of conductance and dielectric constant at high radio frequencies," *J. Am. Chem. Soc. 58*, 2613 (1936).
54. Arnold, O.M. and Williams, J.W., "The resonance and capacity behavior of strong electrolytes in dilute aqueous solution II. The dispersion of electrical conductance," *J. Am. Chem. Soc. 58*, 2616 (1936).
55. Eigen, M. and DeMayer, L., *in* "Techs. Organic Chem. VIII," Part II, 2nd Ed. (S.L. Friess, E.S. Lewis and A. Weissberger, eds.), Interscience, New York, (1963).
56. Persoons, A.P., *J. Phys. Chem. 78*, 1210 (1974).
57. Eigen, M. and Schwarz, G., *J. Phys. Chem. N.F., 4*, 380 (1955).
58. Eigen, M. and Schwarz, G., "Orientation field effects of polyelectrolytes in solutions," *J. Colloid. Science. 12*, 181 (1957).
59. Figure 13 of ref. 31.
60. Schwan, H.P., *Proc. Inst. Radio Engineering,* 1841 (1959).
61. Delbos, G., Bottreau, A., Margat, C. and Salefran, J., *J. of Microwave Power, 13*, 69 (1978).
62. Minakata, A., "Dielectric dispersion of polyelectrolytes due to ion fluctuation," *Ann. N.Y. Acad. Sci. 303*, 107 (1977).
63. Vetter, K.J., "Electrochemical Kinetics," 104, 203, 204, Academic Press, Inc., New York (1967).
64. Kortum, G., "Treatise on electrochemistry," Chapter XII, Elsevier, 2nd Ed., (1965).
65. Pyle, S.D., Nichols, D., Barnes, F.S. and Gamow, E., "Threshold effects of microwave radiation on embryo cell systems," *Ann. N.Y. Acad. Sci. 247*, 401-407 (1975).
66. Fout, Randy, A., Master's Thesis, unpublished, University of Colorado (1979).
67. Webber, M.M., Seltzer, L.A. and Barnes, F.S., "Damage from short microwave pulses on neuroblastoma cells," to be submitted.
68. Fröhlich, H., "Bose condensation of strongly excited longitudinal electric modes," *Phys. Lett. 26A* 402 (1968).
69. Bhaumik, D., Bhaumik, K. and Dutta-Roy, B., "On the possibility of 'Bose condensation' in the excitation of coherent modes in biological systems," *Phys. Lett. 56A*, 145 (1976).
70. Fröhlich, H., "Evidence for Bose condensation-like excitation of coherent modes in biological systems," *Phys. Lett. 51A*, 21 (1975).
71. Prohofsky, E., "Effects of microwave absorption by the DNA helix," Workshop on Mechanisms of Microwave Biological Effects, University of Maryland, College Park, Maryland, May (1979). To be published.

72. Keilmann, Fritz, "Nonthermal microwsve resonances in living cells," Proceedings fo the NATO Advanced Study Institute, "Coherence in Spectroscopy and Modern Physics," Versilia (1977).
73. Webb, S.J. and Booth, A.D., "Absorption of microwaves by microorganismus," Nature *222,* 1199 (1969).
74. Gringauz, K.I., et al., Scientific Session of the Division of General Physics and Astronomy, USSR Academy of Sciences, *Sov. Phys.-Usp.* (Translation) *16*, 568 (1974).